AA003520

VIII International Scientific Conference on New Trends, Strategies and Structural Changes in Emerging Markets (NTSSCEM 2021)

SHS Web of Conferences Volume 114 (2021)

Moscow, Russia
21-22 May 2021

Editor:

V.M. Matyushok

ISBN: 978-1-7138-3472-4

Printed from e-media with permission by:

Curran Associates, Inc.
57 Morehouse Lane
Red Hook, NY 12571

Some format issues inherent in the e-media version may also appear in this print version.

This work is licensed under a Creative Commons Attribution 4.0 International License. License details: http://creativecommons.org/licenses/by/4.0/.

No changes have been made to the content of these proceedings. There may be changes to pagination and minor adjustments for aesthetics.

Printed with permission by Curran Associates, Inc. (2022)

For additional information, please contact EDP Sciences – Web of Conferences
at the address below.

EDP Sciences – Web of Conferences
17, Avenue du Hoggar
Parc d'Activité de Courtabœuf
BP 112
F-91944 Les Ulis Cedex A
France

Phone: +33 (0) 1 69 18 75 75
Fax: +33 (0) 1 69 28 84 91

contact-edps@webofconferences.org

Additional copies of this publication are available from:

Curran Associates, Inc.
57 Morehouse Lane
Red Hook, NY 12571 USA
Phone: 845-758-0400
Fax: 845-758-2633
Email: curran@proceedings.com
Web: www.proceedings.com

TABLE OF CONTENTS

Statement of Peer review

In submitting conference proceedings to *Web of Conferences*, the editors of the proceedings certify to the Publisher that

1. They adhere to its **Policy on Publishing Integrity** in order to safeguard good scientific practice in publishing.
2. All articles have been subjected to peer review administered by the proceedings editors.
3. Reviews have been conducted by expert referees, who have been requested to provide unbiased and constructive comments aimed, whenever possible, at improving the work.
4. Proceedings editors have taken all reasonable steps to ensure the quality of the materials they publish and their decision to accept or reject a paper for publication has been based only on the merits of the work and the relevance to the journal.

Title, date and place of the conference

NTSSCEM 2021 - VIII International Scientific Conference New trends, strategies and structural changes in emerging markets
(Moscow, Russia, May 21 - 22, 2021)

Proceedings editor(s): V.M. Matyushok

Date and editor's signature

28/06/2021

Preface

It is a great honour for me to address you, the conference participants, and our partners, with whose support we are holding this event. Let me express my deep gratitude to everyone who decided to participate in our conference and share the research results. We will try to reflect all valuable thoughts in the materials of the conference "New Trends, Strategies and Structural Changes in Emerging Markets" in the Post-Pandemic Era.

We are witnessing the growth in the world at the global and national levels of the so-called "challenges " i.e. contradictions between the existing state of society and the need for real changes that require an adequate "response" from the society. Among them is the inequality of world development, the transition from natural competition to economic confrontation, the threat of climate change, the aggravation of food security, the growth of migration to developed countries, the decline in the authority of international organizations, etc. The pandemic has only exacerbated these challenges.

The growing instability in the world economy grounds several experts to assert that the world has entered a bifurcation point. Either a long period of chaos lies ahead, or the system will move to a new state of order. It is logical to turn to economic theory and a retrospective analysis of previous development to understand how the world economy will develop and make a scientific forecast as a systematic view of the future. In these conditions, the development of adequate models ("responses") of society to "challenges", the search for new orderliness ("stability"), become more relevant than ever.

Realizing this, the founder and permanent head of the WEF in Davos, Klaus Schwab, developing his ideas about the "Fourth Industrial Revolution" from 2016, last summer put forward the idea of the "Great Restart" of capitalism as a model for the global restructuring of the world economy on a digital basis, clean energy and long-term abandonment of the pursuit of profit and income growth of the middle class. This model became the main topic of the online format of the WEF, held in January 2021. Key messages echo the 2015 UN Sustainable Development Goals and the European Green Deal to achieve zero greenhouse gas emissions in EU member states by 2050. All the ideas put forward are supported by the new leadership of the United States, the IMF, and the heads of the world's largest companies. The outwardly attractive ideas of the “reset” conceal the interests of capital and the world elite, who will get their hands on climate weapons and technologies of total digital control. There will be a redistribution of global wealth through instruments such as fines and taxes on carbon emissions, lower revenues from the export of traditional energy resources, increased spending on the purchase of digital and "green technologies", more environmentally-friendly vehicles.

The world community needs not to build an economy operating for the "golden billion", but models and strategies for the development of national economies as a "new order". We need a model that would direct efforts and resources, new technologies, and care for the environment to improve the level and quality of life of the population of developed and developing countries. I am sure that the work of the conference will make a significant contribution to the development and promotion of such ideas. I wish you success in the work of the conference!

Vladimir Matyushok,

Chairman of the Scientific Committee
Doctor of Sciences (Economics), Full Professor,
Honoured Worker of Higher School of Russia

© The Authors, published by EDP Sciences. This is an open access article distributed under the terms of the Creative Commons Attribution License 4.0 (http://creativecommons.org/licenses/by/4.0/).

An approach to tackle negative effects in project management

Ivan V. Nevolin*, and *Oleg E.* Khrustalev

CEMI RAS, Nakhimovsky pr-t, 117418 Moscow, Russia

Abstract. We introduce a model to support scenario analysis while managing costs of high-tech project. If the project results from joint efforts of firms with one per single stage the model suggests budget redistribution in case of funding shortage or exchange rate volatility. Given negative outer effects for the project the model requires the project to be completed and this constraint forces firms to diminish their profit or to make a loan if needed. The model proposed suggests human-machine (expert) interaction to build scenario for analysis but is simpler compared to method of successive concessions from computational point of view.

1 Introduction

Planning is a standard practice in high-tech projects development: circumstances arise along the time, and they require an appropriate reaction and control. A number of acting bodies, engineering difficulties and complexity of production technology require risk management techniques in the project implementation [1]. Obstacles in these techniques application associate with the number of contractors, competitors, nomenclature, production constraints, etc. – the dimension of the problem design and production grows rapidly when it comes to project planning [2], and the variables in question must be coordinated in an appropriate way. An increasing dimension of the problem entails both quantitative issues (an increase in the number of variables) and structural ones. The decision-maker is forced to deal with indicators of different nature: technical (transportation capabilities, weight of units and assemblies, etc.), financial (underlying budget), production (compatibility of components from different suppliers, equipment requirements and labor conditions) and physical (for example, size or solidity) features. A single expert or even a group of experts in one field is unlikely to cope with the task of project management – generally, solution requires knowledge from various disciplines, as well as an understanding of how variation in one factor affects the values of others. However, special techniques – mathematical models and software – are used to coordinate multidisciplinary efforts [3,4].

Risk management with instrumental support suggests decisions following two paths. First, one could ground on previous experience involving probabilistic methods. Second, one could implement scenario analysis to assess the impact of specific events without sufficient number of observations.

* Corresponding author: i.nevolin@cemi.rssi.ru

© The Authors, published by EDP Sciences. This is an open access article distributed under the terms of the Creative Commons Attribution License 4.0 (http://creativecommons.org/licenses/by/4.0/).

Probabilistic methods result in statistical distribution estimates of project costs and schedule given the random risks. The project analysis starts from the point estimate [5,6]: statistical estimates map technical specifications into costs for every single stage of the project. For example, given the engine development costs one could build a statistical function that maps technical specification into R&D expenditure. Project schedule ensures the planned result and describes the base scenario subjected to risk. The latter affect the cost of single stages both upward and downward. Experts assess statistical distributions of risks and their cross-correlations. Further, project variables as well as relations between contractors form the simulation model to estimate statistical variations in the baseline budget. As a result, the decision-maker receives for analysis the probability distributions of the total project budget and costs of single stages. This information supports additional funding allocation between project stages and contractors.

Scenario analysis allows the decision-maker to assess sensitivity of the project indicators to predetermined distortions and effects of different origin whether external or internal. To simulate the state of the environment – conditions external to the project – it is convenient to operate with the concept of "scenario". Scenario is a set of single events that changes the state of the environment, or – in other words – a function that brings the environment out of its normal state adopted in the project planning. Thus, a demand for reaction arises as a decision-maker's response to the certain scenario.

Every event in any scenario has two characteristics: presence (or absence) and scale. Given this, it is possible to define the main events affecting the project and implement them in as software modules. Combining these events by adjusting the modules the decision-maker builds various scenarios. If one of the project participants gets better off under some scenario, an appropriate share of funding could be redistributed to ensure project completion. Details of this approach are outlined below.

2 Cost management in scenario analysis

Given a sufficient level of abstraction, we discuss a project management model with two key features. First, it supports interconnection of project stages and production processes providing the decision-maker with some kit of building blocks. Second, the model reflects the development stages, production and demand for resources, i.e. includes a cost analysis. An economic result of the model calculation is a project budget corresponding to the production plan. Following general description we focus on the principles of resources redistribution between the project participants while specific production processes only clarify the balance relationships between input and output variables.

Rather often is a task of maximizing result (not in terms of profit, but in terms of coverage area for communication facilities, computing power, drug efficiency, etc.) at minimum costs. In this formulation, budget optimization under the changing environment is possible in the following manner: 1) maximize the benefit while minimizing the cost of the project implementation, or 2) redistribute budget between participants so that the project is implemented, albeit with a reduced profit margin.

The first approach treats the entire life cycle of the project and arising in this context is a classic problem of multi-criteria optimization. To adjust funding of the stages one could apply the method of successive concessions, and human-machine decision support systems supporting this method are known [7]. So, the project implementation in economic terms sounds like a task of effect maximization at minimum costs. Output of the project belongs to a finite and discrete set of options to implement a product or service. The discreteness of a set is determined by two aspects. First, the characteristics of equipment (or substances) do not change continuously: each part or component is an independent product that has specific dimensions, weight, price, etc. Secondly, only some combinations of specific components

© The Authors, published by EDP Sciences. This is an open access article distributed under the terms of the Creative Commons Attribution License 4.0 (http://creativecommons.org/licenses/by/4.0/).

and assemblies give a truly viable device, that is, there is a compatibility problem that needs to be solved. These combinations could be obtained using morphological synthesis [8] and are thus external to the model in question. Further, for analysis purposes the set of options could be reduced to the Pareto frontier – non-dominated project outputs in terms of revenue and budget. The Pareto frontier, therefore, consists of outputs with a minimum budget at fixed profit or those with a maximum profit at fixed costs.

The second approach of cost optimization – the budget redistribution between project participants – is based on the production processes revision at each stage of the project given the fixed layout of a specific product. Let the minimum resource requirements and input-output relations to be known for each stage. Then the decision-maker can vary the cost structure without falling below the threshold value – one that ensures the project output. An optimization begins with an initial budget estimate and follows the less painful – for the project implementation – cost shortage. The most obvious solution to the problem in this case is to assign financing at the threshold level for each of participants. However, the correct solution foresees the following. First, the financing of some participant could be at a level higher than the threshold, but with an insufficient rate of return. This may affect the participant's investment or scientific and technical policy of the. Secondly, in case of funding below the threshold the project has still to be implemented. For this, a participant may require some debt. And then a question arises that can be answered to only by looking at the general state of the participant: is the enterprise capable of paying debt on market conditions, and if not, whether to subsidize the interest rate or develop other supporting mechanism.

Often these are the projects that compete in the high-tech sector as, for example, in the aerospace and defense industry – not the individual enterprises. Leaving the problems of coalitions formation outside the framework, as well as the solution of technical problems associated with the search for the technology layout, we describe a model for drawing up an optimal project budget structure. At the same time, the issues of technological and production links are exogenous to the model, and data about them are provided for calculations as ready to use. Also, issues of options specification for the project are outside the framework and their cost structures are considered to be known. Under this setting, the model makes it possible to answer the question: how to change the base budget of the entire project when the external conditions change for the entire industry or for an individual participating enterprise.

3 Model formalisation

After the general characteristics of the model and its capabilities consider a formal description. Let K firms to implement a project jointly – index them with $k = 1,2,3, K$. The whole project is divided into K stages, and each firm develops only one of them. Firms use standard accounting forms, and we consider one that describes the cost structure of a stage. If the structure contains n fields, then the budget of firm k at time t at discrete time is expressed as $f_k^t = \sum_{i=1}^{i=n} x_i^t[k]$, where $x_i^t[k]$ – the value of field i for firm k at time t. If a period T is required to complete the project successfully, then the total project costs are $\sum_{t=1}^{T}\sum_{k=1}^{K} f_k^t$. Each firm requires a certain threshold funding annually so that nothing threatens the implementation of the project. In the model, this funding is calculated as follows. Let the production matrix $M^t(k) = [m_{i,j}^t(k)]_{i,j=1}^{n}$ of the firm k be known with every element showing how many resources from field i are required to use resource j when implementing production plan of firm k at time t. Matrix elements that correspond to zero costs, i.e. the fields of the standard form of cost structure with no expenditure on the stage, are also considered equal to zero. If $a_i^t(k)$ is the minimum cost at field i for firm k at time t, then $x_i^t[k] \geq a_i^t(k)$ and $x_j^t[k] = m_{i,j}^t(k)x_i^t[k] \geq a_j^t(k)$. Next, define the field of the cost structure reaching its constraint in budget variation. Formally, let the item i_0 to reach its

© The Authors, published by EDP Sciences. This is an open access article distributed under the terms of the Creative Commons Attribution License 4.0 (http://creativecommons.org/licenses/by/4.0/).

threshold while the others lie in interior of feasible region. Then, non-zero costs are $x_{i_0}^t[k] = a_{i_0}^t(k)$ and $x_j^t[k] = m_{i_0,j}^t(k)a_{i_0}^t[k] \geq a_j^t(k)$. Consequently, $a_{i_0}^t(k) \geq a_j^t(k)/m_{i_0,j}^t(k)$, $j = \overline{1,n}, k = \overline{1,K}, t = \overline{1,T}$. Given this the budget of firm k at time t cannot be lower than $\bar{f}_k^t = \sum_{i=1}^{i=n} x_i^t[k] = \sum_{i=1}^{i=n} m_{i_0,j}^t(k)x_{i_0}^t[k] = a_{i_0}^t(k)\sum_{i=1}^{i=n} m_{i_0,j}^t(k)$.

This approach to threshold calculation has the following advantages. First, it is beneficiary in cases of predefined lower values of cost items such as labor, amortization of reserve or temporarily unloaded production facilities, etc. Secondly, from a computational point of view, it is more convenient to determine an item i_0 reaching its threshold. By definition, $m_{i,j}^t(k) = x_j^t[k]/x_i^t[k]$ for nonzero elements, therefore resulting in $m_{i,i}^t(k) = x_i^t[k]/x_i^t[k] = 1$ and $a_{i_0}^t(k) = \max_{i,j}[a_j^t(k)/m_{i,j}^t(k)]$.

After the funding threshold calculation, we proceed to budget distribution at time t between the firms involved. Consider an enterprise k which receiving funding of f_k^t at time t. From a computational point of view, it is convenient to operate with $(f_k^t - \bar{f}_k^t)$ – the excess (or deficient in case of negative value) financing compared to base one ensuring the project implementation. We distribute the difference between costs items in appropriate proportions and the firm's profit in case of excess funding or demand for a debt otherwise. In case of the positive outcome a cost item $x_j^t[k]$ of firm k increases proportionally taking form of:

$$x_j^t[k] = m_{i_0,j}^t(k)\left[a_{i_0}^t(k) + \frac{f_k^t - \bar{f}_k^t(1+\pi_k^t)}{(1+\pi_k^t)\sum_{i=1}^{i=n} m_{i_0,j}^t(k)}\right]$$

where π_k^t is a rate of return for firm k at time t.

If the outcome of $(f_k^t - \bar{f}_k^t)$ difference is negative, the cost item takes a threshold value while the firm makes a debt of $c = (f_k^t - \bar{f}_k^t)$ for τ years with an annual rate of r_k^t. Therefore, cost items take a form of:

$$x_j^t[k] = m_{i_0,j}^t(k)a_{i_0}^t(k)$$
$$x_n^{t+i,t}[k] = \frac{c}{\tau} + \frac{c(\tau - i)}{\tau}r_k^t, i \in \mathbb{N}$$

where $x_n^{t+i,t}[k]$ states for increased costs at time $(t+i)$ due to the debt at time t. With no loss in generality let this cost item be indexed with n. Note, in the above case of $(f_k^t - \bar{f}_k^t) \geq 0$ no debt at time t is made and thus $x_n^{t+i,t}[k] \equiv 0, i \in \mathbb{N}$. The potential payments because of debt entail a revision of the firm's budget. Formally, the costs of the firm under the item n take a form of:

$$\tilde{x}_n^t[k] = x_n^t[k] + \sum_{i=0}^{t-1} x_n^{t,1}[k]$$

Substantially, the following happens. Budget financing of the firm k at time t under some scenario is expected to be f_k^t. If it turns out to be less than the threshold value of $\bar{f}_k^t$, the firm makes a debt. In fact, a sum of the cost items results in firm's budget $\bar{f}_k^t$ for the project with insufficient funding being complemented by loans. But the firm pays for complete budget at time t with increased expenses due to interests in next periods. In case of directive rates of return this means an increase in the total project costs – whether prom public sources or commercial ones – by the amount of credit payments. This growth demonstrates the decision-maker that participation in the project becomes much less profitable compared to the baseline scenario as a debt is needed. And the amount of credit payments denotes withdrawals arise from income that other projects participants have to face in case of joint efforts. These withdrawals though could also be offset from the public sources with the project budget

© The Authors, published by EDP Sciences. This is an open access article distributed under the terms of the Creative Commons Attribution License 4.0 (http://creativecommons.org/licenses/by/4.0/).

adjustment. The letter – increased project developments costs – is rather general in aerospace and defence [9]. Therefore, we arm a decision-maker with a tool supporting budget estimation in case of financing shortage for a single project stage in some time period.

The model suggests the project budget adjustments not only in case of reduced funding, but also under exchange rate fluctuations. Accounting for the budget dependence on imported components requires some modification in the above formulas.

Let the base budget f_k^t of firm k at time t ground on the exchange rate p_0^t, the share of imported components d_k^t in the cost of raw materials $x_1^t[k]$, without loss of generality, indexed with one. Then the change in the currency rate to p_1^t increases the cost on

$$x_1^t[k]\frac{p_1^t - p_0^t}{p_0^t}d_k^t$$

Therefore, a scenario with exchange rate fluctuations the above schema to calculate cost items entails replacements in formulas:

1) of cost items

$$x_1^t[k] \to x_1^t[k]\left\{1 + \frac{p_1^t - p_0^t}{p_0^t}d_k^t\right\}$$

2) of constraints

$$a_1^t(k) \to a_1^t(k) + x_1^t[k]\frac{p_1^t - p_0^t}{p_0^t}d_k^t$$

With calculations remain the same under replacements the model suggests a tool of processing a superposition of events forming scenarios. If a single event leaves the rate to be stable, neither demand for marginal costs nor constraints require modifications. If there is a for debt in a certain period, the cost of the debt becomes an extra funding to the general project budget.

4 Conclusion

The model described processes a set of parameters and variables. The first are budget of the project, the distribution of financing among the stages, production links between the stages, interest rates, resource restrictions, the share of imported components and the exchange rate. An analyst quantifies their values and depending on the settings, the cost structure is calculated as a vector of variables. Feeding the model with a certain set of input values, the analyst initializes a specific scenario, and as a result, she receives the best budget distribution for this particular scenario. Variation in at least one of parameters creates a new scenario. The analysis and selection of a specific financing schema, therefore, relies on the results of the scenarios calculation. Competition in the model doesn't take a form of participants' aspiration for larger share of budget, but looks like alternative solutions contest, which results in the project budget changes, resources redistribution between stages, choice of production processes and financial sources. The formalization of the model imposes requirements on the data: production costs, schedule and description of processes become critical.

Acknowledgment

The article was prepared with financial support by the Russian Science Foundation, grant № 21-78-20001 «Development of the theory and model tools for optimizing the management of

© The Authors, published by EDP Sciences. This is an open access article distributed under the terms of the Creative Commons Attribution License 4.0 (http://creativecommons.org/licenses/by/4.0/).

the diversification of defense production in the context of the economic crisis and growing threats to Russia's national security»

References

1. NASA, NASA cost estimating handbook (2008)
2. D. Motte, R. Bjärnemo, In Proceedings of the 25th International Conference on Design Theory and Methodology - DETC/DTM'13, **5**, (2013)
3. M. Mukelabai, D. Nešić, S. Maro, T. Berger, J.P. Steghöfer, In Proceedings of the 33rd ACM/IEEE International Conference on Automated Software Engineering (2018)
4. A.M. Batkovskiy, E.G. Semenova, A.V. Fomina, E. Iu. Khrustalev, O.E. Khrustalev, Indian Journal of Science and Technology, **9**, 20 (2016)
5. G.S. Levenson, H.E. Boren, D.P. Tihansky, F. Timson, *Cost-estimating relationships for aircraft airframes* (1972)
6. F.D. Harris, M.P. Scully, Journal of the American Helicopter Society **43**, 1 (1998)
7. Golshein E.G., Borisova E.P., Dubson M.S. Economics and mathematical methods, **26**, 4 (1990)
8. A.V. Andreichikov, O.N. Andreichikova, Journal of Computer and Systems Sciences International **45**, 4 (2006)
9. M.V. Arena, R.S. Leonard, S.E. Murray, O. Younossi, *Historical Cost Growth of Completed Weapon System Programs* (2006)

© The Authors, published by EDP Sciences. This is an open access article distributed under the terms of the Creative Commons Attribution License 4.0 (http://creativecommons.org/licenses/by/4.0/).

Assessing the applicability of the CRM concept in a public autonomous institution

A.D. Lagunova , and *A.A.*Pastushkova*

MIREA-Russian technological University (RTU MIREA), 119454 Moscow, Prospekt Vernadskogo 78, Russia

Abstract. In this article the issues of assessing the applicability of Customer Relationship Management in the processes and conditions of the functioning of autonomous state institutions are being examined. The authors consider the outlook of applying the concept of CRM, from the point of view of further implementation of CRM systems in the functional activities of the establishment.

1 Introduction

The CRM concept is an interaction model that is aimed at maximizing the life cycle by building long-term relationships with the client through the formation of loyalty with the help of personal proposals - an issue whose applicability assessment must be identified for use in state autonomous institutions.

2 Materials and Methods

Before considering the issues of customer relationship management of an organization, it is necessary to clarify its status and the restrictions it imposes on its activities. In this study, the authors propose to consider the activities of state autonomous institutions, which, according to Federal Law No. 174 of 03.11.2006, are defined as "a non-profit organization created by the Russian Federation, a constituent entity of the Russian Federation or a municipal formation to perform work, provide services in order to exercise the powers provided for by the legislation of the Russian Federation. Government bodies, the powers of local self-government bodies in the spheres of science, education, health care, culture, mass media, social protection, employment of the population, physical culture and sports, as well as in other areas in cases established by federal laws (including activities to work with children and youth in these areas) ". [1]

To assess the applicability of the CRM concept to an organization of this type, it is necessary to determine the differences between a state autonomous institution and a commercial organization. The study of this issue in various sources, primarily legislative [1,2,3], shows that at the current time there is a confusion of the legal status of an autonomous

* Corresponding author: pastushkovaanya@gmail.com

© The Authors, published by EDP Sciences. This is an open access article distributed under the terms of the Creative Commons Attribution License 4.0 (http://creativecommons.org/licenses/by/4.0/).

state institution as a non-profit organization and commercial organizations engaged in entrepreneurial activity. The main differences between a state autonomous institution and a commercial organization are presented in Table 1.

Table 1. Differences between a state autonomous institution and a commercial organization.

Comparison criterion	State autonomous institution	Commercial organization engaged in entrepreneurial activity
Purpose of creation	To carry out certain functions of the state in the social sphere, and Federal Law No. 174 does not provide for the extraction of profit by autonomous institutions as the goal of their activities.	Extraction of profit
Financing guarantee	According to paragraph 1 of Art. 20 of the Federal Law of 03.11.2006 N 174-FZ "On Autonomous Institutions", the founder is obliged to give a task and a subsidy for its implementation, the cost of providing paid services is carried out at the expense of the institution's own funds.	None. The commercial entity bears its own liability for its obligations.
Distribution of profits between participants	No distribution of profits between participants	Profits are distributed between the participants
Legal capacity of the above types of subjects of civil transactions	Allows the named type of institution as an economic entity to conclude civil law contracts with any counterparties.	Allows the named type of institution as an economic entity to conclude civil law contracts with any counterparties.
Means of earning money	The broadest freedom in raising funds within the established framework, in comparison with a budgetary institution	Complete freedom of action
Profits	They come into its own disposal and are used by it to achieve the goals for which it was created, unless otherwise provided by law. Moreover, the owner of the property of the autonomous institution does not have the right to receive income from the autonomous institution's activities and the use of the property assigned to the autonomous institution (clauses 8, 9, article 2 of Law N 174-FZ), which also determines its special legal regime.	Become available for its own disposal
State subsidies	Allocated to carry out government or administrative assignments from the founder	Not expected (subsidies are possible within the framework of state business support programs under a number of conditions)

It should be emphasized that subsidiary support of autonomous institutions is intended solely to fulfill the task of the founder. The autonomous institution must find the resources necessary for the provision of paid services on its own at its own expense. Thus, in modern

© The Authors, published by EDP Sciences. This is an open access article distributed under the terms of the Creative Commons Attribution License 4.0 (http://creativecommons.org/licenses/by/4.0/).

conditions, the main goal of the functioning of an autonomous institution is to make a profit. [4]

At the same time, the basis for the management of autonomous institutions is a task formed by the founder, which cannot be canceled, and which is aimed at the implementation of work, the provision of services in order to exercise the powers of state authorities provided for by the legislation of the Russian Federation in such areas as: science, education, health care, culture , mass media, social protection, employment of the population, physical culture and sports. [1,5]

In view of the fact that the main activity of the institution is the provision of services, and the largest group of subjects are consumers, we will consider the process of interaction with this target audience.

As noted above, the involvement of the institution in the state and partial funding at its expense should be noted. Most of them are aimed at working with individuals on a budgetary or preferential basis. At the same time, attention should be paid to the work on the provision of services on a paid basis, which is advisable to optimize in order to increase the profit of the institution from this type of service. Considering this activity as interaction with the institution's clients, we will assess the applicability of the CRM concept from the standpoint of the transition to a customer-oriented approach.

At the same time, it is extremely important to separate CRM as a business concept (strategy) and CRM as a project for the implementation of information systems and technologies of the same class. The authors are of the opinion that these concepts are related as "general" and "particular", and it is not advisable to consider the second outside the contour of the first, since such an approach will not allow realizing the key effects of this strategy. This position can be confirmed by a number of authoritative opinions, for example, E. Payne, as one of the recognized researchers in the field of CRM efficiency [6], as well as the collected statistics on the reasons for failures in the implementation of CRM systems (Fig. 1). As you can see, more than a third of all failures (36%) are associated with failures in the development of strategic aspects of the project (setting goals and developing strategies).

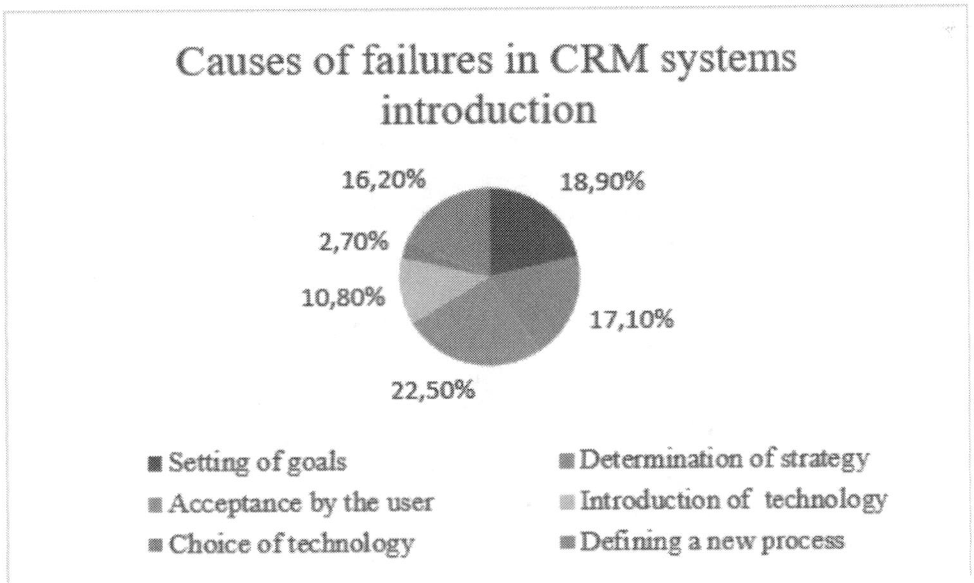

Fig. 1. Causes of failures in CRN systems introduction [7]

The concept of CRM, aka the system of internal standards and business processes for customer service, appeared in the mid-1990s. An engagement model that assumes that the customer is at the center of the entire business philosophy and that the focus is on measures

© The Authors, published by EDP Sciences. This is an open access article distributed under the terms of the Creative Commons Attribution License 4.0 (http://creativecommons.org/licenses/by/4.0/).

to support effective marketing, sales and customer service. Supporting these business goals includes collecting, storing and analyzing information about customers, suppliers, partners, as well as about the internal processes of the company.

A customer-centric approach is a strategy by which a company collects information about its customers at all stages of its life cycle (attraction, retention, loyalty), extracts knowledge from it and uses this knowledge in the interests of its business by building mutually beneficial relationships with them, based on using advanced management and information technologies.

The purpose of applying the strategy is to maximize the full value of the client throughout the entire life cycle, and as a result - to increase the company's competitiveness, since well-built relationships based on a personalized approach to each client and maintaining a balance of values (created by the organization for the client and the client for the organization), allow you to retain customers and help attract new ones.

In the CRM concept, there is a focus not on the product, but on the client with personalization of offers. Often, to increase profitability, customer companies are segmented into different categories (for example, using ABC-, RFM- and other types of analysis) and develop several different methods of serving customers depending on their belonging to a particular group. At the same time, it is extremely important to rationally provide services by tracking the value of relationships for the client and for the organization, and as a result - to maximize the full profit from the client throughout the entire life cycle.

Let us assess the possibility of applying the CRM concept in a state autonomous institution according to the key features of customer relationship management in Table 2, highlighted on the basis of an assessment of implementation in a state autonomous institution.

Table 2. Assessment of the application of the CRM concept in a state autonomous institution

Key features of the CRM concept	Assessment of implementation in a state autonomous institution
Focus on customer retention and increasing their total bottom line as part of a targeted customer retention strategy	Let's consider the state autonomous institution clients from the point of view of receiving services: - budgetary basis. the state autonomous institution is interested both in re-applying on a preferential basis at the expense of the state budget and for a commercial service (implementation of a cross-sale approach due to the complex implementation of techniques, but pay special attention to the sales matrix, the method of communicating with customers, as well as employee motivation). - commercial basis. Institutions are interested not only in attracting and retaining customers, but also in increasing the amount of sale (implementation of the up-sale principle).
Arranging relationships with a range of stakeholders that contribute to the market efficiency of the company	Let's highlight the following stakeholders and their states according to OMG Essence - Kernel and Language for Software Engineering Methods [8]. This standard provides a description of the life cycle of each stakeholder, as well as their interactions. - founder (involved) - as a rule, this is a representative of a constituent entity of the Russian Federation or a higher-level institution. Interaction is based on the receipt and control of the fulfillment of a state or administrative assignment, assessment of the effectiveness of the institution's work; - regulatory authorities (represented). It is extremely difficult to influence the loyalty of a relationship. All interaction takes place with various representatives of regulatory bodies and is formed on the basis of unilateral relations by providing accounting documents. - the customer of the service (involved, in agreement. Loyalty is formed due to satisfaction with the service provided; - state autonomous institution employees (involved, in agreement) Employee loyalty is formed by external motivation (bonuses, benefits, career growth) and internal (team atmosphere, relationships at work, personal growth).
Emphasis on cross-functional marketing approach	Currently, lower-level institutions (district/municipal) are focused on a functional approach, but there are opportunities for implementing a cross-functional (mixed) approach, as in higher-level organizations (city/federal district or state level).

© The Authors, published by EDP Sciences. This is an open access article distributed under the terms of the Creative Commons Attribution License 4.0 (http://creativecommons.org/licenses/by/4.0/).

3 Results and Discussion

Thus, the data presented in the table allows us to conclude about the applicability of the CRM concept in an autonomous institution. Regardless of the institution's involvement in the state or its independence, it is advisable to apply the CRM concept when working with clients. This will increase the loyalty of both buyers and other contact audiences, which will allow you to count on re-applying for the service.

Managing customer relationships in an autonomous institution means: implementing the procedure for attracting new customers, transforming neutral buyers into loyal customers, and making regular

customers your business partners (agents). Such a scheme is valid in any market, and only competent customer relationship management allows to work according to it. And although the main regulations usually affect the front office of the organization, namely three areas (sales, marketing and service), the degree of customer satisfaction and loyalty also largely depends on the activities of the back office: the quality of the service provided, the timing of its implementation, and the organization of mutual settlements.

Customer service schedules are a chain of small elements that are often overlooked. But it is these elements that play a key role, here are examples of some of them:

- script for answering a phone call;
- speed of reaction to the client's request;
- structure of the commercial proposal;
- loyalty program scheme;
- an offer to purchase related products, etc.

Like any project, a CRM project should begin with a survey of the current situation, it is necessary to conduct a CRM audit - an audit of the current level of customer relations. The overall strategic structure of CRM can be defined by five main processes, such as: the strategy development process; value creation process; multichannel integration process; information management process; the process of evaluating the effectiveness [4]. The procedure for applying this scheme in a state autonomous institution remains the same as for commercial enterprises, with some adjustment in terms of customer segmentation: that part of customers who are served within the framework of an assignment from the founder for the provision of services in order to exercise the powers of state bodies provided for by the legislation of the Russian Federation authorities do not have to go through the process of assessing the value received by the institution (stage of the value creation process). The process of evaluating effectiveness in terms of results for shareholders also requires adjustments.

It should be noted that the issues of adapting the strategic structure of CRM in a government agency are beyond the scope of this article.

4 Conclusions

Thus, the CRM concept is applicable to government agencies and includes an orientation towards personalized offerings.

As mentioned earlier, it is advisable to classify clients by different types of analyses and provide services in accordance with this. When implementing CRM systems, it is necessary to rely on the experience of successful projects. It is important to understand that the system is only part of the tool for attracting and retaining customers, and the main focus is on the company's strategy.

© The Authors, published by EDP Sciences. This is an open access article distributed under the terms of the Creative Commons Attribution License 4.0 (http://creativecommons.org/licenses/by/4.0/).

Acknowledgements

The article was prepared within the framework of writing a master's thesis on information systems and technologies together with the diploma supervisor.

References

1. Federal Law "On Autonomous Institutions" dated 03.11.2006 N 174-FZ
2. Federal Law "On Amendments to Certain Legislative Acts of the Russian Federation in Connection with the Improvement of the Legal Status of State (Municipal) Institutions" dated 08.05.2010 N 83-FZ
3. Federal Law "On Non-Commercial Organizations" of 12.01.1996 N 7-FZ
4. E. Polotovskaya, Ezh-Lawyer, 47 (2011)
5. BC RF Article 69.2. State (municipal) task
6. Adrian Payne. *Handbook of CRM: Achieving Excellence in Customer Management*
7. CRM Landmark, http://www.crmlandmark.com/
8. Object Management Group / Essence – Kernel and Language for Software Engineering Methods, 10 (2018)

© The Authors, published by EDP Sciences. This is an open access article distributed under the terms of the Creative Commons Attribution License 4.0 (http://creativecommons.org/licenses/by/4.0/).

SHS Web of Conferences **114**, 01003 (2021) https://doi.org/10.1051/shsconf/202111401003
NTSSCEM 2021

Automotive industry: trends in the implementation of modern corporate strategies

Oleg Zhilkin*, and *Svetlana* Kiryukhina

Peoples' friendship university of Russia, Miklukho-Maklaya str.6, 117198 Moscow, Russia

Abstract. The article attempts to identify new directions for the strategic development of automotive enterprises, taking into account current trends, using the example of both the largest German concerns, such as BMW Group and Volkswagen AG, and actively developing Chinese manufacturers - BAIC, Chery, Geely, JAC, JAV. The emphasis is on the production of hybrid and electric vehicles. Special attention is paid to the prospects for the development of the Russian segment of the production of cars with hybrid and electric drives and identifying problems that hinder its advanced development and the reasons that hinder the expansion of the segment of using cars with hybrid and electric drives, both Russian and imported

1 Introduction

Corporate strategy acts as the main tool for strategic management and represents a pre-planned reaction of the company to changes in the external environment, the trajectory of its behavior, developed to achieve the set goals. Analysis of foreign experience in the formation of corporate strategies showed that the main attention in their development is paid to four aspects of business strategy:

• selection of product markets and investment volumes;

• determination of possible volumes of production and sales of products;

• identification of assets and competencies necessary to ensure sustainable competitive advantage;

• development of functional strategies.

The main component of developing a strategy for commodity-market investment is the choice of a business area and an analysis of the dynamic processes occurring within it.

When defining investment sectors, the corporation's management primarily proceeds from several circumstances:

• out of the need to gain a foothold in those commodity markets where there is a steady growth in sales;

• from accelerating the introduction of the most promising innovative products into serial production;

• from the tasks of continuous development of their own research and development work.

To ensure a sustainable competitive advantage, an economically sound choice of strategic asset allocation is required, i.e. resources that surpass in their characteristics the resources of

*Corresponding author: zhilkin_on@pfur.ru

© The Authors, published by EDP Sciences. This is an open access article distributed under the terms of the Creative Commons Attribution License 4.0 (http://creativecommons.org/licenses/by/4.0/).

competitors, and strategic competencies - strategically important for the corporation areas of activity, the development of which is based on special knowledge or competencies. Consequently, in the process of developing a corporate strategy, it is necessary to provide for the costs of creating and maintaining strategic assets and competencies designed to provide the appropriate technical capabilities for the production of competitive products.

The final stage in the formation of a corporation's business strategy is the development of functional strategies and programs. Typically, functional strategies cover the areas of production, distribution, communications, informatization, etc.

Along with the global market strategy, a market segmentation strategy is being developed. Functional strategies are complemented by specific programs (quality, customer relations, branding, etc.).

The main approaches to the development of an organization's strategy boil down to the fact that a strategy is a combination of the results of theoretical analysis and the intuition of developers, which, first of all, should be the subjects, who then detail and implement the strategy. It is also important that a strategy can never be thought out and calculated to the end, and its adjustment as external and internal conditions change is a necessary procedure.

Based on the foregoing, it should be emphasized that there is no universal method of developing a strategy suitable for all occasions.

2 Literature review

Many scientists were involved in the formation of corporate strategies, however, the generally recognized leaders are K. Andrews (Andrews KR, 1971), M. Porter (Porter M., 2013), G. Hamel and K. Prahalad (Prahalad K.K., Hamel G., 2011).

In particular, the Association for the Development of Electric, Unmanned and Connected Transport and Infrastructure and the Analytical Agency "AUTOSTAT" are engaged in analyzing the trends of what is happening in the modern car production market.

3 Theoretical foundations for the formation of corporate strategies

Harvard Business School is considered to be the leader in the development of procedures for the formation of strategies. K. Andrews (Andrews K. R., 1971), M. Porter (Porter M., 2013), G. Hamel and K. Prahalad (Prahalad K.K., Hamel G., 2011) developed in the XX century. the main approaches to the formation of such strategies as "Economic Strategy", "Business Strategy" and "Leadership Strategy".

K. Andrews proposed a strategy based on the correspondence between the existing market opportunities and the organization's capabilities at a given level of risk. Approaches to developing business strategy based on the competitive position of the organization and the competitive strategies themselves were developed by Porter, and the concept of core competencies belongs to G. Hamel and K. Prahalad. Analysis of the model of five forces of competition makes it possible to determine the strengths and weaknesses of the organization in the market and in the areas, strategic changes in which, in accordance with the forecast, can give maximum results for business development.

The different state of the external and internal environment of organizations explains the diversity of the organizations themselves and their actual state.

The multifactorial nature of the parameters that determine the position of each organization does not allow to systematize these parameters and formalize the process of developing an image of the existing and desired state of the organization.

© The Authors, published by EDP Sciences. This is an open access article distributed under the terms of the Creative Commons Attribution License 4.0 (http://creativecommons.org/licenses/by/4.0/).

Practice shows that the best results are achieved by organizations that, when choosing goods and services, not only focus on meeting the needs of customers, but can also outstrip the needs of consumers or, even better, form these requests. For example, the need for mobile phone communication has been generated by mobile telesystem companies. There is a relationship between the amount of demand and the price of a commodity, which is called the law of demand. The following categories of demand should be considered:

• primary demand, which is understood as the minimum demand determined by the needs of the population (basic food products, consumer goods, etc.);

• secondary demand - needs of a higher order, which appear with a time lag in relation to the primary demand (delicacies, etc.);

• realized demand, which is accompanied by the actual purchase of goods;

• rush demand caused by the desire of more and more layers of consumers to make a purchase of goods purchased by other buyers;

• unsatisfied (or deferred) demand caused by the lack of the required product on sale;

• emerging demand, which is focused on new products that are being prepared for release.

However, knowing the demand does not yet guarantee success in the market. It is necessary to structure the process of transforming resources into finished products in such a way that the value of resources is maximally preserved and increased due to the synergy effect. Therefore, it is required to analyze the potential of the organization in terms of the ability to provide effective processes within the organization itself. As a result, it is not only the stage of the product life cycle that matters, but also the life cycle stage at which the organization itself is located.

An important condition for the formation of an innovative strategy is to take into account, through a formalized toolkit, the potential opportunities and prospects for the use of certain innovations in connection with the development of the internal and external environment of the enterprise. The lack of such accounting may lead in the future to technological or market inadequacy of currently relevant innovative products and services.

4 The main strategic directions for the development of automobile transport

The main strategic directions for the development of transport from a historical point of view, cars with internal combustion engines appeared later than electric cars. However, due to the lack of sufficiently developed technologies for storing electricity, the large weight of batteries, which reduced the payload of the car and high operating costs, from an economic point of view, cars with internal combustion engines turned out to be more practical.

However, in modern conditions, when it is widely said that the reserves of explored deposits of natural resources from which fuel for cars with internal combustion engines is produced is steadily decreasing, that oil and oil products have recently become an instrument in the "games" of big politics, that cars with internal combustion engines are one of the main sources of environmental pollution, and the level of development of modern technologies makes it possible to create cars on alternative traction with comparable costs, the world leaders of car manufacturers increasingly began to pay attention to the creation of cars, the movement of which is provided by electrical devices.

Fig. 1 shows the dynamics of sales of electric vehicles and plug-in hybrids in the world for 2017-2019. The data presented clearly shows that there is a stable trend of growth in demand for cars with electric and hybrid drive, and in 2018 the growth is taking on an exponential nature, which suggests that the demand for cars with electric and hybrid drive will continue to grow steadily in the future.

© The Authors, published by EDP Sciences. This is an open access article distributed under the terms of the Creative Commons Attribution License 4.0 (http://creativecommons.org/licenses/by/4.0/).

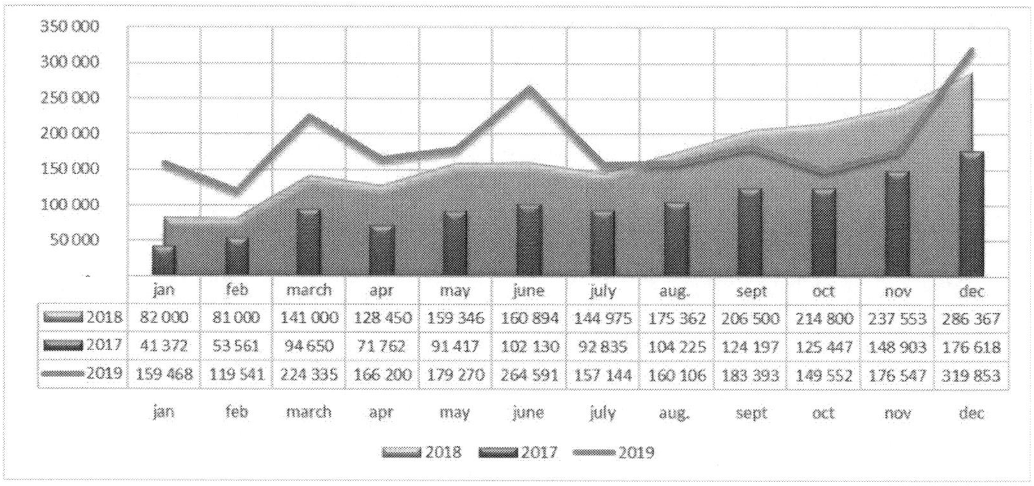

Fig. 1. Dynamics of sales of electric vehicles and plug-in hybrids in the world [5].

Figure 2 shows the comparative dynamics of sales of electric cars and cars with internal combustion engines in the world for 2012-2020

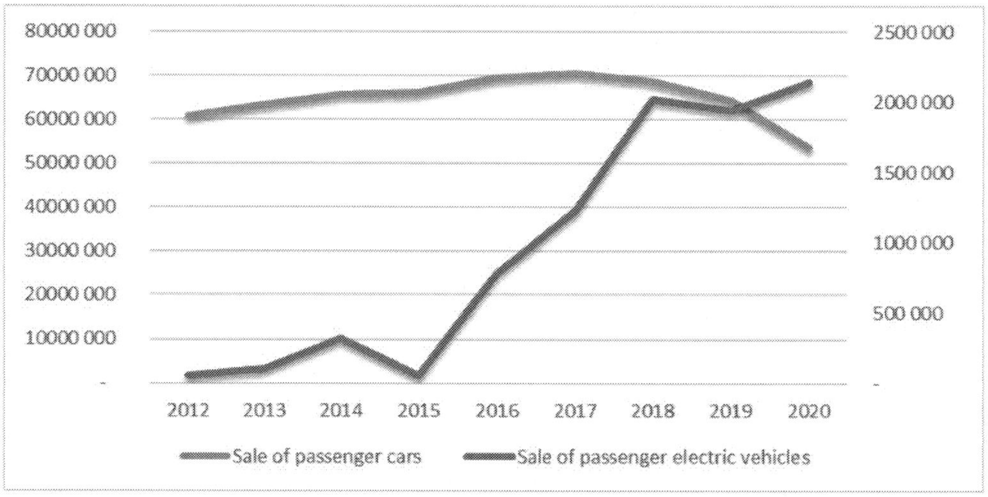

Fig. 2. Comparative dynamics of sales of passenger cars and passenger electric vehicles for 2012-2020

Analysis of data on the dynamics of sales of electric vehicles and rechargeable hybrids in the world in 2017-2019. (Table 1) shows a steady increase in demand for cars with engines that do not use refined petroleum products.

Table 1. Sales of electric vehicles and plug-in hybrids in the world in 2017-2019. (thousand pcs).

Month / year	01	02	03	04	05	06	07	08	09	10	11	12	Total
2017	41,3	53,5	94,6	71,7	91,4	10,2	92,8	10,4	12,4	12,5	14,8	17,6	12,2
2018	82,0	81,0	14,10	12,8	15,96	16,8	14,4	17,5	20,6	21,4	23,7	28,6	20,1
2019	159,4	119,5	224,3	166,2	179,2	264,5	157,1	160,1	183,3	149,5	176,5	319,8	22,6

© The Authors, published by EDP Sciences. This is an open access article distributed under the terms of the Creative Commons Attribution License 4.0 (http://creativecommons.org/licenses/by/4.0/).

Analysis of the strategies of the largest car manufacturers at the present stage has shown that almost all companies include projects for the production of non-electric and hybrid vehicles as priorities.

For example, one of the world leaders in the global automotive industry, the BMW Group, intends to become a leader in autonomous driving technology and driver assistance functions by introducing new electric and hybrid models. Sales of the company's electrified vehicles - electric vehicles and hybrids - grew immediately by 65% in 2017, reaching 103,080 units.

In July 2019, representatives of the company's top management announced ambitious goals for the electrification of their car fleet:

• in 2021 - to double sales of electric models and hybrids;

• by 2023 - two years earlier than planned - to increase to 25 electrified models on the market. More than half will be fully electric.

• by 2025, sales of electrified vehicles will grow by an average of over 30 percent per year.

There is growing interest in innovative solutions for electric vehicle technology and auto giant Volkswagen AG, which is about to change plans for the development of electric vehicle technology. This is due to the tightening of standards for carbon dioxide emissions by the EU authorities. If earlier the concern planned to invest 30 million euros in the development of technologies for electric vehicles by 2022 and in 10 years to produce 50 models of electric vehicles, now the company intends to significantly increase the production of new models of electric vehicles by 2028 by increasing their number to 70 and increasing the number assembled electric cars from 15 million to 22 million units. The concern entered the TOP-10 in terms of sales among electric car models in 2020 (Table 2).

Table 2. Top sellers among electric car models in 2020 (in thousands of rounded units)

Tesla Model 3	365
Wuling HongCuang Mini EV	119
Renault Zoe	100
Tesla Model Y	80
Hyundai Kona Electric	65
Volksvagen ID.3	57
Nissan Leaf	56
SAIC Baojun E-Series	48
Audi e-Tron	48
Great Wall ORA R1/ Black Cat	47

From the data presented, it can be seen that the Tesla Model 3 model takes a confident leading position. In 2017, she was ranked 106th in the ranking.

Particular attention should be paid to the strategic programs of car manufacturers in the Asian region. For example, China is the largest car market in the world. New energy vehicle sales have continued to grow at a rapid pace over the years. The country was in third place among the countries with the largest share of electric vehicles in the global transport market, its share was about 1%. The leading positions here were occupied by such companies as BAIC, Chery, Geely, JAC. But according to the results of the "pre-crisis" 2019 and 2020, the year of the pandemic, one can notice a significant decline in sales.

Fig. 3 shows the dynamics of sales of electric vehicles and plug-in hybrids of 14 Chinese companies in 2018-2020, which were leaders in this area in 2018.

© The Authors, published by EDP Sciences. This is an open access article distributed under the terms of the Creative Commons Attribution License 4.0 (http://creativecommons.org/licenses/by/4.0/).

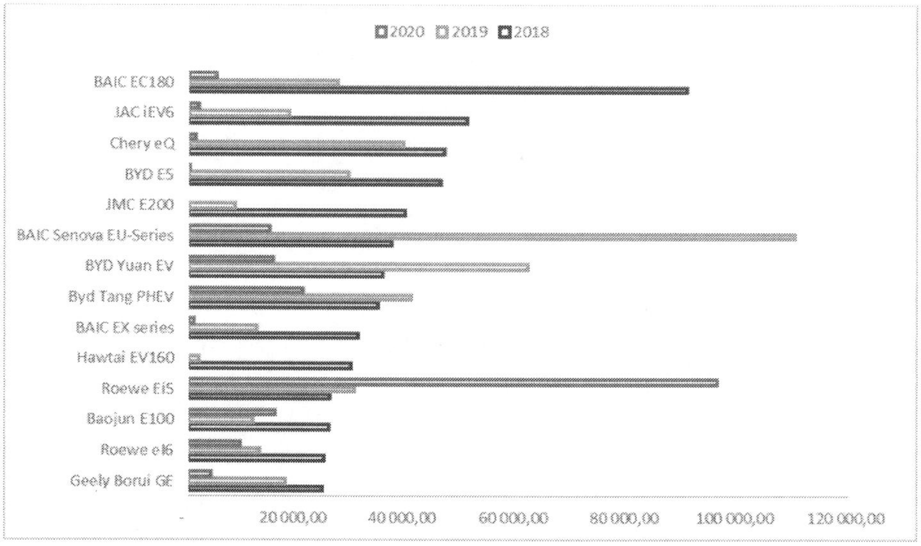

Fig. 3. Comparative dynamics of sales of passenger cars and passenger electric vehicles for 2012-2020

Norway is currently the world leader in the consumption of electrically driven vehicles (about 40%). The vice-leader in the use of electrically driven vehicles is Sweden with an indicator of more than 5%.

The past tough year of the COVID-19 pandemic has proven extremely successful for the electric vehicle market.

EV-Volumes has released preliminary sales figures for BEV (battery electric vehicles) + PHEV (Plug-in hybrid electric vehicles), which are classified as plug-in.

If in the "pre-crisis" 2019, sales of electric vehicles practically did not increase, and according to some estimates, there was even a decrease in volumes in key markets - in the PRC and the USA, then in 2020, despite all the quarantines, an impressive increase in sales of electric vehicles was recorded.

Thus, at the end of 2020, BEV + PHEV sales grew by 43% compared to 2019, and their share in global sales exceeded 4%.

The growth of the electric vehicle market in Europe for the year was 137% (!), and sales of BEV + PHEV reached 1.4 million units and this is in the context of the crisis in the automotive market as a whole, where sales fell by 20%. For the first time since 2015, sales of electric vehicles in Europe exceeded sales in China.

The share of electric cars in European sales reached 10.2%, while in China the share of NEV was 5.5% (5.1% in 2019).

Germany, where not so long ago electric cars were not popular, came to the second place after China in terms of sales.

5 Russian electric vehicle market: state and problems

Russian electric vehicle market: state and problems Russia is one of the largest players in the global car consumption market, but, unfortunately, has recently lost its position among manufacturers.

The first serial Russian car on electric traction was produced by AvtoVAZ - Lada Ellada. 10 million euros were invested in the development of the model. In this case, the cost of the production model is 1,250,000 rubles. In addition, AvtoVAZ has developed and launched another electric model - Vesta EV.

© The Authors, published by EDP Sciences. This is an open access article distributed under the terms of the Creative Commons Attribution License 4.0 (http://creativecommons.org/licenses/by/4.0/).

One of the largest manufacturers of commercial vehicles in Russia, GAZ Group, produces a serial Russian electric car - Gazelle Next Electro.

In general, electric modes of transport in the Russian Federation have good prospects. In 2020, 687 electric vehicles were sold in the country in absolute terms, compared with 353 in 2019.

In the regional rating of new electric vehicles, Moscow is the leader (240 units) for 2020, followed by St. Petersburg (69 units) and the third position is the Moscow region (60 units). The Krasnodar Territory is the leader among non-capital regions (36 units), followed by the Primorsky Territory (31 units), the Sverdlovsk Region (21 units), and the Novosibirsk Region (20 units). More than 10 (pcs.) Electric cars were purchased in 4 constituent entities of the Russian Federation: Perm Territory, Irkutsk, Chelyabinsk and Tyumen regions.

Today, the main limiting factor is the price of electric vehicles, which is still too high. In addition, it should be noted that an insufficiently developed network of electric filling stations plays a role in the development of electric transport.

Namely, it depends on the rate of development of this infrastructure when Russia will switch to electric vehicles.

For comparison, in the United States there are now more than 50 000 charging stations, in Amsterdam there are about 20 000 charging stations. While in Russia, according to the plugshare.com service, which marks all charging points for electric vehicles, today there are less than 1 000. According to the forecasts of the Association for the Development of Electric, Unmanned and Connected Transport and Infrastructure (AETI), in the coming years the number of charging stations in Russia should increase two to three times. Russian developers are actively installing charging stations for electric vehicles in the construction of housing and shopping and entertainment centers.

Taking into account the fact that the number of electric cars in the country is growing steadily, there is an urgent need to create a regulatory document that would oblige construction and management companies to ensure the availability of electric charging stations in the yards of apartment buildings, shopping centers, hotels, etc.

So, for example, in St. Petersburg on January 1, 2019, a local regulatory act began to operate, obliging developers to install charging stations in new residential complexes.

If we talk about what the Russian fleet of electric vehicles is, then according to the analytical agency AUTOSTAT, 10 836 electric cars were registered in the Russian fleet of passenger cars as of January 1, 2021. At the beginning of the year, the electric car market is represented in our country by 18 models of 14 different brands.

About 80% of all electric cars registered on the territory of the Russian Federation belong to one model - Nissan Leaf, which has a fleet of more than 9 thousand units. The second place in the rating is taken by Tesla (models 3, S, X, Y - about 700 units in total). This is followed by the Mitsubishi i-MiEV (400 units).

6 Conclusions

Based on the study, it was revealed that the production of electric and hybrid vehicles, after many years of neglect, has now become one of the most important areas of strategic planning both at the global and Russian levels in the automotive industry.

It was revealed that all the theoretical positions of K. Andrews, M. Porter, G. Hamel, K. Prahalad on obtaining competitive advantages in the development of long-term strategies in the automotive industry have the right to be directly applied. As an example, the main strategic directions of development of both leading Western European, Chinese, and Russian companies - car manufacturers are considered. A stable growth in demand for cars that do not use internal combustion engines for both personal and commercial use has been revealed.

© The Authors, published by EDP Sciences. This is an open access article distributed under the terms of the Creative Commons Attribution License 4.0 (http://creativecommons.org/licenses/by/4.0/).

In the course of the study, bottlenecks were identified that hinder the spread of electric vehicles, in particular, in the territory of the Russian Federation, and recommendations were given to solve this problem.

References

1. V. V. Iosifov, E. E. Bobylyov, Financial Analytics: Problems and solutions, **10-11,** 1273 (2017)
2. A. V. Pas`ko, E-Management, **2,** 16 (2019)
3. D. A. Plexanov, S. M. Zaverskij, N. M. Churkina, Russian Entrepreneurship, **16(24),** 4375 (2015)
4. M. Porter, *Competitive strategy*, 301 (M.: Al`pina Biznes-buk, 2013)
5. K. K. Praxalad, G. Xe`mel, *Core competencies of the Corporatio*n, 117 (SPb: Piter, 2011)
6. S. V. Ratner, S. S. Maslova, Finance and Credit, **23(22),** 1281 (2017)
7. Boucar Diouf, Ramchandra Pode, Financial Analytics: Problems and solutions, **10-12,** 1380 (2017)
8. T. S. Remizova, Financial Analytics: Problems and solutions, **10-12,** 1380 (2017)
9. K. O. Tarxanovskij, Accounting in budget and non-profit organizations, **17,** 34 (2017)
10. E. D. Chirkina, Management of electronic economic systems, **8,** 21 (2017)
11. K. E`ndryus, *Corporate strategy Concept*, 252 (2010)

© The Authors, published by EDP Sciences. This is an open access article distributed under the terms of the Creative Commons Attribution License 4.0 (http://creativecommons.org/licenses/by/4.0/).

Belarusian case study of P2P lending market digitalization: state-of-the-art, needs and perspectives

Joanna Koczar[1], *Yury* Karaleu[2,*], and *Aliaksandr* Dudkin[2, 3]

[1]Wroclaw University of Economics and Business, 53-345, Komandorska Str., 118/120, Wroclaw, Poland
[2]School of Business of Belarusian State University, Oboinaya Str., 7, 220004 Minsk, Belarus
[3]JSC BSB Bank, Pobediteley ave., 23/4, 220004 Minsk, Belarus

Abstract. The paper investigates the present state and prospects of development of the Belarusian market of mutual lending, the business model and the results of the activity of KUBYSHKA crowdlending Internet platform, the possibilities of organizing cooperation between Belarusian banks and crowdlending platforms, obstacles to the development of the market for mutual lending.

1 Introduction

Since the early 2000s and with the development of information technology, the increasing level of globalisation, the anonymous nature of the flow of money, a new innovative element of the money market (in the non-banking sector) had emerged – the market for mutual lending or crowdlending. The uniqueness of the crowdlending or marketplace lending (in contrast to other segments of the non-banking sector of the financial market) is the absence of a traditional financial intermediary (primarily a bank) in the process of transferring money from an individual (lender or investor) to the borrower. The latter could be:

- legal entities (basically SMEs) – also called person-to-business lending (P2B lending) – business lending by a private person;
- another individual – also called peer-to-peer lending (P2P lending) – lending by one private person to another [1, 2].

The crowdlending world market is basically presented by the Asian market (84% with a turnover of USD420 billion in 2018), the American market (12% market share with a turnover of USD60 billion) and the European market (3% market share with a turnover of USD15 billion). The remaining balance of 1% is split between Africa and Australia [3].

The P2P lending market, which is the object of this survey, is not fundamentally a new stream but it has indeed recently become relatively cheap and easy to find sufficient investors to finance the amount requested by the borrower through the virtual crowdlending platforms. After a loan agreement is concluded between parties (often via the platform), the crowdfunding platform transfers the invested loan amount from the investor to the borrower. In return, the borrower has to repay the loan amount to the lender over a predefined period,

*Corresponding author: yykorolev@sbmt.by

© The Authors, published by EDP Sciences. This is an open access article distributed under the terms of the Creative Commons Attribution License 4.0 (http://creativecommons.org/licenses/by/4.0/).

plus interests. So, credit check, rating, brokering, processing, operation, etc. are handled by the crowdlending platform. Even any payment delay or borrower's inability to pay are usually handled by the platform, without the investors themselves having to intervene, in exchange for a fee (mostly paid by the borrower) [3, 4]. So, crowdlending platforms "epitomise the idea of disintermediated 'middle-man-free' access to finance, especially insofar as platforms are conceived as technological means that allow borrowers to directly access available funding on the one hand, and lenders to invest in specific loans on the other" [5].

The first-ever worldwide launched crowdlending platform named ZOPA (Zone of Possible Agreement) was founded in the UK in 2005. It still remains one of the largest and most authoritative in the world. Over the past 15 years, half a million borrowers have been loaned through ZOPA for a total amount of over GBP5 billion; investors of the site (lenders) received interest income in the amount of more than GBP250 million [6].

In 2006, similar crowdlending platforms PROSPER and LENDING CLUB appeared in the United States. Since that moment, the practice of P2P lending has become the most widespread in the world, both in high-income economies such as UK, USA, Sweden, Australia, Canada, China as well as in emerging economies such as India and Brazil.

Since 2012, similar services have become and appear in Russia. ZAYMIGO, LOANBERRY and Вдолг.ру are the most prominent today [7, 8].

Summing up the aforesaid, we consider that interaction between individuals on the mutual landing based on the global Internet, remotely, online, is an innovative digital technology in the field of finance (fintech project), fully correlates with the digital transformation trend in all sectors of the global economy.

2 Belarusian market of mutual landing

Belarusian financial market includes the banking system (the main and largest market participant), as well as the non-banking sector (leasing companies, forex companies, insurance organizations, professional participants in the securities market, microfinance organizations, etc.). The National Bank of the Republic of Belarus (National Bank) and the Belarusian Government strongly encourage the development and competition between various participants in the financial market. The intensification of competition and even the preferential development of the non-banking sector were declared in the Strategy for the Development of the Financial Market of the Republic of Belarus until 2020, approved by a joint resolution of the Council of Ministers and the National Bank of March 28, 2017, No. 229/6, as one of the strategic objectives.

Despite the declared goals, the Belarusian market for mutual lending is still in its infancy. Until now, the issues of granting loans between individuals in the Belarusian legislation has not been properly regulated [9]. The work on the preparation of the draft Decree regulating the activities of such services in the Republic of Belarus, initiated by the National Bank back in 2017, has actually been suspended today. At the same time, in the report of the First Deputy Chairman of the Board of the National Bank Sergei Kalechits it is proposed to pay special attention to the development of online borrowing services [9].

3 Business model and operational results of KUBYSHKA crowdlending Internet platform

Today, the Belarusian market of mutual lending is represented by a single player – professional intermediary KUBYSHKA crowdlending Internet platform, founded by Financial and Analytical Bureau LLC (Holding Company).

© The Authors, published by EDP Sciences. This is an open access article distributed under the terms of the Creative Commons Attribution License 4.0 (http://creativecommons.org/licenses/by/4.0/).

The activity of KUBYSHKA crowdlending Internet platform was launched in December 2016. The National Bank, as the main regulator of the national financial market, was promptly informed by the founder about the start of activities and about the business model of the platform based on online Internet resource www.kubyshka.by [10].

Today 1,300 individuals (borrowers and lenders) are registered as users of the crowdlending Internet platform. Over the period since the beginning of the activity through the Internet site, more than 3,000 loans were issued totalling USD900 thousand (in equivalent). As shown in Fig. 1, the current portfolio of 900 loans by reference to the principal outstanding is equal to USD183 thousand.

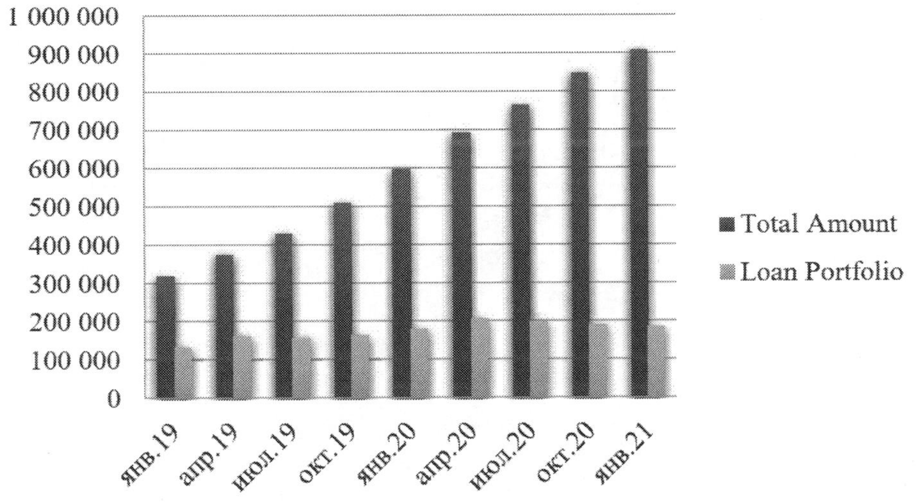

Fig. 1. The dynamic of the loan portfolio and total loan amount, USD (in equivalent).

According to the business model of KUBYSHKA crowdlending Internet platform, the registration of borrowers is carried out after they have signed, in the presence of a representative of the company, a loan adhesion agreement, a user agreement and when the individual's consent to process his/her personal data from the Credit Register is obtained. Registration of lenders is carried out remotely.

Despite the fact that the Holding Company is not a party to the treaty and is not responsible for any financial risks of non-repayment of debts, it assumed the responsibility for assessing the solvency of borrowers and guarantors (if any), assigning a credit rating (basic (from 0.0 to 5.0) or advanced rate (from 0.0 to 10.0)) and setting a personal credit limit. As practice has shown, most borrowers undergo an initial assessment of solvency, which does not require supporting documents and certificates, as well as the presence of guarantors, which, as a result, ensures that borrowers receive a basic credit rating. The average ratings of borrowers entering into loan deal in local (BYN) and foreign currency are 3.2 and 3.4, respectively (Fig. 2).

© The Authors, published by EDP Sciences. This is an open access article distributed under the terms of the Creative Commons Attribution License 4.0 (http://creativecommons.org/licenses/by/4.0/).

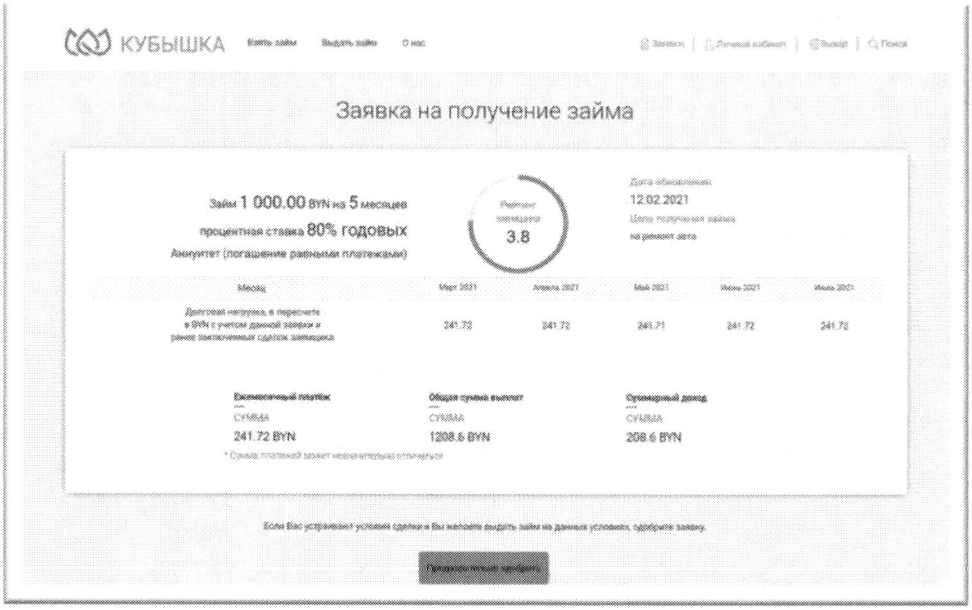

Fig. 2. The interface of loan application in KUBYSHKA crowdlending Internet platform with the attributed credit rating (personal data of the borrower are not presented).

The specific feature of KUBYSHKA crowdlending Internet platform is the assignment of the borrower to set the loan terms and conditions (such as the size of the loan, the loan terms, interest rate, the loan currency) when presenting loan applications to the crowdlending platform. If the lender accepts the submitted loan application, he transfers the money to the borrower. By doing so, the lender joins the loan adhesion agreement and since that moment the loan deal is considered concluded.

The procedure (schedule) for the repayment of loan debt and accrued interests are established automatically when concluding a loan deal using the annuity method.

Settlements between parties for issuing and repaying loans and paying interests are carried out remotely making instant card-to-card money transfers between parties using the "virtual terminal" of the partner bank (OJSC Belgazprombank), which is implemented into KUBYSHKA crowdlending Internet platform.

4 Preconditions for the development of the market for mutual lending

The political and socio-economic uncertainty in Belarus, enhanced since the second half of 2020, has become the main reason for the large-scale outflow of deposits of individuals from the banking system. As shown in the Analytical Review of the National Bank "Survey on the implementation of the banking principle "Know your customer" (KYC approach) in relation to attracting term deposits of individuals, in 2020 deposits in foreign currency decreased from USD6.3 billion to USD4.7 billion or more than 25%; deposits in BYN – from BYN5.2 billion to BYN4.6 billion or by 11.6% [11] (Fig. 3).

© The Authors, published by EDP Sciences. This is an open access article distributed under the terms of the Creative Commons Attribution License 4.0 (http://creativecommons.org/licenses/by/4.0/).

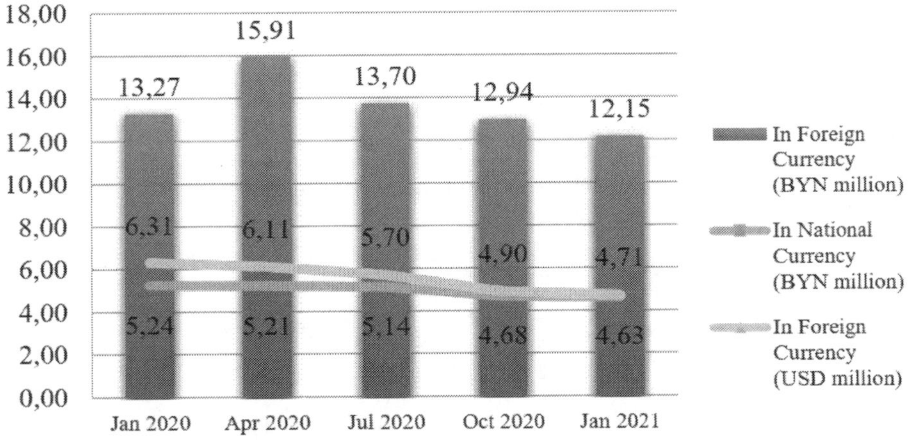

Fig. 3. Dynamic of term deposits in Belarusian banks in 2020.

This, in turn, has led to unstable capital markets, a significant decrease in the liquidity of the banking sector and a reduction of available borrowings for individuals (or caused the total suspension of lending) with a conjoined growth of the borrowing value [12] (Fig 4).

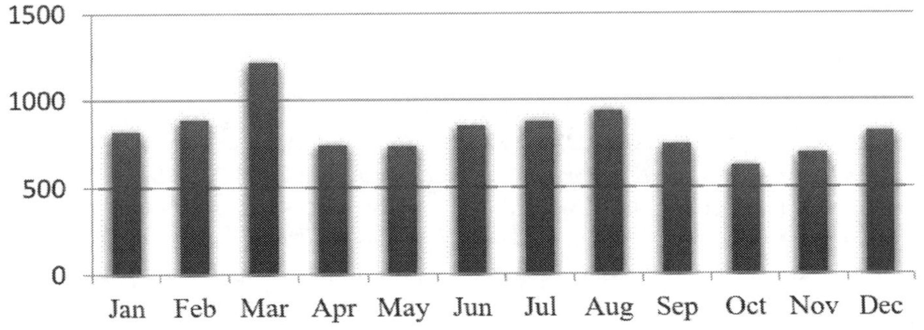

Fig. 4. Dynamic of bank lending to individuals in Belarus in 2020, BYN million.

Such a large-scale decrease in the volume of lending to individuals by banks stimulates individuals to seek alternative sources of financing for their own current needs. Only an insignificant part of the banking customers became clients of KUBYSHKA crowdlending Internet platform and a significant part of them apparently turned to participants of the shadow banking system, qualitative and quantitative evaluation of which is extremely difficult in Belarus due to a lack of information required for monitoring.

At the same time, banking customers with limited access to bank resources, who became users of KUBYSHKA crowdlending Internet platform, as well as previously registered borrowers, fully demonstrated an urgent need for credit resources that lead to rising interest rates for new loan applications.

Fig. 5 below shows the dynamic of interest rates for new loan applications over the past two years. The weighted average interest rates reached a maximum by January 2021: 243% per annum – for transactions in BYN and 97% per annum – for transactions in foreign currency.

© The Authors, published by EDP Sciences. This is an open access article distributed under the terms of the Creative Commons Attribution License 4.0 (http://creativecommons.org/licenses/by/4.0/).

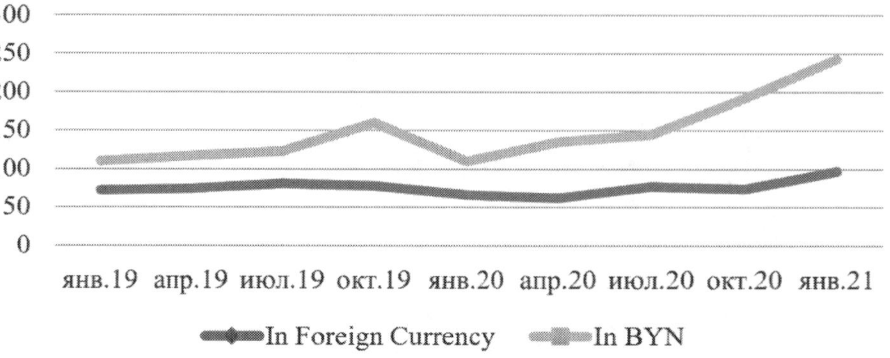

Fig. 5. The dynamic of interest rates for new loan applications over the past two years.

The described situation would not have been possible in a balanced market environment, providing free access of individuals with a higher credit rating to credit resources through crowdlending Internet platforms with a reasonable personal credit limit and sensible interest rates.

On the other hand, such a rise in interest rates in the absence of obvious factors of a sharp deterioration of borrowers' solvency (proved by rational level of resulting risk-reward ratio) is quite attractive for another group of KUBYSHKA crowdlending Internet platform – lenders. The profitability of lenders' investments is much higher than the interest rates on bank deposits and investments in securities on the Belarussian stock market. Simultaneously, lending of investments on crowdlending platforms is less than the risk of operations in the Forex market and, most importantly, it is controlled and managed by the lender himself.

The attractiveness of such a model of personal investments is illustrated by the statistics of overdue and toxic debts. As of February 1, 2021, according to KUBYSHKA crowdlending Internet platform data, more than 81% of the funds invested by creditors were repaid in full amount including interest payments, and in some cases – with penalties and fines. The remainder of the current portfolio is classified as standard loans – 7.8%, loans with insignificant (often technical) overdue – 1.8%, toxic loans with more than 90 days overdue 9.4% [10].

In order to reduce the risk of non-repayment of issued loans, the Holding Company has assumed the responsibility of assisting creditors in the event of defaults or delay. In particular, the proceedings of disputes between borrowers and lenders are carried out in the Arbitration Court, established at the Financial and Analytical Bureau LLC and registered in the manner established by the legislation.

Despite the attractiveness of the model of mutual lending for both borrowers and lenders, as described earlier, a significant part of the personal funds is withdrawn from financial turnover and transferred to the unregulated shadow borrowing market and criminalization.

5 Needs and Perspectives of the development of Belarusian P2P lending market

The KUBYSHKA crowdlending Internet platform extensive operation experience in the real Belarusian economic environment has revealed a number of problems that demand the immediate solution for the further development of P2P technology. For example, the cooperation between banks and crowdlending platforms, in addition to the infancy of the market for mutual lending and the absence of a large number of participants, is seriously hampered by the fears of banks related to their joined cross-services. Some experts raised

© The Authors, published by EDP Sciences. This is an open access article distributed under the terms of the Creative Commons Attribution License 4.0 (http://creativecommons.org/licenses/by/4.0/).

concerns about ambiguous interpretation by the National Bank of the Republic of Belarus of the possibility and legality of providing cross-services by banks together with crowdlending platforms. In our opinion, such fears are not justified, because, in a similar regime, Belarusian banks have been selling services of insurance companies at their points of sale for several years. This kind of cross-sales does not raise doubts because it is an established practice that has never been questioned. In addition, international experience demonstrates only the positive effects of the proposed collaboration.

Among other urgent problems in the development of P2P technologies, the following can be distinguished:

- Organization of monitoring and control of crowdlending platforms as professional financial intermediaries;
- A legal regulation of lending deals between individuals;
- The implementation of personal bankruptcy;
- A legal regulation of the activity of collection agencies;
- The inclusion of the history of borrowings on crowdlending platforms in the credit history of borrowers;
- The development of risk insurance of non-recovery funds, etc. [1].

For example, to ensure enforcement in the event of borrower's default, it seems appropriate to amend Law On economic insolvency (bankruptcy) dated July 13, 2012, No. 415-3 in the part concerning the introduction of the institution of personal bankruptcy. A similar institute exists in many countries of the world: in the USA, Germany, England, Sweden, Denmark, Spain, and not so long ago it appeared in the Russian Federation [13].

To qualify the issues of granting loans between individuals as an entrepreneurial activity, it is necessary to clarify the legality of actions to provide no more than two loans by an individual without registration as an entrepreneur, which, in our opinion, can be regarded in the situation under consideration as a definition of the concept of 'on a regular basis' [9].

Each of the highlighted problems is a range of measures and activities, new legal mechanisms that deserve separate consideration and are just beyond the scope of this article.

6 Conclusions

According to the crowdfunding-platforms.com, the European crowdlending market had in 2013-2018 yearly growth of +70% on average. Even despite the expected slowdown of its growth over the next years because of tightening of monetary and credit conditions in response to COVID-19 and increased inflationary pressures, specialists consider the value of the European crowdlending market equal to USD38 billion in 2023 representing a forecasted yearly growth of +16% on average [3].

Taking into account such prospects for the development of the P2P crowdlending market, an incentive for its development at the national level could be the development and enactment of a new special legal act regulating the activities of crowdlending platforms (services) as a professional intermediary, as well as amending the Law On economic insolvency (bankruptcy) and a number of other regulatory legal acts. The creation of a full-fledged regulatory legal framework will fully legalize mutual lending, ensure proper state control over this market, and increase budget revenues by including the corresponding income of individuals in the taxable base.

Integration of Belarusian banking system with crowdlending Internet platforms and collaboration between them will allow maintaining and increasing the client base, more effectively using the intellectual potential of the staff and the existing territorial network of points of sale. All this will ensure high quality, speed and convenience in managing the personal finances of individuals, and will significantly increase the availability of modern financial services for them.

© The Authors, published by EDP Sciences. This is an open access article distributed under the terms of the Creative Commons Attribution License 4.0 (http://creativecommons.org/licenses/by/4.0/).

Acknowledgements

The article has been prepared as a part of the project financed by the Ministry of Science and Higher Education in Poland under the programme „Regional Initiative of Excellence" 2019–2022 project number 015/RID/2018/19.

References

1. Y. Y. Karaleu, A. B. Dudkin, *Finance architecture: new solutions in the digital economy*, 272 (2019)
2. V. A. Kuznetsov, Money and credit, **1**, 65 (2017)
3. Crowdlending Guide: What is it and how to invest? https://crowdfunding-platforms.com/
4. A. B. Dudkin, Business. Innovation. Economy, **1**, 52 (2017)
5. V. Bavoso, J Bank Regul, 21, 395 (2020)
6. ZOPA, https://www.zopa.com
7. K. Treskova, P2P lending: what is it, https://brobank.ru/
8. A. B. Dudkin, Legal world, **6**, 78 (2017)
9. D. L. Kalechits. Ensuring financial stability in 2020 and targets for 2021, https://www.nbrb.by/
10. KUBYSHKA, https://kubyshka.by/
11. Analytical review of the National Bank of the Republic of Belarus "Survey on the implementation of the banking principle "Know your customer" (in relation to attracting time deposits (deposits) of individuals), https://www.nbrb.by/
12. Statistical Bulletin of the National Bank of the Republic of Belarus, https://www.nbrb.by/
13. Y. Y. Karaleu, *Innovative development of the economy: entrepreneurship, education, science* (2017)

© The Authors, published by EDP Sciences. This is an open access article distributed under the terms of the Creative Commons Attribution License 4.0 (http://creativecommons.org/licenses/by/4.0/).

Changes in the role of digital services in retail in the context of a crisis

Natalia Usova*, *Mikhail* Loginov, and *Andrei* Drevalev

Ural State University of Economics, 8 Marta Str., 62, 620990 Ekaterinburg, Russia

Abstract. The article studies the transformation of the role of digital services in the retail trade in the context of the economic crisis that started in 2020. The subject of the research is digitalization of retail trade services. The study aims to identify the trends in the retail digital services development in the context of the economic crisis as well as to identify the priorities of their development in the near future. Research methods used in the scope of the study include the following ones: retrospective analysis, comparison, grouping, and synthesis.
The main results of the study:
1. The authors explore specific features of digitalization of services in the retail trade in the national market.
2. The study identifies the market development trends of the retail digital services in the context of a crisis.
3. The authors propose measures aiming digital transformation of the national market of retail trade services in the face of the current situation in the national economy.
The obtained results constitute the scientific novelty of the study as the authors identify the market trends and propose specific activities to ensure development of digital services in retail in the context of a crisis in the national economy.

1 Introduction

There are two segments that can be identified in the national economy of the Russian Federation, which, on the one hand, are complementary to each other, and on the other hand, can be treated as competitors. Namely, there is the traditional economy, referred to as the "analog" economy, as opposed to the digital economy which represents an innovative vector of economic development and increasing efficiency of the national economy.

The economic crisis that began in 2020 had a significant impact on all industries, markets and forms of business activities, retail trade being among them.

The economic context of 2020 became a catalyst for the introduction and development of digital technologies in various business activities. Retail trade, often associated with the traditional analog economy, is not an exception, on the contrary, it has become one of the major areas of digital transformation.

* Corresponding author: nata-ekb-777@yandex.ru

© The Authors, published by EDP Sciences. This is an open access article distributed under the terms of the Creative Commons Attribution License 4.0 (http://creativecommons.org/licenses/by/4.0/).

The study aims to explore theoretical aspects of digital services in retail and identify priorities for their development in the context of the economic crisis.

The aim has determined the tasks solved in the article, including the following ones: studying specific features of digitalization of services in the national retail market, identifying the trends in the development of the digital services in the retail market in the context of the crisis in order to propose activities and measures facilitating the digital transformation of the national market of retail services taking into account the current situation in the national economy.

2 Methods and implications

The development of digital technologies in the national economy belongs to the priorities among the key activities of the state authorities. Moreover, the importance and the need for digital services in retail has increased during the coronavirus pandemic and the subsequent economic crisis. On the one hand, the economic crisis has had a negative impact on the retail segment and the services it provides. It resulted in a decrease in the number of retailers and diminished economic efficiency of their activities in the national market. However, at the same time, the business model of existing retailers has changed and evolved. There is a clear trend towards digitalization of activities, which allows companies to cut down the costs in comparison to traditional retail outlets.

In the scope of the study, the authors relied on the methods of retrospective analysis, comparison, grouping, and synthesis.

The research is based on the results of the previous studies by the authors and Nedorostkova E. [1-3] as well as approaches of the following researchers: Perepechaeva D.M., Matosyan V.A. and Arslanov R. [4], Bostoganashvili E.R. [5], Kochieva A.K. and Dalakova A.N. [6], Nikolaeva M.A. and Kamolova D.S. [7], Pilé Ya.E. [8], Shapovalova A.R. and Ralyk D.V. [9].

A lack of the scientific analysis should be noted as a result of the literature review. The development of digital services in the retail trade in the context of the economic crisis has not yet obtained the consideration that this issue deserves, which became a starting point and the basis for the research conducted by the authors.

3 Results and discussion

Most companies have faced a lack of financial resources needed to pay salary and obligatory payments, a decrease in demand for goods and services and a corresponding downturn in the company's financial performance, as well as an increase in accounts receivable.

The economic crisis was caused by non-economic factors in 2020. It led to a number of consequences.

Firstly, a decrease in the revenues of enterprises and companies, as well as bankruptcy of organizations in various sectors due to the drop in demand and the lack of financial resources for business activities. The share of bankruptcies of small businesses in various segments of the service sector has reached 35%.

Secondly, layoffs and release of the unused labour force due to the staff optimization and business closure led to an increase in the unemployment rate with a subsequent decrease in the purchasing power of the population.

Thirdly, the state authorities recognized the need to implement additional measures to support enterprises and organizations in various sectors and industries in order to preserve the integrity of the economic structure at the national level.

© The Authors, published by EDP Sciences. This is an open access article distributed under the terms of the Creative Commons Attribution License 4.0 (http://creativecommons.org/licenses/by/4.0/).

Retailers had to reconsider the business model and implement digital tools to maintain interaction, both at the corporate level and with their target audience. Due to the changes in the external environment, 2020 became a period of active development of digital services in the retail trade.

Resulting from the anti-pandemic restrictions, the population had to start using and master digital technologies, including choosing services online, ordering and paying for the orders while staying at home. Most likely, many will not return to the previous pre-pandemic consumption patterns. The consumers benefit from getting the opportunity to choose services at any time and with the use of any device with the Internet access. For example, a company in the foodservice industry not having online presence and not offering delivery services have faced sharp decrease in the level of its competitiveness.

According to the Central Bank of the Russian Federation, the decrease in the volume of paid services in April and May 2020 was more significant than that in the retail trade turnover: -37.9% in April and -39.5% in May YoY. Operational indicators of consumer activity indicate that in the context of partial reopening of shops, cafes and service providers, the decline in consumer activity slowed significantly in June 2020 YoY [10].

The Russian Association for Electronic Communications in cooperation with the National Research University "Higher School of Economics" (HSE University) and Microsoft have carried out a research of the impact of the COVID-19 pandemic on the development of national business [11]. According to the results of the study, the most popular tools to be in demand in the near future include "tools for online communications" (44% of respondents) and "tools for automated communication with clients" (9%). The contribution of digital channels to total revenues will also increase by the end of 2021, the context of the events of spring 2020 being the major catalyst for the digital transformation.

It should be noted that digital services of retail trade used to be considered previously as a supplementary channel and format of interaction with the target audience, but in the today's context these services have become a forefront and are in high demand among the customers.

Offering digital services in the retail sector is becoming a priority for development in a crisis because it allows companies to cut down the costs of business activities and at the same time maintain continuous interactions with the target audience.

The development of e-commerce ensures that the goods and services are available for the end consumer and territorial differences are almost neutralized regarding the availability of traditional offline services. For retailers, the transition to the Internet is especially relevant as it allows companies not only to maintain a certain sales volume, but also to increase their competitiveness. The market success of a retailer is no longer a function of the size of the company itself and the associated economies of scale, as the costs of opening and managing a digital retailer are significantly lower. The changes are an incentive for the rapid development of retail segment of the national market of digital services.

Apart from that, in the context of the pandemic, the companies have faced the impact not only of the mobile presence, but also that of a mobile application, social media pages as well as the official website or an online store – the role of these channels has significantly increased. Many organizations have maintained their competitiveness mainly due to the implementation of online communication policies.

Before the pandemic, a buyer had the opportunity to choose the way and the channel of making a purchase of a good or a service, but in the context of a lockdown situation has changed dramatically. The closure of stores and the transition of interactions to the Internet led to the sharp decrease in the profitability of business activities. There had been a positive trend in the development of digital services in retail before the pandemic of 2020, however it couldn't offset the negative impact of the lockdowns which resulted in strong negative trends. In particular, according to the Federal State Statistics Service, the retail trade

© The Authors, published by EDP Sciences. This is an open access article distributed under the terms of the Creative Commons Attribution License 4.0 (http://creativecommons.org/licenses/by/4.0/).

turnover in Russia in May 2020 fell by 19.2% YoY [12]. At the same time, compared to April 2020, the retail trade turnover increased by 6.7% in May 2020. Trade in non-food products in May 2020 decreased by 29.2% YoY, but it increased by 12.3% compared to April same year.

A number of factors can be identified that slow down the development of digital services in the retail trade sector.

Firstly, the inability to touch the product, which a consumer potentially considers to purchase. For some of the potential clients, this drawback is one of the important ones.

Secondly, there are doubts and uncertainty about the quality of the goods purchased online.

Thirdly, a potential customer may have certain fears associated with an online store regarding the contacts: it is not always clear who to contact in case of problems with the purchase and difficulties may arise in case of the need to return the goods and get a refund.

At the same time, the share of this segment is growing steadily. According to experts, if the retail market in general experienced decline in 2020, the e-commerce, in turn, have demonstrated the opposite trend.

Thus, according to the results of a meeting with representatives of the e-commerce and the market of online trade, it is estimated that the share of digital trade may have increased by up to 30% [13].

A consumer experiencing the restrictions and the effects of the pandemic, prefers to purchase goods either in a convenience store, corner shop or online. Shopping malls and large stores are not popular and relevant in the context of a pandemic. Digitalization becomes a key factor in the development of the retail trade. It allows companies to create new business models, change the behaviour and shape the needs of customers.

The following trends in the development of digital services in retail can be identified as a result of 2020 [14].

Firstly, there has been an explosive growth in online trade and the popularity of marketplaces. Moreover, the trend will be in effect in the future as well, due to favourable prices compared to traditional stores, alongside with the possibility of relatively cheap or free delivery of the purchased goods.

Secondly, the growth of e-commerce has been distinct, as evidenced by the results of numerous studies. In particular, a research on customer preferences in purchasing New Year's gifts has found out that the number of respondents who prefer digital retail services exceeds the number of supporters of traditional offline ones. In this regard, we can identify growing popularity and relevance of omnichannel retailing, which provides continuous interaction with a customer at any time and with the use various platforms while keeping the complete history of interaction.

Thirdly, the transformation of consumer habits has occurred. Before the COVID-19 pandemic, the representatives of millennials and generation Z had constituted the main target audience of digital services in the retail trade, but then other groups of the population began using digital services in 2020 as well. At the same time, the degree of involvement of different groups in purchasing and consuming digital services still varies and is not stable.

Fourthly, the structure of demand for various goods and services has changed. In particular, several categories should be mentioned as the new leaders of growth: household goods, health and beauty products (+ 151%), games (+ 129%), pet supplies (+ 115%). At the same time, the demand for clothes and accessories has decreased, which is attributes to self-imposed isolation and lockdowns.

Fifthly, the changes also affected the demand for mobile applications. For example, according to the results of 2020, the demand for mobile applications had the following structure by the share of consumers mentioned the corresponding category: food delivery 36%, social media 33%, retail shopping 33%, podcasts and music 28%, finance, video

© The Authors, published by EDP Sciences. This is an open access article distributed under the terms of the Creative Commons Attribution License 4.0 (http://creativecommons.org/licenses/by/4.0/).

streaming accounted for 27% each, games 25%, video conferencing 24%, news 22%, at-home workouts 13%, education for children 11%, education for adults 9%, stress relief and productivity 7% each, and finally 16% didn't mention anything of the above.

In addition, during the 2020 pandemic, many customers discovered non-grocery shopping online (30%) and food items online shopping (21%). Buying foodstuffs with the delivery service using mobile applications attracted 21% of the respondents [15].

In general, 53% of the users admitted that they had discovered at least one type of online shopping, which they would continue using in the future.

The further development of digital services in the retail trade and the e-commerce will be influenced by the following factors identified in Table 1 presenting authors' own elaboration based on market analysis [16].

Table 1. Drivers contributing to the growth of digital services in the retail trade.

Factor	Specific characteristic
New online shoppers	Three months of lockdowns brought at least 10 million customers into online trade, most of them will continue to buy products and order services online
Switching to teleworking	Five to seven million people will keep teleworking. Some of the remote workers will move to distant suburbs or smaller cities – where the product range available offline is limited
Isolation at home: March-May	Forced reduction of frequency of offline shopping. Increased demand for online services in March-May 2020 has created a high base effect, reducing the growth rate in 2021
Reducing purchasing power of the population	Decrease in general consumption. Effect is limited to the ostentatious / premium consumption of the middle class
Increase in the frequency of online purchases	Decrease in the number of visits to shops which will have effect for several months
E-Grocery sector growth	A sharp increase in spring 2020: expansion of geography, market power and audience of the market leaders. Increase in investments and economies of scale will facilitate further market penetration

The analysis of the factors shows that there are the prerequisites for further development of digital services in the retail trade and opportunities not only to maintain the achieved level of demand, but also to advance, although consumers have faced an economic crisis that affects the level of income and the structure of consumption.

The payment method is another important factor, which also affects the demand for digital services in retail. The trends characterizing consumer behaviour in the period prior to the crisis of 2020 are reflected in Table 2 (authors' own elaboration based on [17-19]).

Table 2. Payment methods used by the population for online orders of goods and services (as percent of the respondents aged 15 – 74 who ordered goods and services online)

No	Payment method	2015	2016	2017	2018	2019
1	Bank card	60.2	68	75.9	80	85.2
2	Cash on delivery	38.3	37	33.9	35.5	36.3
3	With an ATM or a self-service terminal	13.7	13	14.2	11	10.3
4	At the Russian Post office	13.5	13	13.6	11.8	11.9

© The Authors, published by EDP Sciences. This is an open access article distributed under the terms of the Creative Commons Attribution License 4.0 (http://creativecommons.org/licenses/by/4.0/).

Table 2. Continued

5	Mobile phone payments	7.1	8	10.2	9.9	10.6
6	Using electronic money	10.4	9	8.0	5.4	6.0
7	At a bank branch	4.9	5	5.9	4.8	4.8
8	At the organization – distributor of goods	1.8	2.4	2.6	2.4	3.4
9	At the organization - producer of goods (provider of services)	1.3	1.8	1.5	1.7	2.1

As it can be noted from the table above, payments with a bank card showed increasing popularity among the population (the indicator increased from 60.2% of respondents in 2015 to 85.2% in 2019) as well as payments with the use of a mobile phone (7.1% and 10.6% respectively). The trends signify the growing demand for digital financial services. Payment in cash on delivery is losing its popularity, although the decline is not so significant (from 38.3% in 2015 to 36.3% in 2019). The fact that the decrease in popularity of cash payments was so modest can be attributed to relatively low level of financial literacy of the population and can signify the need for involvement of the state authorities and timeliness of financial literacy programs and activities.

The demand for digital financial services substantially increased in 2020. The study on the use of electronic payments by the Russians citizens in 2020 [20] identified the following trends.

Among the services paid for online by the residents of large Russian cities mentioned the most prominent ones are mobile communications mentioned by 82% of the respondents, orders in online stores (81%) and utilities (72%). At the same time, the share of customers who pay online for mobile communications has decreased compared to the previous year: the indicator was 86% in 2019. It can be partly attributed to introduction of services bundling and family plans for which only one family member is charged a regular fee for the set of services. Apart from telecommunications services the plan can include advanced financial services or social media benefits resulting from emerging digital ecosystems.

The Russian citizens have more often paid online for the ready-made food delivery, taxi services, online gaming as well as increased the frequency of online money-transferring operations. Not surprisingly, there was a significant decrease in the share of those who pay online for transport tickets, tickets for cultural events, as well as book hotels.

The leader of online payment services in Sberbank Online used by 83% of the respondents. Yandex.Money was the runner-up with 55% of respondents using it. At the same time, Google Pay was most often used (29%) for contactless and mobile proximity payments. It should be noted that contactless payments continue to gain popularity - 53% of the Russians already pay for their online purchases with the help providers of corresponding digital financial services. At the same time internet-banking and bank cards remained the most popular solutions for internet payments in 2020: each of them was used by 90% of the respondents who pay online. Electronic money was mentioned by 78% of the respondents.

All the trends prove the growing demand for digital services, both in the retail trade and in the financial services sector.

3 Conclusions

© The Authors, published by EDP Sciences. This is an open access article distributed under the terms of the Creative Commons Attribution License 4.0 (http://creativecommons.org/licenses/by/4.0/).

The development of the market of digital services of retail trade has a number of positive aspects for the consumers in the national economy.

The focus of competition shifts from the traditional, or “analog”, business activities to the digital ones, which results in the increasing role of individual online stores and dominating positions of marketplaces. Amazon and Alibaba are vivid examples of the companies that have transformed into most popular digital platforms enjoying demand at the global level.

The development of the digital services in retailing allows companies to customize and personalize offers for consumers, and alongside with that neutralize the negative influence of location-related issues.

The rise of innovative retail start-ups is also associated with the opportunity to improve the efficiency of existing businesses by providing digital retail services. The costs of opening and managing a digital online store are significantly lower than those of the traditional outlet.

The improving competitiveness of national retailers on the one hand, is interlinked on the other hand with the improving quality of life of the population thanks to the twenty-four-hour availability of digital services provided by retailers. It should also be noted that the availability of services is not affected by the degree and level of materialization of a good or a service.

Further improvements in the sales system and the development of digital services in the retail trade will allow retailers neutralize the influence of the aspects of geography, ensure the development of multi-format marketplaces and provide possibility of choice for the consumer. The most important factors for a buyer in the traditional framework include the location of the seller, the overall service quality and the quality of specific services provided by the company. The context of the forced self-isolation and the transition of interactions to the Internet have made other factors more important: the seller’s market profile as perceived by customers, brand awareness, the quality of products, the terms of payment and delivery of purchased goods and provision of services. The sellers benefit from introducing digital services in retailing business activities as they allow companies to maintain their market share and continue interaction with the target audience by using digital technologies. On top of everything, producers and consumers can interact online with the use of various platforms and channels including online stores, marketplaces, social media, etc.

The globalization of digital technologies has already changed consumer needs and the models and patterns of consumer behaviour. In the context of the digital transformation of the national economy, these processes will be intensified and will be expressed in an increasing share of e-commerce and digital trade, as well as the shift of the focus from traditional retail outlets and activities to the digital ones.

References

1. M. Loginov, N. Usova, E. Nedorostkova, Adv. Ec. Bus. Man. R., **486**, 161 (2020)
2. M. Loginov, N. Usova, E. Nedorostkova, SHSConf, 94, 1 (2020)
3. M. Loginov, N. Usova, A. Drevalev, Sm. Inn. Sys. Tech., **227**, 1 (2021)
4. D. Perepechaeva, V. Matosyan, R. Arslanov, Ec. Bus. Th. Pr, **61**, 160 (2020)
5. E. Bostoganashvili, Ec. Yest. Tod. Tom. **8**, 155 (2018)
6. A. Kochieva, A. Dalakova, Ec. Sust. Dev. **38**, 54 (2019)
7. M. Nikolaeva, D. Kamolova, Rus. J Ret Man **4**, 81 (2017)
8. Ia. Pilé, Ec. Yest. Tod. Tom. **9**, 126 (2019)

© The Authors, published by EDP Sciences. This is an open access article distributed under the terms of the Creative Commons Attribution License 4.0 (http://creativecommons.org/licenses/by/4.0/).

9. A. Shapovalova, D. Ralyk, Sc. XXI Act. Dir. Dev. **1-2**, 114 (2019)
10. Economy **53** (Bank of Russia, Moscow, 2020) www.cbr.ru
11. New accents of digital transformation: how spring 2020 influences Russian business (RAEC, Microsoft, HSE University, Moscow, 2020) raec.ru
12. Ministry of Industry and Trade: retail trade in the Russian Federation will recover as stores open in the regions (TASS, Moscow, 2020) tass.ru
13. Denis Manturov hosted a meeting with representatives of the e-commerce market (Ministry of Industry and Trade, Moscow, 2020) minpromtorg.gov.ru
14. Online consumption in 2020: trends and forecasts (Retail.ru, 2020) www.retail.ru
15. Loyalty in self-isolation: how customers discover new online stores (Criteo, 2020) URL: www.criteo.com
16. eCommerce: growth forecast for 2020-24 (Data Insight, 2020) datainsight.ru
17. G. Abdrakhmanova, L. Gokhberg, M. Kevesh et al., *Digital Economy Indicators in the Russian Federation: 2017: Data Book* (HSE University, Moscow, 2017) www.hse.ru
18. G. Abdrakhmanova, K. Vishnevskiy, L. Gokhberg et al. *Digital Economy Indicators in the Russian Federation: 2019: Data Book* (HSE University, Moscow, 2019) www.hse.ru
19. G. Abdrakhmanova, K. Vishnevskiy, L. Gokhberg et al. *Digital Economy Indicators in the Russian Federation: 2020: Data Book* (HSE University, Moscow, 2020) www.hse.ru
20. Electronic payments of the Russian citizens in 2020 (ICT.Moscow, 2020) ict.moscow

© The Authors, published by EDP Sciences. This is an open access article distributed under the terms of the Creative Commons Attribution License 4.0 (http://creativecommons.org/licenses/by/4.0/).

Climate change scenarios as a driver for "Race to Zero" campaign

Leonid Sorokin[1,2,*]

[1]Peoples' Friendship University of Russia, Miklukcho-Maklay Str., 6, 117198 Moscow, Russia
[2]Atmospheric Physics Laboratory, LLC, Moscow, Russia

Abstract. The Race to Zero is the largest credible alliance aiming to reduce twice emissions by 2030, with the main goal of moving towards a carbon-zero economy by 2050. Zero emission technologies can reduce the greenhouse gas concentrations in Atmosphere that can help to slowdown the Global Warming but for making the Earth's climate system more stable we need implement Negative emission technologies. Negative emission technologies can significantly reduce the greenhouse gas concentrations in Atmosphere that can getting down the global average temperature to the pre-industrial level and prevent Global Warming and future Sea Level Rise. The currently observed climatic changes cannot be completely explained by the increase in the concentration of carbon dioxide in the atmosphere. Taking into account the impact of methane emissions, including the Polar Regions and the World Ocean, it will be possible to explain rapid changes in the Earth's climate. For getting the Earth's climate system stable and predictable we need to reduce greenhouse gas concentrations in the atmosphere and the global average temperature to the pre-industrial levels. If this aims cannot be achieved then Climate Change and associated with Global Warming future Sea Level Rise in the nearest future would be the most important Risk factors in the Global World and World Economy, that can provide the world massive losses and economic crisis.

1 Introduction

In the present time the greenhouse gas (GHG) emissions from humanity exceed natural ones. This dramatically affects climatic conditions and increases the number and severity of natural disasters.

The World Bank's Disaster Risk Management data [1] reveals the huge increase of economic losses from natural disaster during the past four decades: "Since 1980, more than two million people and over $3 trillion have been lost to disasters caused by natural hazards". Over this period, total annual damage has increased six times from $23 billion in the 1980s to $150 billion in the last decade.

One of the new global trends is urbanization. The concentration of the population close to the transport infrastructures as roads, railways, airports, seaports, rivers and costal lines can cause the significant problems for Disaster Risk Reduction. According to the World

*Corresponding author: sorokin-lv@rudn.ru

© The Authors, published by EDP Sciences. This is an open access article distributed under the terms of the Creative Commons Attribution License 4.0 (http://creativecommons.org/licenses/by/4.0/).

Bank's report Aftershocks: Remodeling the Past for a Resilient Future [2] by Global Facility for Disaster Reduction and Recovery (GFDRR) in the nearest future by 2050 the models show that:" population growth and rapid urbanization alone could put 1.3 billion people and $158 trillion in assets at risk to river and coastal floods". The previous World Bank's report Investing in Urban Resilience, remodeling by 2030, underline that without significant investment into making cities more resilient, natural disasters may cost cities worldwide $314 billion each year [3].

Under the threat by climate change and enormous economic losses from natural disasters the UN Secretary-General António Guterres called on all leaders with reducing greenhouse gas (GHG) emissions by 45 per cent over the next decade, and to net zero emissions by 2050.

The global campaign "Race to Zero" was launched at the UNSG's Climate Action Summit 2019 by the President of Chile, Sebastián Piñera. The Race to Zero is the largest credible alliance aiming to reduce twice emissions by 2030. The main goal of moving towards a carbon-zero economy by 2050 looks very attractive. But one can put a question about the reversibility of climate change and how the proposed measures to reduce greenhouse gas emissions could affect the Earth's future climate.

Race to Zero the global campaign is mostly based on the reduction of CO_2 emissions, which reflects maximum effort toward or beyond a fair share of the 50% global reduction in CO_2 by 2030 identified in the IPCC Special Report [4] on Global Warming of 1.5 °C. The most of Race to Zero actions focused on the regulation of oil and gas consumption and other carbon related products. Looking on the Race to Zero Lexicon we can see the carbon related definitions: Climate neutral(ity); Carbon neutral(ity); Carbon negative; Like for Like. It also contains definitions related to greenhouse gas emissions: Net zero; Absolute zero; GHG neutral(ity); Climate positive (net negative); Offsetting; Insetting; Neutralization; Compensation; GHG reductions; GHG removals.

UN Secretary General António Guterres targets global community to reduce greenhouse gas emissions and this is a very precise definition. One of the Race to Zero Criteria is: "Pledge at the head-of-organization level to reach (net) zero GHGs as soon as possible".

The main measures of limiting Global Warming were working out in United Nations Framework Convention on Climate Change (UNFCCC): Conference of the Parties, Adoption of the Paris Agreement, 12 December 2015 [5]. With the coming into force of the Paris Agreement following the ratification of countries responsible for over 55% of global emissions, increased attention is focusing on the means by which Contracting Parties can implement the Agreement and achieve its objective of limiting global warming to 'well below 2 °C above pre-industrial levels' and 'to pursue efforts to limit the temperature increase to 1.5 °C above pre-industrial levels' [5, Article 2a]. As the primary means of achieving this, the Parties agreed to 'aim to reach global peaking of greenhouse gas emissions as soon as possible … so as to achieve a balance between anthropogenic emissions by sources and removals by sinks of GHG in the second half of this century' [5, Article 4.1], at which time net emissions would be zero.

2 Zero or Negative greenhouse gas emissions

For getting the Earth's climate system stable and predictable we need to reduce greenhouse gas concentrations in the atmosphere to the pre-industrial levels. It is very important to understand that the World Ocean could not be no longer a deposit of carbon dioxide (CO_2) and methane (CH_4) during the warming of ocean water layers, just the opposite it can be the main source of GHG emissions in the atmosphere in the period of Global Warming.

On the aim of reducing greenhouse gas to net zero emissions by 2050 the Global Warming do not stop at 1.5 °C and it looks like a lowest possible estimation concerning for the best

© The Authors, published by EDP Sciences. This is an open access article distributed under the terms of the Creative Commons Attribution License 4.0 (http://creativecommons.org/licenses/by/4.0/).

scenarios. The different scientific groups provide various estimations from 1.5 °C up to 4 °C by 2050. We can look on it in details.

The Global Warming (GW) is estimated as aced of the year's globally averaged temperature (Earth's global annual temperature) above the levels since the Industrial Revolution (pre-industrial levels the baseline 1951-1980 mean). The rate of Global Warming is in relation of greenhouse gas concentrations. The M. Previdi et al. (2013) find the relation [6] of the climate sensitivity for doubled CO_2 concentrations of about 3 °C. Including in the model all other factors as reflection from ice sheet and vegetation albedo feedback the maximum level of GW can rise up to 4-6 °C [6].

The difference between the World Ocean surface temperatures in different parts can cause an increase in extreme weather related events. One of the leading US climatologist J. Hansen (2016) [7] stressed that the 2 °C global warming above the pre-industrial level is highly dangerous and he discussed a possibility to stabilize the Earth climate in the case of GHG emissions reduction [8]. Cooling Ocean surface in the North Atlantic with the simultaneously warming Southern Ocean surface provide the increase in wind speed and the Atmosphere kinetic energy that can provide the better conditions for more powerful storms [7].

Max Planck Institute for Chemistry (MPIC) in Germany provides a research on long-term meteorological datasets and climate modeling for socioeconomic and population projections. The leader of MPIC climatologists group J. Lelieveld et al., (2016) in his model predict the 4 °C global warming above the preindustrial level by 2050, this will provide the conditions for rising the maximum temperature during the hottest days up to 50 °C for more then one month [9]. This will cause the difficult climate conditions for North Africa and Arab states 500 million populations and can lead to the climate migration. These events can happen two times faster and two times higher (2 °C) than it was planned by COP21 in Paris 2015.

According to the NASA's Goddard Institute for Space Studies the year 2020 was the second (1.02 degrees Celsius) after the warmest on record 2016, exceeded both the Global Warming +1 °C level.

The paleoclimate data can be used for estimations of Sea Level as a function of the global annual temperature over the Globe [10-13].

Heat content accumulated in the World Ocean and Atmosphere after the relaxation period of the climate system will provide melting of the relevant to it ice deposits and as a result increasing the Sea Level. This model [12-16] based on logistic equation provides both the time series and the stationary solution for the future Sea Level as a reaction on global annual temperature increase.

The model stationary solution on Global Warming of one degree Celsius temperature increase above pre-industrial levels corresponds to the future Sea Level Rise on +7.5 meters [14-16]. This is our current perspective if in the nearest time the accumulated Heat content due to the GW will melt the relevant ice deposit.

In the case "to pursue efforts to limit the temperature increase to 1.5 °C above pre-industrial levels" [5, Article 2a] will be successful in the nearest future the Sea Level will rise on +11.25 meters [14-16].

If not, the limiting global warming to "well below 2 °C above pre-industrial levels" [5, Article 2a] it will provide the Sea Level Rise up to +15 meters [14-16].

This research was done for the different scenarios of Global Warming including the cases of +4 °C corresponding to SLR +30 meters and +8 °C accordingly to SLR +60 meters [14-16].

As an example [14] we provide an estimation for the influence of the SLR due to GW on the European airport infrastructure, including reduction of air traffic, passenger traffic and cargo transportation. In the case of "EU's adaptation strategy to the climate change of 2 °C, in the long run, this temperature will lead to rising sea levels up to 15 meters. Consequently,

© The Authors, published by EDP Sciences. This is an open access article distributed under the terms of the Creative Commons Attribution License 4.0 (http://creativecommons.org/licenses/by/4.0/).

this involves the flooding of 172 airports (19.9%), a reduction of 20.4% in passenger traffic and cargo by 14.7% " [14].

We discuss the future Climate Change due to the GHG emissions and it's possible consequences on Global Warming and related to them Sea Level Rise during the International Conferences in: UAE (2012) [17]; Coimbra (Portugal, 2012) [18]; Belgrade (Serbia, 2013) [12]; Nice (France, 2000-2018) [14]; Belgrade (Serbia, 2016) [15]. The main proposals were: "Making efforts for the getting down of the global average temperature to the pre-industrial level that will reduce the number of Natural Disasters in times and stop the Sea Level Rise. For getting the Earth's climate system stable and predictable we need to reduce greenhouse gas concentrations in the atmosphere to the pre-industrial levels".

The proposals on implementing vitally needed Zero or Negative emissions of GHG for preventing Climate Change vs. Global Warming and future Sea Level Rise were widely discussed and published in Belgrade (Serbia, 2016) [15] and during IGU conferences in: Moscow (Russia, 2015); New Delhi (India, 2016) [16, 19]; Hyderabad (India, 2017).

In 2016 L. V. Sorokin was the only delegate from Russian Federation on the UN conference: "UNISDR Science and Technology Conference on the Implementation of the Sendai Framework for Disaster Risk Reduction 2015-2030" (27-29 January, 2016 | Geneva, Switzerland), with the report on "Weather Modification Application for Disaster Risk Reduction". During this event the suggestions on implementing Zero or Negative emissions of GHG and preventing methane emission from Arctic for stopping Climate Change were represented on the session Work stream 2 "Understanding disaster risk, assessment and early warning", Group 2 "Vulnerability and exposure", Chair R.B. Singh.

The next year the risks of economic losses from Climate Change, extreme weather related events and future Sea Level Rise (Thematic Area: National and local disaster risk reduction strategies integrated with climate change and sustainable development plans) were the topic of proposals made by L. V. Sorokin to the Leaders forum of the United Nations Global Platform for Disaster Risk Reduction (Global Platform, 22-26 May, 2017, Cancun, Mexico).

If the targets of getting down the greenhouse gas concentrations in the atmosphere and the global average temperature to the pre-industrial levels cannot be achieved then Climate Change and associated with Global Warming future Sea Level Rise in the nearest future would be the most important Risk factors in the World and World Economy, that can provide the world economic crisis and massive losses due to flooding of the coastal and low land areas of the Earth's land.

So to slow down the Global Warming and future Sea Level Rise the humanity need significantly reduce the concentrations of GHG in the atmosphere that will launch the era of Negative GHG emission technologies.

The importance of implementing the negative emissions of GHG is on mainstream of the leading climatologists. So T. Gasser et al., (2015) in the paper [20] write: "To limit global warming to <2 °C we must reduce the net amount of CO_2 we release into the atmosphere, either by producing less CO_2 (conventional mitigation) or by capturing more CO_2 (negative emissions)". The leading US climatologist J. Hansen et al., (2017) in his paper [21] devoted to the future generation stressed the requirement of implementing negative CO_2 emissions. European Academies Science Advisory Council (EASAC) prepares special policy report [22] dedicated to Negative emission technologies and their role in meeting Paris Agreement targets.

3 Link to Arctic Methane emissions

The currently observed climatic changes cannot be completely explained by the increase in the concentration of carbon dioxide in the atmosphere. Taking into account the impact of

© The Authors, published by EDP Sciences. This is an open access article distributed under the terms of the Creative Commons Attribution License 4.0 (http://creativecommons.org/licenses/by/4.0/).

methane emissions, including the Polar Regions and the World Ocean, it will be possible to explain rapid changes in the Earth's climate.

Using the NOAA satellite data of the GHG distributions we can see the inequality over the North and South Pole regions. The higher concentration of GHG was observed in the North hemisphere. This can be in connection with the Polar amplification. The Arctic amplification is polar amplification in the region of the Earth's North Pole only and Antarctic amplification is concerning the region of the South Pole. At the present time the Arctic amplification is stronger then the Antarctic amplification.

Taking into account the Arctic amplification we can see in Russian Federation the current rise of the polar region annual temperature two times higher then over the Globe.

From the NOAA satellite data in August 2015 the Arctic Ocean has a sea surface temperature anomaly up to +8 °C, at the same time the North part of Atlantic and Pacific Oceans has a surface temperature anomaly of +11.9 °C. This is an example of heating the sea surface water in the Earth's North Pole region. All this can affect the deep layers of ocean water, heating the layers of bottom sediments containing deposits of Methane hydrate.

Methane hydrate ($4CH_4 \cdot 23H_2O$) also known as methane clathrate is stable at low temperatures and high pressure typical for ocean depths of more than 300 m. Methane hydrate deposits in the World Ocean thousands of times more then the current content in the Earth's Atmosphere.

Clathrate gun hypothesis [23] is based on the stronger evidence of the uncontrolled runaway methane clathrate breakdown during the Global Warming periods. On our opinion this is most credible scenario of the Climate destabilization due to the Global Warming. In the paper [24] Carolyn Ruppel and John Kessler (2016) provide the analysis of the interaction of Climate Change and methane hydrates stability in the Ocean sediments. This research can be very important for modeling of methane clathrate breakdown during the current warming of deep layers ocean water.

One of the first research [25] on aquatic sources and sinks of CO_2 and CH_4 in Russian Polar Regions was done by I. P. Semiletov (1999). This was the base for recent discovery [26] of massive methane emissions from the Eastern Arctic Seas (EAS) into the atmosphere done by Shakhova and Semiletov et al., (2010). They were found that the amount of methane entering the atmosphere from the EAS shelf is higher than the methane emission from the Entire world ocean [27]. After that in 2011 they discovered in the Laptev Sea the mega-seeps of bubble methane, reaching 1000 meters or more across. The abnormal bubble fluxes can produce up to 100g of methane per square meter per day or more [28]. They found the high probability of a catastrophic release of methane into the atmosphere [29-31] in the Eastern Arctic Seas.

Another source of methane is the Polar Siberia region with melting huge permafrost areas. In research [32] G. E. Oblogov et al., (2020) provide an analysis of methane content and emission in the permafrost landscapes of Western Yamal, Russian Arctic region. Huge deposits of organic materials can be a source of methane emissions during the melting Siberian permafrost. The degradation of the permafrost landscapes can also be explained by an increasing the temperature in it's the upper layers.

In total the new evidences of the huge methane content and emission from Polar Siberia with melting huge permafrost areas and the mega-seeps of bubble methane in the Eastern Arctic Seas, together with methane emission from the World Ocean can explain the huge gap in GHG World balance.

4 Conclusions

1. Making efforts for the getting down of the global average temperature to the pre-industrial level that will reduce the number of Natural Disasters in times and stop the Sea

© The Authors, published by EDP Sciences. This is an open access article distributed under the terms of the Creative Commons Attribution License 4.0 (http://creativecommons.org/licenses/by/4.0/).

Level Rise. For getting the Earth's climate system stable and predictable we need to reduce greenhouse gas concentrations in the atmosphere to the pre-industrial levels.

2. Zero emission technologies can reduce the GHG concentrations in Atmosphere that can help to slowdown the Global Warming.

3. Negative emission technologies can significantly reduce the GHG concentrations in Atmosphere that can getting down the global average temperature to the pre-industrial level and prevent Global Warming and future Sea Level Rise.

4. The currently observed climatic changes cannot be completely explained by the increase in the concentration of carbon dioxide in the atmosphere. Taking into account the impact of methane emissions, including the Polar Regions and the World Ocean, it will be possible to explain rapid changes in the Earth's climate.

5. Taking in account the Arctic amplification we can see in Russian Federation the current rise of the polar region annual temperature two times higher then over the Globe. That can explain the heating the sea surface water in the Earth's North Pole region and can affect the deep layers of ocean water, heating the layers of bottom sediments containing deposits of Methane hydrate. The deposits of Methane hydrate in Ocean sediments can become unstable and provide a catastrophic release of methane into the atmosphere.

6. Clathrate gun hypothesis is based on the stronger evidence of the uncontrolled runaway methane clathrate breakdown during the Global Warming period. It can explain the overheating the Earth's atmosphere in the Arctic Polar region and Arctic amplification.

7. The main goal of preventing the worst-case scenario of Global Warming is preventing methane emission from Arctic region and World Ocean.

8. The Global Warming and future Sea Level Rise will terminate the interglacial period (warmer global average temperature lasting thousands of years with comfortable climate conditions).

9. Global Warming is the precursor of falling in the Next Glacial Period, which can last from 80 thousands years up to 120 thousands years (Last Glacial Period). The prerequisites for the beginning of the Next Glacial Period are the sea level rise, changing the trajectories of Ocean Currents and flooding of the coastal and low land areas of the Earth's land.

10. The strategy on adaptation to climate change does not consider the possibility of a Sea Level Rise at the upper level +1 m. In the case of the EU's adaptation strategy to the climate change of +2 °C in the nearest future the SLR can exceed 15 times the level of +1 m. This +1 m limitation on adaptation to climate change could lead to catastrophic economic losses due to the flooding of the coastal and low land areas with huge infrastructure.

11. Climate Change and associated with Global Warming future Sea Level Rise in the nearest future would be the most important Risk factors in the Global World and World Economy, that can provide the world economic crisis and massive losses due to flooding of the coastal and low land areas of the Earth's land.

References

1. World Bank's Disaster Risk Management, https://www.worldbank.org
2. Aftershocks: Remodeling the Past for a Resilient Future, World Bank's Global Facility for Disaster Reduction and Recovery (GFDRR) (2018) https://www.gfdrr.org
3. Investing in Urban Resilience: Protecting and Promoting Development in a Changing World. World Bank, Washington, DC. (2016) https://openknowledge.worldbank.org
4. IPCC Special Report on Global Warming of 1.5 C, https://www.ipcc.ch
5. United Nations Framework Convention on Climate Change (UNFCCC): Conference of the Parties, Adoption of the Paris Agreement (2015)

© The Authors, published by EDP Sciences. This is an open access article distributed under the terms of the Creative Commons Attribution License 4.0 (http://creativecommons.org/licenses/by/4.0/).

6. M. Previdi et al., Quarterly Journal of the Royal Meteorological Society. Wiley (2013)
7. J. Hansen et al., Atmospheric Chemistry and Physics **16**, 3761 (2016)
8. J. Hansen et al., PLoS ONE, **8(12),** e81648 (2013)
9. J. Lelieveld et al., Springer Netherlands, 1 (2016)
10. J. Hansen et al., Phil. Trans. Roy. Soc., **A371,** 20120294 (2013)
11. J. C. Zachos, G. R. Dickens, R. E. Zeebe, Nature **451**, 279 (2008)
12. L. V. Sorokin, The Cost of Adaptation for Climate Change: Future Sea Level Estimation, International Scientific Conference Economic Sciences on the Crossroad, 557 (2013)
13. L. V. Sorokin, Bulletin of PFUR. Series Mathematics, Information Sciences, Physics, **4**, 18 (2015)
14. L. V. Sorokin, G. Mondello, GREDEG Working Paper, **2015-46** (2015) http://www.gredeg.snrs.fr
15. L. V. Sorokin, G. Mondello, Europe and Asia: Economic Integration Prospects. CEMAFI International, Chapter 21, 349 (2016)
16. L. V. Sorokin, G. Mondello, Climate Change, Extreme Events and Disaster Risk Reduction. Sustainable Development Goals Series. Springer, Cham, 181 (2018)
17. L. V. Sorokin, The Malaysian View on Drought Consequences and Water Crisis. *Emerging Economies: Development Challenges and the Innovative Approach Solutions,* 257 (2012)
18. L. V. Sorokin, European Integration Process in Western Balkan Countries, Chapter 11, 208 (2012)
19. L. V. Sorokin, Climate Change, Extreme Events and Disaster Risk Reduction. Sustainable Development Goals Series. Springer, Cham, 233 (2018)
20. T. Gasser et al., Nat. Commun. 6:7958 (2015)
21. J. Hansen et al., Earth System Dynamics, **8**, 577 (2017)
22. *Negative emission technologies: What role in meeting Paris Agreement targets?,* EASAC policy report, **35**, 37 (2018)
23. M. Maslin, M. Owen, S. Day, D. Long, Geology, **32** (2004)
24. C. Ruppel, J. Kessler, Reviews of Geophysics, **55** (2016)
25. I. P. Semiletov, J. Atmos. Sci., **56**, 286 (1999)
26. N. Shakhova, I. Semiletov, A. Salyuk, V. Joussupov, D. Kosmach, Ö. Gustafsson, Science, **327**, 1246 (2010)
27. N. Shakhova et al., Nature Geoscience, **7**, 64 (2014)
28. N. Shakhova et al., Phil. Trans. R. Soc. A, **373,** 20140451 (2015)
29. N. Shakhova, I. Semiletov, I. Leifer, P. Rekant, A. Salyuk, D. Kosmach, Journal Geophys. Res., **115** (2010)
30. I. Leifer et al., The Cryosphere, **11**, 1333 (2017)
31. N. Shakhova et al., Nature Communications, **8**, 15872 (2017)
32. G. E. Oblogov et al., Russian Arctic. Geosciences **10(10),** 412 (2020)

© The Authors, published by EDP Sciences. This is an open access article distributed under the terms of the Creative Commons Attribution License 4.0 (http://creativecommons.org/licenses/by/4.0/).

Development of the sectoral system of remuneration of medical workers of state healthcare institutions of the constituent entities of the Russian Federation

Irina Omelchenko[1], *Oleg* Dozortzev[1], *Marina* Danilina[1,2,3,*], *Alexander* Vas'kov[1], and *Olga* Zabelina[1]

[1]Research Institute VNII of Labor, Ministry of Labour of Russia, Zemlyanoy Val, 34, 105064 Moscow, Russia
[2]Finance University under the Government of the Russian Federation, Leningradsky prosp., 49, 125993 Moscow, Russia
[3]Russian Economic University, Stremyanny per., 36, 117997 Moscow, Russia

Abstract. The article analyses the Development of a sectoral system of remuneration of medical workers of state healthcare institutions of the constituent entities of the Russian Federation. It was determined, the PM differentiation, as well as the differentiation of the cost of a fixed set of consumer goods and services, is too low for use in the OSS, while the value of the coefficient of variation for the median wage is much higher, and at the same time, if this indicator is used in the OSS, with a significant decrease in the differentiation between the most the richest and poorest budgets will not experience the problems. Thus, the most appropriate indicator, taking into account the existing resources of the federal budget, for use in order to take into account the economic differentiation of a constituent entity of the Russian Federation is the median salary of employees of organizations.

1 Introduction

The determination of the minimum tariff rate (minimum salary) is the central point for the development of a remuneration system, regardless of the level at which it is supposed to be applied. The minimum wage rate is the rate of payment for the labor standard (standard), having the lowest characteristics for all factors that could form a tariff. The level of remuneration of an employee engaged in such work, provided full performance of his work duties must provide him the consumption level sufficient to reproduce its ability to labor.

The size of the minimum wage rate (official salary) must be at a level not less than the minimum wage, established at the state level. From a scientific point of view, the most justified criterion for ensuring the share of salaries in the wage structure at a level of at least 50%, according to the Unified Recommendations, it is advisable to preserve the use of the PCG (modified taking into account the current labor functions of medical workers), but with

*Corresponding author: marinadanilina@yandex.ru

© The Authors, published by EDP Sciences. This is an open access article distributed under the terms of the Creative Commons Attribution License 4.0 (http://creativecommons.org/licenses/by/4.0/).

the use of basic minimum salaries, the establishment of which is enshrined in Article 144 of the Labor Code of the Russian Federation, while linking the size of the minimum wage of a worker of the 1st category to the size of the value of the concept of "minimum consumer budget".

2 Materials and methods

To date, there are no guidelines for setting salaries in budgetary institutions, with the exception of the minimum wage, which is equal to the subsistence minimum, but is not mandatory in terms of setting minimum salaries. In turn, the PM is formed on the basis of the consumer basket and reflects the value of the minimum set of food products, non-food goods and services, as well as mandatory payments and fees, necessary to preserve human health and ensure his life.

At the same time, the minimum permissible level of satisfaction of the material and spiritual needs of a person employed in labor, as well as the basic value and structure of consumption of goods and services necessary for a person for normal life, are more accurately and reasonably characterized by the concept of “minimum consumer budget”.

The consumer budget is a broad concept that characterizes the structure and balance of incomes and expenses of people (families and single people, employed and unemployed, etc.), as well as the structure of needs that determines the directions of using income for the purchase of various goods and services, for investments, savings and other purposes.

The consumer budget expresses the value of a set of food and non-food goods and services. Consumer budgets differ in the degree to which basic physiological social and spiritual needs are met. The expenditure side of the consumer budget is determined by the following factors:

1. Physiological needs of people (biological factor).
2. Type and nature of work (professional factor).
3. The level of cultural development of a person (general cultural factor).
4. Working conditions: the level of mechanization, automation, hygienic conditions, working hours and rest (production factor).
5. Habitual lifestyle and standard of living (socio-historical factor).
6. Age composition of the family (demographic factor).
7. Natural and climatic conditions (geographic factor).
8. National customs (ethnic factor).
9. Intentions and moods of people (psychological factor).

There are a large number of combinations of the above factors that determine the needs and form consumer budgets for various categories of people. At the same time, the minimum consumer budget is one of the main system-forming basic budgets (along with the subsistence minimum). The minimum consumer budget is a criterion for a socially acceptable level of reproductive consumption (calculated at the lowest level of welfare).

Taking into consideration the fact that the level of wages should contribute to the reproduction of the labor force and correspond to the cost estimate of the restoration of the ability to work, the sociocultural needs of a person, it is the minimum consumer budget that should be used as the base salary (official salary). An example of the formation of the structure and calculation of the cost of the minimum consumer budget of a man of working age engaged in light physical labor (with a minimum level of labor complexity) is presented in the previous report on the research work "Development and scientific substantiation of proposals to improve wage systems ...". The cost of the consumer budget in March 2019 prices amounted to 13,432.51 rubles, which exceeds the minimum wage set as of 01.01.2020.

Another scientifically grounded approach to setting the minimum wage rate is to determine the ratio of the minimum consumer wage relative to the average wage for the

© The Authors, published by EDP Sciences. This is an open access article distributed under the terms of the Creative Commons Attribution License 4.0 (http://creativecommons.org/licenses/by/4.0/).

relevant period. If this ratio is at a level exceeding 45-50%, then the corresponding value cannot be taken as the minimum tariff rate, since there will be no funds left for normal differentiation by other tariff-forming factors (complexity, working conditions), as well as for incentives for over-fulfillment of norms labor costs and bonus indicators. If this ratio is in the range of 35-45%, then this calculated value can be taken as the minimum tariff rate.

In practice, taking into consideration the tight deadlines and the lack of flexibility of domestic legislation, it was decided to accept the minimum wage as of 01/01/2020 as the minimum value that can correspond to the minimum wage rate (salary). The use of the minimum wage, taking into consideration the complete reduction of its separation from the minimum necessary physiological conditions for the reproduction of labor (PM), can be considered a reasonable decision, but not optimal.

At the same time, the OSS takes into consideration the differentiation in the size of wages in the constituent entities of the Russian Federation due to the application of the coefficient of regional economic differentiation. The second no less complex and difficult task in the development of a wage system is to determine the composition of tariff-forming factors, their relationships with each other and the principles of differentiation of tariff payment.

Taking into consideration a wide range of compensation and incentive payments, taking into consideration the impact on wages of various working conditions, including the establishment of compensations for the negative impact of unfavorable environmental conditions on the body of workers, the main tariff-forming factor will be the complexity of labor. The complexity of labor, expressed both in the direct physiological zeal of the employee and in the qualification requirements for the work performed, as well as, in general, labor functions are described in detail in professional standards, and the differentiation by levels of complexity is reflected when the positions of medical workers are divided according to the PCG.

Taking into consideration the large number of meetings of the working group on the development of the OSHP, which includes specialized experts from the Ministry of Health of the Russian Federation, expressing their opinion on the complexity of work (in particular, on the level of responsibility in certain positions), it was decided to update the existing CAP in order to form more objective differentiation by the complexity of job positions. The PKG "Medical and pharmaceutical personnel of the first level" (2 qualification levels), "Nursing and pharmaceutical personnel" (6 qualification levels), "Doctors and pharmacists" (6 qualification levels) were expanded. The number of qualification levels in the PCG "Heads of structural units of institutions with higher medical and pharmaceutical education (specialist doctor, pharmacist)" has not been changed.

Taking into account the complexity coefficients agreed with the experts of the Ministry of Health of the Russian Federation during the development of the model, reflecting fair differentiation in the remuneration of medical workers, the most relevant grid of complexity coefficients for use in OHS when excluding qualification categories from salaries is the UTS dated 01.12.2001, which has 18 digits and the ratio of the 18th to the first digit is 4.5: 1. Taking into account the fact that in the new PCG (developed by the Ministry of Health of Russia and the Ministry of Labor of Russia) differentiation by 16 qualification levels is used, it seems justified to use the first 16 categories of the proposed UTS, taking into account the decrease in coefficients in proportion to a certain influence on the complexity of labor of qualification categories to the value " no category ".

3 Results and discussions

The most controversial issue in the course of determining the complexity coefficients, reflecting the differentiation between the qualification levels of medical workers, was the question of including the complexity coefficients of qualification categories in the grid. On

© The Authors, published by EDP Sciences. This is an open access article distributed under the terms of the Creative Commons Attribution License 4.0 (http://creativecommons.org/licenses/by/4.0/).

the one hand, employees who confirm the category can perform more complex work, on the other hand, all the work that employees must perform in one position, regardless of the qualification category, are described in job descriptions. Thus, it is not clear how the confirmation of the qualification category affects the complexity and quality of the services provided by health workers.

If the accounting of qualification categories is included in the coefficient grid, it is necessary to revise and update absolutely all job descriptions, which should reflect the specific duties that the employee must perform, taking into consideration the confirmation of a particular qualification category. As a result, it was decided not to include qualification category coefficients in the grid.

A distinctive feature of the post-Soviet period of development of the economic system of the Russian Federation is an extremely high degree of stratification of the constituent entities of the Russian Federation according to the socio-economic level of development, which is expressed in all aspects of society, including wages. In the Russian Federation, there is a high differentiation of wages, and it is impossible not to take into consideration the economic component when determining the fair amount of wages, since the purchasing power of money in the constituent entities differs significantly. In this regard, as well as taking into consideration the need to reduce the differentiation of wages of medical workers in different constituent entities of the Russian Federation, when developing an OSHP, it is necessary to reasonably take into consideration economic differentiation.

For this purpose, it is proposed to take into consideration economic differentiation when calculating the official salary of medical workers using the concept of "coefficient of regional economic differentiation".

The basis for determining this coefficient at various stages of the study of the problem was:

- the cost of a fixed set of consumer goods and services;
- living wage;
- average monthly accrued wages of employees of organizations;
- average monthly accrued wages of employees in organizations, individual entrepreneurs and individuals (hereinafter - ZNR);
- the median value of the accrued wages of employees of organizations.

For interregional comparisons of the purchasing power of the population in the Russian Federation and individual constituent entities of the Russian Federation, the cost of a fixed set of consumer goods and services is calculated on the basis of consumption volumes that are uniform for all constituent entities of the Russian Federation, as well as regional average prices. The set includes 83 items of goods and services, including 30 types of food products, 41 types of non-food products and 12 types of services. The cost of the set is given per month.

Taking into account the task of objectively assessing economic differentiation in the OSS, the cost of a fixed set is the most appropriate indicator for the formation of official salaries.

However, the financial capabilities of the state cannot provide such a rapid reduction in the gap between the wages of workers in various constituent entities of the Russian Federation, the consequences of using this indicator in determining official salaries will lead to very high costs for OSH, to an imbalance in the labor market in the public sector, to protest moods of employees of state and municipal organizations in other social spheres.

Therefore, as a basis for determining the coefficient of regional economic differentiation, alternative options were proposed, assessed by the coefficient of variation of the sample (for the constituent entities of the Russian Federation).

Preparation of proposals for the formation of lists and conditions for the establishment of compensation and simulation payments.

For a number of compensatory and incentive payments, the amounts and conditions for implementation are determined by the Labor Code of the Russian Federation and decrees of

© The Authors, published by EDP Sciences. This is an open access article distributed under the terms of the Creative Commons Attribution License 4.0 (http://creativecommons.org/licenses/by/4.0/).

the Government of the Russian Federation (payments for regional regulation of wages, bonuses for detecting oncological diseases and payments for birth certificates, etc.). For the rest, it is proposed to establish the size and conditions of implementation separately by the Government Decree.

In accordance with Article 147 of the Labor Code of the Russian Federation, the minimum wage increase for employees engaged in work with harmful and (or) hazardous working conditions is 4% of the tariff rate (salary) established for various types of work with normal working conditions. It is proposed to establish the amount of additional payments depending on the class of labor conditions from 4 to 12% (3.1 - 4%; 3.2 - 6%; 3.3 - 8%; 3.4 - 10; 4 - 12%), which will ensure optimal differentiation of these payments by degrees of hazard.

In accordance with the Law of the Russian Federation of July 2, 1992 No. 3185-1 "On psychiatric care and guarantees of the rights of citizens in its provision", Federal Law of June 18, 2001 No. 77-FZ "On the prevention of the spread of tuberculosis in the Russian Federation" [49] and Federal Law of March 30, 1995 No. 38-FZ "On the prevention of the spread in the Russian Federation of a disease caused by the human immunodeficiency virus (HIV)", the order and size of an increase in wages for work with harmful and (or) dangerous working conditions for medical workers of state institutions of the constituent entities of the Russian Federation involved in the provision of psychiatric care, directly involved in the provision of anti-tuberculosis care, carrying out diagnostics and treatment of HIV-infected, as well as persons, whose work is related to materials containing the human immunodeficiency virus are established in the manner determined by the executive authorities of the constituent entities of the Russian Federation.

In order to unify the approaches to the establishment of these compensation payments, the Project proposes the establishment of uniform amounts of compensation payments to medical workers directly involved in the provision of anti-tuberculosis care in the amount of 40%; medical workers who diagnose and treat HIV-infected people, and persons whose work is associated with materials containing the human immunodeficiency virus in the amount of 60%; medical professionals involved in the provision of mental health care 30%.

In accordance with Article 154 of the Labor Code of the Russian Federation, the increase in wages for night work is established by collective agreements, but not lower than the minimum amount established by the Government of the Russian Federation.

According to the Decree of the Government of the Russian Federation No. 554 of July 22, 2008, the minimum wage increase for night work is 20 percent of the hourly wage rate (salary (official salary) for each hour of night work).

At the same time, in a number of regions, in the remuneration systems introduced after 2005, the rates of increase in the payment for hours of night work of medical workers, previously established within the framework of regional tariff scales, were retained - 50%.

In this connection, it is proposed to establish a single general requirement for compensation payments in all regions for work at night in the amount of 50% of the official salary calculated per hour of work.

It is also proposed to include in the list of incentive payments a supplement for work in an ambulance in the amount of - 50%, in the same amount to establish extra charges to the hourly rates of consultant doctors (specialist doctors) attracted from other institutions; an allowance for medical specialists for medical care provided on an outpatient basis in the amount of 30% and an allowance for medical workers of feldsher-obstetric points (heads of feldsher-obstetric points, feldshers, obstetricians (midwives), nurses, including nursing nurses) for the provided medical care on an outpatient basis in the amount of 30%; allowances for the precinct: to district doctors - district therapists, district pediatricians, general practice doctors (family doctors), district nurses, district therapists, district pediatricians and nurses of general practitioners (family doctors) for medical care provided on an outpatient basis at a rate of 40%; bonus for work in mobile teams.

© The Authors, published by EDP Sciences. This is an open access article distributed under the terms of the Creative Commons Attribution License 4.0 (http://creativecommons.org/licenses/by/4.0/).

These payments are offered instead of previously paid within the framework of the national project "Health", as well as in exchange for payments for work experience or seniority.

In accordance with article 105 of the Labor Code of the Russian Federation, in those jobs where it is necessary due to the special nature of work, as well as in the production of work, the intensity of which is not the same during the working day (shift), the working day can be divided into parts so that the total the duration of the working time did not exceed the established duration of the daily work.

In order to unify payments for the allowance for dividing the working day into parts, it is proposed to establish a single amount of the said allowance at a rate of 30%.

At present, in most regions, payments are made "bonus for work in rural areas" in the amount of 25%. It is envisaged to preserve it in the systems of remuneration of employees of institutions, since it is important for attracting qualified medical personnel to the village.

Incentive payments in the form of bonuses for work in hospitals for war veterans and bonuses for work in children's homes are established in the current remuneration systems of the regions. In connection with the social significance of the institutions, it is also envisaged to preserve the specified payment in the amount established in the existing systems of remuneration of labor in the regions.

The payment for mentoring evaluates the contribution of specialists to the accelerated development of practical skills of young specialists, as well as their professional adaptation, consolidation and development of professional skills among young specialists. Improving the efficiency of employment of graduates who have passed the accreditation of specialists, including those who have completed targeted training. In this connection, it is proposed to establish the specified mark-up at a rate of 15%.

It is proposed to establish the remuneration based on the results of work according to indicators and criteria at a minimum amount (10%) in order to exclude employers' abuse in the appointment of these payments and to establish the transparency of their calculation.

4 Conclusion

As it was determined, the PM differentiation, as well as the differentiation of the cost of a fixed set of consumer goods and services, is too low for use in the OSS, while the value of the coefficient of variation for the median wage is much higher, and at the same time, if this indicator is used in the OSS, with a significant decrease in the differentiation between the most the richest and poorest budgets will not experience the problems described above.

Thus, the most appropriate indicator, taking into account the existing resources of the federal budget, for use in order to take into account the economic differentiation of a constituent entity of the Russian Federation is the median salary of employees of organizations.

References

1. Allegretto, Sylvia, Arindrajit Dube, Michael Reich, and Ben Zipperer, *Credible Research Designs for Minimum Wage Studies* (Working paper, Institute for Research on Labor and Employment, UC Berkeley, 2013)
2. P. Bastos, N. Monteiro, Journal of Economics and Management Strategy, **20(4),** 957 (2011)
3. Belman, Dale and Paul Wolfson, *What Does the Minimum Wage Do?* (Kalamazoo, MI: W.E. Upjohn Institute for Employment Research, 2014)

© The Authors, published by EDP Sciences. This is an open access article distributed under the terms of the Creative Commons Attribution License 4.0 (http://creativecommons.org/licenses/by/4.0/).

4. Belser, Patrick and Uma Rani, *Minimum Wages and Inequality, Janine Berg (Ed.) Labour Markets, Institutions and Inequality: Building Just Societies in the 21st Century* (Edward Elgar Publishing, Cheltenham, UK, 2015)
5. V. Gimpelson, R. Kapelyushnikov, A. Lukyanova, Comparative Economic Studies, **52(4),** 611 (2010)
6. S. Kapelyuk, Economics of Transition, **23(2),** 389 (2015)
7. A. Lukiyanova, N. Vishnevskaya, The Decentralization of Minimum Wage Setting in Russia. Working papers by NRU Higher School of Economics. Series WP BRP "Economics/EC", **90** (2015)
8. A. Muravyev, Europe-Asia Studies, **66(8),** 1270 (2014)

© The Authors, published by EDP Sciences. This is an open access article distributed under the terms of the Creative Commons Attribution License 4.0 (http://creativecommons.org/licenses/by/4.0/).

SHS Web of Conferences **114**, 01008 (2021) https://doi.org/10.1051/shsconf/202111401008
NTSSCEM 2021

Digital challenges for creative industries: case of opera

Nina Trubnikova*, and *Severyan* Tsagareyshvili

RUDN University, 117198, Miklukho-Maklaya str.6, Moscow, Russia

Abstract. Digital technologies today are seen as a powerful driving force of creative industry growth. The authors reveal the development of operatic theatre as a significant entity of a new market reality - the economy of impressions. Notwithstanding the fact that the majority of opera houses are state funded, the issue of economic efficacy is gaining relevance for theaters; increasingly more attention is paid to promoting theatrical product. Despite the fact that the eliteness of operatic art and loyalty to tradition poses restrictions on implanting new instruments of theater branding. active digitization of the opera product is the only plausible means of attracting the attention of the new generations who are used to this format of representing cultural content. This dilemma is analyzed in the article on the basis of the data yielded by a field study in the form of a survey involving Russian and Italian opera lovers.

1 Introduction

The processes of digitization today have concerned creative structures and industries which in the past remained practically unaffected by new technological shifts and, for quite a while, underwent zero digital development. This brings to the fore the problems of introducing digital technologies in the products of classical theater, including musical theater with its synthetic nature. Operatic art, among the others, has to face the challenge. The genre that for centuries remained a highly conservative one is now confronted with the vital necessity of looking more attractive for the new generation of listeners. The intense competition with the colorful multimedia content of the mainstream culture along with the massively emerging alternative kinds of entertainment make the issue of financial health especially relevant for opera houses [1, p. 1].

The key driver of technology-related growth for the industry of performing arts can be regarded as their ever-growing need to employ informational/communicational technologies for increasing the spectacularity and uniqueness of their content [2, p. 413]. The potential level of monetization of their product (ticket price, overall revenue of the theater etc.) directly depends on the extent to which the product meets the expectations of the audience.

The restricted offline communication brought about by the COVID-19 pandemic urged the opera to set out on a qualitative digital reform [3, p. 45]. The reorganization, in fact, had begun long before the said events, but for a while it did not receive a wide spreading, nor, more importantly, support of the loyal operatic art consumers. Due to the recent

*Corresponding author: trubnikova_nv@rudn.ru

© The Authors, published by EDP Sciences. This is an open access article distributed under the terms of the Creative Commons Attribution License 4.0 (http://creativecommons.org/licenses/by/4.0/).

communicational lockdown the theaters that used to regard direct live interaction with the spectator as their main channel, now have found themselves under the threat of a financial and HR collapse. In consequence, they have been forced to urgently adapt opera products to digital placement. Online resources, which were normally just a complementary means, have suddenly become the only channel of communication with the consumers.

Overall online sales of cultural products across the industry doubled in 2020, reaching the record $12.4 bn, while the market share of the sector increased to 25% compared to 9% the year before [4]. The merging of the technological and creative spheres is only expected to intensify in the nearest future [5, p. 2].

The problem of the research is rooted in the fact that, in the digital era, the traditional marketing paradigm which extended onto opera as an art and an elite show business, undergoes an inevitable transformation.

Marketing, which in the pre-digital economic system existed as an outer auxiliary function in the sphere of operatic art, never used to affect the creative core. Now, however, it is expanding its frontiers. Marketing today is integrating with theater, while beginning to actively form a new opera product, seeking new points of contact with the consumers of different ages, including the younger generation. Formation and transformation of the operatic product by use of digital instruments is not always unilaterally understood and unanimously accepted by the audience. This is why researching the audience's attitude towards the potentials and limitations of introducing such instruments in the process of creating and promoting opera is an important academic and practical task for effective management of the modern opera theater and for shaping its communication strategy. Digital technologies have provided the industry of PR and marketing with new complex methods for evaluating demographic and behavioral characteristics of the consumers who leave informational traces of their presence [6, p. 25], as well as for using this information to form marketing strategies [7, p. 172].

Today, one of the main drivers of transformational processes in cultural economy in the conditions of digitization is the steady growth of "culture of participation" [8, p. 86] presupposing interaction of the "networking public" with the media space [9] which optimizes dialogue between the audience and a cultural organization. Kaiser defines the use of digital technologies by cultural organizations as a special strategy able to fundamentally change the consumers' perception of a given cultural entity [10, p. 132]. It completely reorganizes the traditional creative industries of design, production, consumption, service and plays an important innovational role for their development. On the one hand, creative entities can receive inspiration, financial backing and talent support through of audience analysis, crowdfunding and crowdsourcing with the help of the Internet. [11]. On the other hand, creative institutions can keep in touch with their consumers at a distance on the basis of online platforms. They can quickly register the dynamics of creative demand and select creative products for the consumers with precision and accurate timing.

The so-called virtual agglomeration of creative industries is quickly expanding as it positively influences the life of creative industries, including the development of opera.

The existence of digital technologies generate new commercial opportunities for operatic institutions - such that extend beyond the boundaries of traditional merchandising and published products (programs, billboards etc.) [12]

The authors of this research single out three types of digital impact as applied to operatic art:

– Acting on the operatic product per se – introducing technologies in the show itself (digital lighting, sound, effects, props)

– Optimizing distance demonstration of the operatic content through airing and online streaming, as a compensation for the absent live sound. This same purpose is also served by digital libraries of opera houses and halls as well as by professional recordings and

© The Authors, published by EDP Sciences. This is an open access article distributed under the terms of the Creative Commons Attribution License 4.0 (http://creativecommons.org/licenses/by/4.0/).

demonstrations involving technologies of virtual and augmented reality; by digitizing and archiving best performances and concerts.

– Engaging new technologies as means of promotion for operatic art and for drawing a wider audience. This concerns such formats as advertising in social networks, digital PR communications, and some non-standard instruments (advergaming, virtual excursions to opera halls etc).

A special place in this regard is occupied by representing opera organizations in social media through application of SMM (social media marketing) instruments. Operatic institutions use social networks for spreading information about their activities and shows given for real and potential audiences [13]. Filling social media with relevant content has a whole range of functions, like informing the audience about forthcoming events, building up loyalty via effective direct marketing (communicating with the audience in comments and direct messages), targeted advertising, drawing new consumers by means of influencer marketing (live streamings and online interviews with famous singers, producers, conductors).

2 Materials and methods

The aim of the research is to reveal the attitude of the consumers (opera lovers) in Russia and abroad towards engaging digital technologies in operatic art and towards digitized promotion of opera products in and outside Russia. The authors also aim to evaluate the effectiveness of such tools in terms of monetizing and optimizing the activity of opera halls in the "Covid-19"- and "post-Covid-19" periods.

In connection with the overall digitization tendency of cultural industries and the necessity of revealing its specifics in regard to opera theaters, the authors have formulated several key questions of the research:

How do operatic product consumers in Russia and abroad feel about involvement of digital instruments in opera?

To what degree are the consumers aware of the directions and technologies of digital transformations in operatic industry; do they use digital resources of musical organizations?

Can digital analogues of opera products eventually replace the traditional formats?

The hypothesis of the research is that the audience regards digital technologies as an effective instrument of popularizing opera, a mark of its adaptability to the realities of the digital society, and, consequently, a token of opera theaters' economic thriving.

For conducting the research the authors have chosen the empirical method of surveying on Google Forms, which is one of the most optimized platforms for this type of study. The research engaged a standardized questionnaire and was done exclusively in the online medium, as this was in complete conformity with the topic of the survey. In the course of the work the authors applied such research methods as comparing and generalizing. The research mainly focused on respondents from two countries - Russia and Italy, as one of the world opera centers. The motivation to take the survey was modeled by the corresponding motivational post Surveys in the Russian and Italian languages which were circulated from 17 December 2020 to 10 January 2021 via two large opera-dedicated groups in social networks. Representativity of the respondents selection is ensured by the narrow thematic character of the groups, all the members of which are opera lovers, i.e. the target audience of the research.

As the platform for the Russian survey the researchers have chosen the group "Voci dell'Opera" [14] based in the most popular Russian social network VKontakte. The group is of an open character; the number of subscribers as of 13 January 2021 amounted to 55606 persons, making Voci dell'Opera one of the widest spread Russian thematic opera-oriented communities. The group is described as a sort of art-featuring magazine whose aim is to

© The Authors, published by EDP Sciences. This is an open access article distributed under the terms of the Creative Commons Attribution License 4.0 (http://creativecommons.org/licenses/by/4.0/).

reveal the opera world to a wide circle of followers, as well as acquaint the public with ballet, academic vocal and classical music in general. The community covers opera music and theatre, old-time and today's singers, musical shows and related events; makes and publishes musical compilations along with advertising music festivals, opera schools and academies, airings, books featuring and covering opera etc. The group's Instagram account and Internet site - http://www.vocidellopera.com - republish most of the information posted by the community.

The survey in Italian was conducted in Facebook, the reference group being called L'Opera lirica dal Loggione & Musica in Opera [15]. This community has fewer subscribers than Voci dell'Opera (13330 persons as of 13 January 2021, over 9000 of this amount being native Italians). Just like the Russian group, this one, too, publishes information concerning operatic shows and airings. The subscribers have a tradition of exchanging observations and views. The comments feature active discussions and parallels drawn between the modern and retro worlds of opera.

The questions of the survey are based on the following:

- Social and demographic data of the participants (gender, age, place of residence);
- The participants' overall regard of digital technologies as applied to operatic art and the process of advertising opera;
- The participants' level of awareness of the fact that digital technologies are used in the world of opera (if they themselves have often encountered examples of this);
- The participants' involvement in digital platforms of musical organizations as well as social networks; if the content of those is useful to them;
- The potentiality of a live show being substituted for by a digital equivalent online.

The overall amount of respondents in the survey is 325 people (210 Russians and 115 Italians).

The Russian survey comprised 57,7% female and 42,3% male participants; the Italian one - 50,4% female and 49,6% male respectively. The participants were divided into groups according to age - a traditional grouping for this kind of sociological study: 15-24, 25-44, 45-64 и 65+.

Most Russian participants (57,7%) fall in the 45-64 age group; next come 23,7% of those aged 25-44. Those are followed by 11,3% aged 15-25, and 7,3% - 65+. The majority of the Italians occur in the 45-64 age group (49,5%), followed by the 25-44 (27,3%). The respondents 60+ amount to 17,2%, those under 20 - to 6%.

It can be thus inferred that both in Russia and Italy the most opera lovers are 45-64 years old, while respondents under 20 are almost not featured at all in this survey. Meanwhile, Italy yields notably more 65+ participants, whereas in Russia they amount to under 10%.

The age grouping of the participants is representative and largely corresponding to other studies of target audiences in the realm of classical music. The Statista study [16] showed the majority of classical music listeners to be 55+, whereas 35% of the classics' followers are under 35.

The choice of an online platform and the thematic particularity of the study account for the low amount of senior participants (65+), while the age of many respondents occurs just in the "representative" slot between 55 and 64.

The Russian participants were represented by residents of many different regions: the majors like Moscow, Saint-Petersburg, Novosibirsk, Voronezh, Omsk, Samara, Perm and Volgograd; and medium-sized towns like Vladimir, Kursk, Krasnodar, Habarovsk, Ryazan and others. The featured overseas territories are Ukraine (Odessa, Lugansk, Novorossiisk), Belarus (Minsk), Israel (Holon, Modiin), Germany (no city/town specification), Austria (Graz).

All the participants of the Italian survey, too, hail from different towns and provinces of Italy: the major centers like Milan, Rome, Naples, Turin, Venice, Genoa, Palermo, Florence,

© The Authors, published by EDP Sciences. This is an open access article distributed under the terms of the Creative Commons Attribution License 4.0 (http://creativecommons.org/licenses/by/4.0/).

Verona, as well as the medium-sized ones - Ferrara, Padua, Modena, Brescia, Rimini, Bergamo, Taranto, Livorno, Lecce, Salerno. The Italian survey also featured residents of small towns like Sant'Agata di Militello.

3 Results and discussion

The respondents' evaluation of the new stage of operatic art development with application of digital formats has confirmed the idea that, far from being in decline, the industry of performing arts in the segment of opera is gradually developing in spite of the financing problems. This idea has also been corroborated in the course of the expert interviews with participants of the industry, who state that, although traditionally the sphere of performing arts is one of the most difficult objects for result-oriented management, digital technologies facilitate the task.

The hypothesis that introducing digital technologies (in the show itself as well as in promotional instruments) on the whole improves the perception of this art has been supported by the respondents. Almost half of the participants in the Russian group (44,9%) gave the response "rather yes than no" to the question about approving the use of digital instruments.

20,4% gave a unilaterally positive answer ("definitely yes"). The Italian opera lovers appeared a bit less enthusiastic: 40,8% more or less positively evaluate the change in question, and 17,1% give a unilaterally positive reply. Notwithstanding the fact that the development of Italy in the main digitization spheres is highly active, such a rather chilly attitude towards digital instruments in opera is accounted by the fact that opera is a cult in the country, and the classical boundaries of this art are carefully preserved.

The survey has also measured the level of the respondents' awareness of concrete digital instruments that are being integrated in opera today.

Next question is connected with particular innovational technologies in the fields of opera which the participants have observed in real life. The frequency of encounter with certain innovational formats reflects the degree of integration of these technologies in the modern opera, as well as how noticeable they are for the consumers, and if they have become part of the modern opera world, an effective means of its promotion and popularizing.

The respondents were asked to chose between several options: mobile applications of theaters and opera houses, viewing a show in 3D glasses, video games with opera-based plots, digital billboards in town.

The results (figure 1) showed that, on the whole, the Italian audience is much better informed about different opera digitization formats than the Russian counterparts. The Russian respondents who have never encountered digital technologies in the opera are twice more numerous than the Italian ones.

It is also worth noting that such an instrument as AR- and VR-technologies which is receiving an increasingly wider spreading in western theaters is hardly familiar to the Russian audience and barely in demand of opera producers and decorators in Russia (only 12,2% of the Russian respondents happened to be aware of it).

Table 1. The awareness of digital instruments in opera as demonstrated by the Russian and Italian audiences.

Innovational technologies in opera	Russian respondents	Italian respondents
Mobile apps of theaters and halls	52,3%	61,5%
Viewing performances in 3D glasses, AR- и VR-technologies	12,2%	56,6%
Video games with operatic plots	12,2%	13,8%
Digital billboards	36,6%	44,4%
Never observed	16,2%	7,7%

© The Authors, published by EDP Sciences. This is an open access article distributed under the terms of the Creative Commons Attribution License 4.0 (http://creativecommons.org/licenses/by/4.0/).

This question presupposed adding individual answers. Among the variants of digital technologies listed by the Russian respondents were:

- online airings, online concerts and streams;
- interviews and coaching online;
- introducing video clips in the show;

The Italians listed the following digital-instruments:

- online broadcasting and streaming;
- online concerts featuring animation films.

It also appeared relevant to obtain information whether the respondents used social networks of theaters and philharmonics, for, as was noted above,

SMM today is one of the most important digital instruments for opera organizations. More than

40% of the respondents in both cases (42% и 44,9% in the Russian and Italian groups respectively) are subscribed operatic social media and regularly follow their contents. Meanwhile there are also respondents who do not use social networks (33% in the Russian group and 25,5% – in the Italian). 25% и 29,6% Russian and Italian opera goers respectively are subscribed to opera theaters' accounts but do not follow the content.

Among the social networks used by respondents, the most popular are:

Vkontakte (46,9%) in the Russian group and Facebook (50%) in the Italian. Instagram, YouTube и Twitter are more popular among the Italian respondents, although they are also actively used by the Russians.

Apart from other aspects, the research analyzed the participants' opinion whether live operatic sound could be replaced by online streamings.

More than a half of the surveyed (54,5% Russians and 56,5% Italians) believe that online broadcastings are in no way able to compete with a real performance. 39,4% of Russian- and 36,5% Italian respondents accept the possibility of replacement as a partial thing. Less than 10% in both cases are willing to replace a live performance with its online version.

Situational marketing changes the communicational picture of the world according to the concrete informational call. Such a call has recently become the COVID-19 pandemic which has proved the toughest check for the opera industry. The life of the theater was completely stopped, and the only thing left to do for opera houses around the globe was show online broadcastings of their best recent years performances.

According to the data on frequency of viewing online broadcastings during the pandemic, this form of representing operatic art in the digital medium was not in high demand even in the time of the pandemic. 41,8% of the Russian- and 42,9% Italian respondents watched the online shows "several times". 37,8% and 29,6% participants respectively did not watch them at all, while only 20,4% and 27,5% watched them regularly.

The final stage in the survey was the question what the respondents would choose in case of not being able to go to the opera house: watch broadcastings, listen to old recordings, or complete abstain from listening to opera music. Most respondents in both cases chose to refuse the digital equivalent of a live performance, and preferred classical recordings to modern broadcasts. 32,3% of Russian respondents are willing to watch online streams, against only 39% of Italian counterparts. 16,2% and 16,5% Russian and Italian participants respectively choose to live without listening to opera at all if real theater is inaccessible. Having analyzed the results of the online survey, we have come to the following conclusions:

- Most participants from both Russia and Italy feel that introducing digital technologies in operatic art is a positive factor, and believe that these technologies can improve the perception of opera today.
- A wide circle of the respondents have encountered digital technologies in opera and in its promotion, which reflects active presence of this technologies in the communicational

© The Authors, published by EDP Sciences. This is an open access article distributed under the terms of the Creative Commons Attribution License 4.0 (http://creativecommons.org/licenses/by/4.0/).

field and proves that they are visible to the consumers, while the Italian audience is a lot more aware them than the Russian

– Most respondents are subscribers and regular followers of operatic social media

– Meanwhile, a significant amount of the participants refuse to accept replacement of live theater with its online version regarding such a change as impossible

– Online broadcastings which replaced real theatrical live in the time of the pandemic were only viewed by a limited number of the consumers: most of them would choose to listen to old classical recordings.

4 Conclusions

The process of digitization of opera is not only a musical and cultural, but also a marketing and economic issue, for it directly affects stability of the theater, the size of investment in it, its creative projects. From the point of view of integrating new instruments in the operatic show and using them for promoting the cultural product, the respondents see these instruments as a positive influence on modern perception of opera. This concerns innovations in the opera content itself (unusual, spectacular technologies attracting the audiences' attention) as well as interactive channels of informing the consumers, such as social networks. The practice of online streamings as an additional activity of operatic institutions successfully services their brand, expands their geography, intensifies the interest of potential viewers. With such an approach, digital formats and platforms do not lead to the crisis of live performing, denaturation of the art and loss of its meaning [17, p. 4].

The moot point, however, remains the possibility of opera "going online" as a new phenomenon which is equalized to traditional theater, and where live sound is replaced with digital "formula". On the one hand, a number of respondents view this a chance to see events that would be inaccessible offline, on the other - the majority feel that it is important that the theater should remain authentic and that its real acoustic core should remain intact.

The research had shown that the opera theater tradition remains firm and, even in the conditions of a pandemic, "switching online" is not always received with a welcome, or proves a successful promotional move.

Thus, it can be concluded that, in today's reality, digital technologies are important, but only additional tools for promoting live opera. They must not change the traditional artistic essence of opera, but harmoniously augment it, while also meeting the requirements of modernity. This is the main difficulty of adapting a conservative genre like opera to the digital reality of today. Time alone will show if, while preserving the integrity and traditions of the operatic theater, digital technologies will be able to attract new audiences as well as increase the loyalty of the existing ones.

References

1. A. Modlinski, L. M. Pinto, Management-journal of Contemporary Management Issues, **25,** 1 (2020)
2. I. Lacasa-Mas, I. Villanueva-Benito, Profesional de la Informacion, **5(4),** 413 (2012)
3. O. Obradovic, M. Senicic, IPSI BGD Transactions on Internet Research, **17(2),** 45 (2021)
4. Worldwide art sales fall by 22% in 2020, Interfax, https://www.interfax.ru
5. I. Sedykh, Industry Performing Arts, National Research University Higher School of Economics (2018) https://dcenter.hse.ru

© The Authors, published by EDP Sciences. This is an open access article distributed under the terms of the Creative Commons Attribution License 4.0 (http://creativecommons.org/licenses/by/4.0/).

6. I. Dickey, W. Lewis, *Handbook of research on digital media and advertising: User generated content consumption,* 1 (2011)
7. Sh. Rodgers, E. Thorson, Routledge, *Digital advertising: theory and research (3rd edition)* (2017)
8. G. Crawford, V. Gosling, G. Bagnall, B. Light, Information Communication Society, **17(9),** 1072 (2014)
9. M. Ito, *Introduction* (Networked publics London: MIT Press, 2007)
10. M. Kaiser, *The Cycle: A Practical Approach to Managing Arts Organizations* (Brandeis University Press, 2013)
11. X. Chen, C. H. Liu, C. C. Gao, Y. Jiang, Sustainability, **13(14)** (2021)
12. F. Avanzini, A. Baratè, G. Haus, L. A. Ludovico, S. Ntalampiras, HCI, **12515,** 325 (2020).
13. L. Wroblewski, *Culture Management: Strategy and Marketing Aspects* (Berlin: Logos Verlag Berlin GmbH, 2017)
14. Voci dell'Opera' VKontakte, https://vk.com/operavoci
15. L'Opera Lirica dal Loggione & Musica in Opera', Facebook, https://www.facebook.com/groups/musicainopera
16. Distribution of classical music listeners worldwide as of the 4th quarter of 2018, by age group, https://www.statista.com
17. C. Pernelet, European overview of the use of digital media for opera, music and dance education, European Network for Opera and Dance Education (2015) https://www.reseo.org

© The Authors, published by EDP Sciences. This is an open access article distributed under the terms of the Creative Commons Attribution License 4.0 (http://creativecommons.org/licenses/by/4.0/).

SHS Web of Conferences **114**, 01009 (2021) https://doi.org/10.1051/shsconf/202111401009
NTSSCEM 2021

Digital transformation and its role in the reproduction of innovative development of the modern banking institution of Russia

Sergey Petrovich Sazonov, *Irina Aleksandrovna* Ezangina, *Alena Aleksandrovna* Polianskaia*, and *Aleksandr Ivanovich* Chunakov

Volgograd State Technical University, Lenin Ave., 400005, Russian Federation

Abstract. The institutional mechanism of the financial services market puts forward new requirements to the banking community, determining the need for innovations in products, technologies, services, and encouraging credit institutions to transform the declared strategy and implemented financial policy. In the present, the innovative factor is decisive in supporting the competitiveness of the bank, and the development of banking activities is determined by the degree of activity in promoting innovations. The article reveals the role, directions, tools of digitalization of the innovative development of a modern commercial bank, strengthening its competitive position.

1 Introduction

The Russian banking sector, which has been developing at an accelerated pace since its market formation, has clear process advantages: most of the stages of its development occurred in the era of digitalization, and the integration of modern technologies into the business system occurred at the first stages. On the other hand, the process of introducing innovations into the activities of Russian banks has its own history, features, certain trends, under the influence of which credit institutions develop new products and services, making their own development more modern. The well-known marketing research of the banking services market, carried out both by credit institutions themselves and by specialized companies, confirms the fact that the existing range of traditional services and products is currently not able to fully satisfy the transforming needs and expectations of customers.

The pandemic actualized the acceleration of innovation reproduction in the banking environment through its digitalization, including through the spread of a new mechanism, strengthening the universality of operational use, and the development of a banking remote application as the basis of an ecosystem characterized by a large-scale partner network and seamless authorization. At the same time, objectively, the trend towards digitalization will not end with the epidemic: banking innovations are necessary to attract new customers and improve financial performance. On the other hand, the larger the bank's financial resources, the better the customer needs; the greater the financial result from the introduction of

* Corresponding author: helenderbet@mail.ru

© The Authors, published by EDP Sciences. This is an open access article distributed under the terms of the Creative Commons Attribution License 4.0 (http://creativecommons.org/licenses/by/4.0/).

innovations, the more vigorously the financial institution reproduces innovative products and services to meet the needs of the market. The importance of the development and implementation of new services dictates the need for their elaboration, the creation of new methods of provision, taking into account the maximum amount of banking risks and the impact of innovative solutions on bank profitability.

The objective of the research is to reveal the role, directions and tools of digitalization in the innovative development of a modern commercial bank, and to strengthen its competitiveness.

Materials and research methods. The research was based on fundamental and practical materials of domestic and foreign authors on competition in financial markets, the competitiveness of banking institutions in the market of innovative services and technologies, open materials of news agencies, and actual data on the activities of a number of domestic banking institutions that actively promote the digital trend in the financial services market. The following research tools were used: methods of observation, analysis and synthesis, induction and deduction, a systematic approach, methods of grouping and comparisons, etc. The research methodology was based on the collection, processing of an array of data, the formulation of relevant conclusions, and their justification.

2 Study results and discussion

At the present, the development of the world banking system is taking place in the conditions of intense and increasing competition, crisis processes and cataclysms in the financial markets. The authors recognize the tactics of permanent innovations as the most important factor in the successful development of banking activities. It is about financial and technological innovations as conditions for stability, competitiveness, ensuring sustainable economic growth of banking structures [6].

Banking innovation in this context is a product, service, technology, the promotion of which means a qualitative transformation of banking activities, as a rule, bringing the bank directly or indirectly an increasing net financial result [1]. At the same time, it should be understood that banking innovation, integrating the creative actions of the internal environment of the bank, implements the principles of:

- novelty — the activity is associated with the creation of a fundamentally new or a significant change in the functional characteristics, consumer qualities of well-known products, services, operations and tools for offering them to consumers;
- consistency (synergetic effect): banking innovations are accompanied by reengineering of the management structure, the mechanism for organizing activities, the process of providing services and implementing banking products, improving banking technologies;
- the maximum possible satisfaction of market demand: innovation is aimed at the customer, his needs, improving the quality and comfort of the received financial and credit services;
- commercial viability.

The modern world practice recognizes that the catalysts of banking innovation are:

- globalization of global financial and banking markets;
- rapidly growing competition between the banking and non-banking sectors in the financial services market;
- global financial crisis [3]. Objectively, there are a number of trends in the Russian financial market that mediate the most effective promotion and development of innovations (Figure 1).

© The Authors, published by EDP Sciences. This is an open access article distributed under the terms of the Creative Commons Attribution License 4.0 (http://creativecommons.org/licenses/by/4.0/).

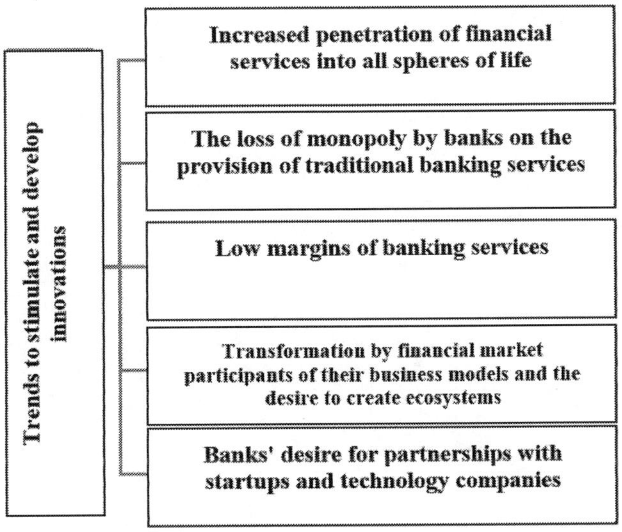

Fig. 1. Trends in stimulating and developing innovations in the banking sector [11]

Among the areas of banking innovations, the Bank of Russia updates: biometric identification of customers; artificial intelligence (AI), mobile technologies, robotic process automation (RPA), distributed registry, cloud technologies (Figure 2).

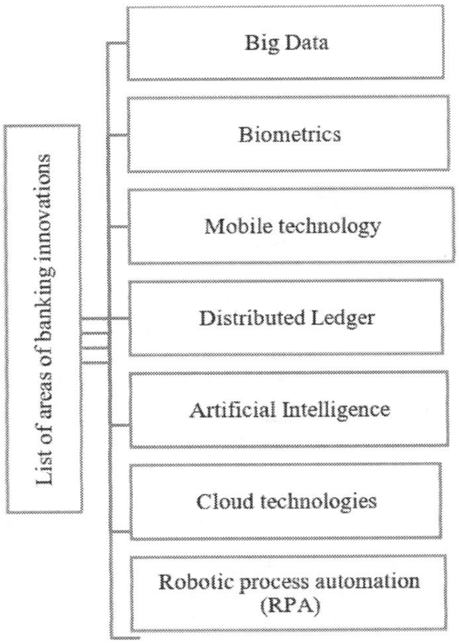

Fig. 2. List of areas of banking innovations [11]

Indeed, the modern world of remote banking dictates the need for constant replenishment of analytical and scoring products that provide real-time customer evaluation and are able to form a behavioral forecast of the customer based on big data technology. In 2021, the Bank of Russia, together with credit market participants, is working on the possibility of launching a digital mortgage using distributed ledger technology. This project has serious prospects in the market, making it possible to make the processes "seamless", to abandon the "routine old technological solutions" [16].

© The Authors, published by EDP Sciences. This is an open access article distributed under the terms of the Creative Commons Attribution License 4.0 (http://creativecommons.org/licenses/by/4.0/).

Comparative characteristics of different types of biometric authentication, as well as all the advantages and disadvantages of each of its methods are shown in Figure 3.

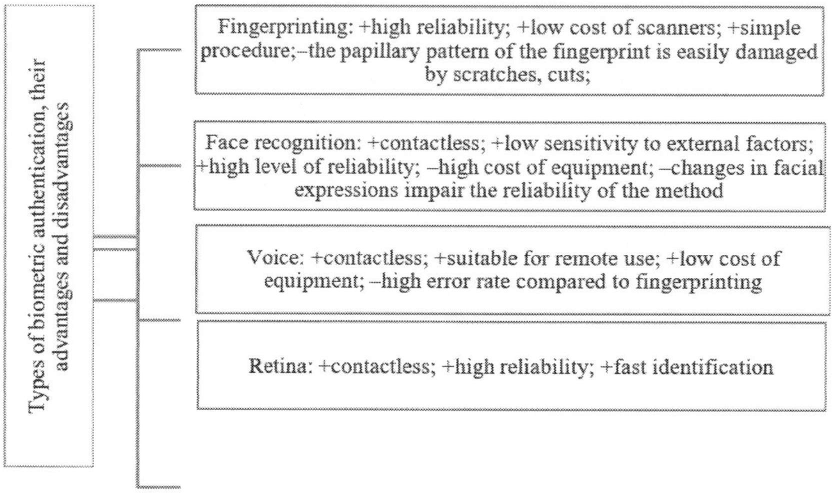

Fig. 3. Types of biometric authentication in a modern commercial bank [17]

In continuation, Table 1 shows examples of the use of chat bot functionality by Russian banks when servicing individuals [9]. At the same time, among the advantages of the latter, the following popular functionality is noted: to inform about the features of products and services; to provide contact information; to make payments, including from card to card; to give financial recommendations to the client; to inform about exchange rates and currency; to keep records of personal finances; to send requests for trade and Internet acquiring and check the counterparty by INN (ORGN (IE); to answer questions of users.

Table 1. The well-known chat bot functionality implemented in the practice of commercial banks of Russia [9]

Chat bot functionality	Name of the bank
Processing the simplest queries (currency exchange rates, etc.), testing the personal finance accounting service, which includes various types of bank income and expenses, flower delivery and ordering goods.	PAO Sberbank, AO Alfa-Bank, AO Raiffeisenbank
Refinancing of loans to the population, with the help of which customers can calculate the terms of this product	PAO VTB
Information about news in the bank, promotions and special offers, reminders of contact information and bank details, addresses of ATM machines, etc.	AO Raiffeisenbank, AO Tochka
The use of chat bots for internal needs of the IT department Creation of community in Telegram, in which IT specialists of the bank work with malfunctions	PAO Home Credit
Checking the account balance, loan debt, making a money transfer to the counterparty in instant messengers	AO Tochka, AO Russian Standard
Formation of an application for a loan (message to the bot of the desired loan amount and sending him a scan of the passport).	PAO Sovkombank

Among the directions of using AI in commercial banks, the following are the most popular: credit scoring; detection of fraudulent transactions; remote identification of customers, monitoring of compliance with norms and regulations; active use of AI in bank marketing; computer vision. Through the use of AI tools such as Big and Machine

© The Authors, published by EDP Sciences. This is an open access article distributed under the terms of the Creative Commons Attribution License 4.0 (http://creativecommons.org/licenses/by/4.0/).

Learning, some banks have successfully applied machine learning in other areas of the banking business. These include, in particular:

- personal advising or an individual financial manager;
- automation of decision-making based on the analysis of big data;
- cybersecurity with the use of blockchain technology;
- personalization of customer offers;
- automation of basic operations and standard business processes: from back office, accounting and audit to legal control of document flow [9].

Using the tools mentioned above, as well as the well-known classifications of banking innovations, modern banks implement various types of innovation strategy: process, technological; strategy of pursuit, double, subordination; hidden, leading, consolidated [18].

Thus, according to the development method, PAO Sberbank implements planned innovation strategies with their division into two large groups: process and technological strategies.

In the first case, it is about focusing on the systematic introduction of innovations into the operational (technological) management system of PAO Sberbank, including changes in the organizational structure of PAO Sberbank;

In the second case, the implementation of innovative measures is practiced in the content component of the services provided by PAO Sberbank, the banking products offered.

At the same time, process strategies are more primary than technological ones, with an emphasis on offering customers fundamentally new banking products and services [20].

According to the degree of integration with the external environment, PAO Sberbank uses the strategy of a leader, and the priorities of its activities are in the plane of constant systematic reproduction of innovations on a flagship basis. The introduction of innovative technologies, products and services on the credit market is focused on the leadership in obtaining "primary" profit; in turn, while competitors evaluate the innovation, the bank keeps going further, developing and commercializing the following "innovations" [21].

Thus, by implementing innovative strategies, the participants of the modern Russian financial sector are forced to constantly take into account the time risks and requirements of technological evolution. It is about such external factors of a stimulating nature as: "viral" marketing; multi-channel availability; personalization; loyalty programs; assistance in managing personal property assets; use of networks; intuitive interface; public pricing [10]. As a result, the concept of interaction with the target community is transformed: open innovation ecosystems replace the well-known technologies of B2B and B2C communications.

The implementation of the client-oriented principle is embodied in the VTB Bank ecosystem, which includes: building a Housing ecosystem (Rebranding of the site "Square Meter", Proptech solution "Domiland", Marketplace for the sale of seized property "CommissiON", Service "Digital Mortgage"); B2B business content (Marketplaces, "Digital accounting for small and medium-sized businesses", "Purchases", "Export Manager" service); Lifestyle Banking ("My Smart City", Multibonus loyalty program, Bicycle rental, VTB-mobile); Big data platforms; Fiscal data operator. At the same time, the VTB ecosystem is built on the "open" principle together with partners, instead of absorbing other players.

Describing the ecosystem of Tinkoff Bank, created since 2013, we note as its peculiarity the concentration of resources on projects not related to classical banking services. The bank has its own global services in the field of insurance, investment, projects for servicing the segment of small and medium-sized businesses, a virtual mobile operator. The ecosystem also includes integrations with partners, including "Rambler. Cashier" and "Afisha-Restaurants". The existence of a super-application, which is a single window of

© The Authors, published by EDP Sciences. This is an open access article distributed under the terms of the Creative Commons Attribution License 4.0 (http://creativecommons.org/licenses/by/4.0/).

access to the services of the bank, should be recognized as a feature of the bank's ecosystem [2].

Table 2 and Figure 4 show the components of PAO Sberbank's progressive ecosystem, which is equally focused on corporate clients and private consumer audience.

Table 2. Directions of development of the PAO Sberbank ecosystem for the private consumer segment as of 2020 [12]

Category	Services	
1	2	
1. Food:	delivery club, samokat, Kukhnya Na Rayone, sbermarket	level kitchen, performance food, my food, sberfood
2. Health:	sberhealth	
3. Goods:	sberprime, sbermobile, sberlogistics	
4. Entertainment:	okko, sberzvuk, afisha	
5. Transport:	sberauto, setelem, citymobil, osago online, yudrive, 2GIS	
6. Realty:	domclick, sovtech	
7. Carrier:	rabota.ru	
8. Education:	sberclass, sberuniversity	
9. Finances:	investments, deposits, loans, yandex.money, sberspasibo	given loans, sberbank online, sberpay
10. Technologies:	bi.zone, visionlabs, sbercloud, sberdisk, r-keeper, sberkorus, cognitive pilot	STC group of companies, sberdevices, smart market and smart apps, sber id
11. Media:	soyuzmultfilm, segmento, sbermarketing	
12. For business:	registration of an IE or LLC online, settlement and cash services, my accounting outsourcing, lawyer for business, rabota.ru, employee check	online cashier and acquiring in one device, acquiring, delivery club - connecting restaurants, sbercloud.advanced, sberservice, sberfactoring

© The Authors, published by EDP Sciences. This is an open access article distributed under the terms of the Creative Commons Attribution License 4.0 (http://creativecommons.org/licenses/by/4.0/).

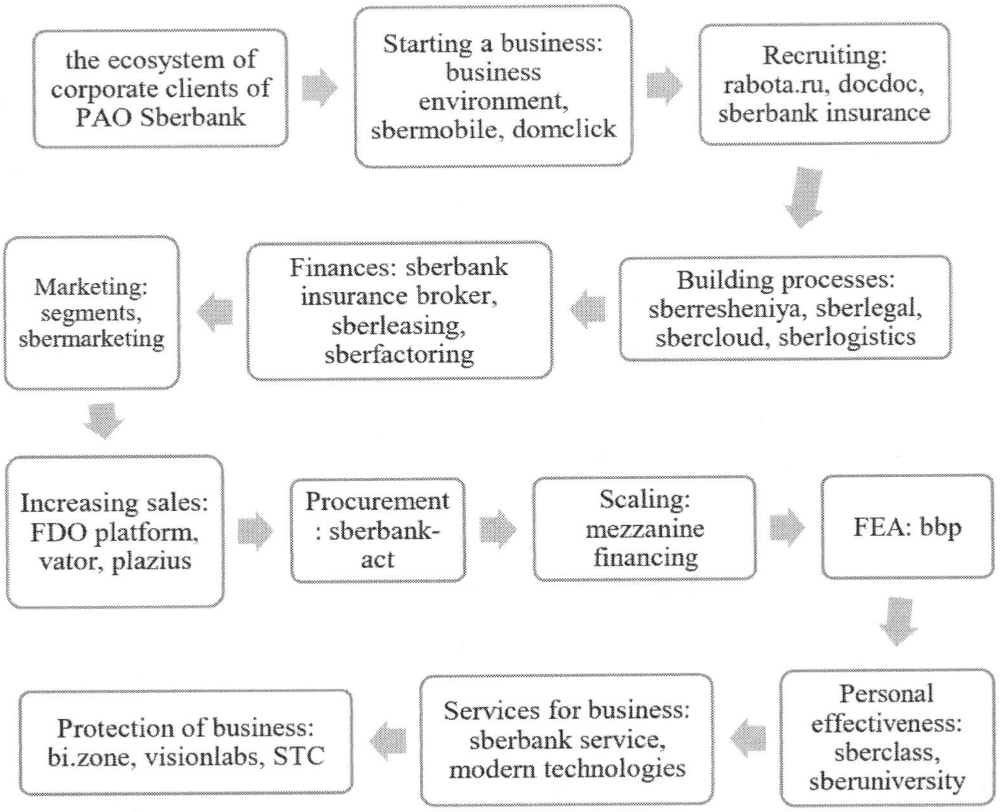

Fig. 4. The ecosystem of PAO Sberbank's corporate clients as of 2020.

The result of the long-term development of the ecosystem of PAO Sberbank, in accordance with the postulates of Strategy 2023, should be a financial business that combines the needs of individuals and legal entities in three categories: e-commerce, healthcare, education (Table 3).

Table 3. Combining the needs of individuals and legal entities on the basis of the financial business of PAO Sberbank [12, 13]

Category	**The needs of individuals**	**The needs of legal entities**	**Services, combining individuals and legal entities**
1	2	3	4
1. E-commerce:	1. Any offer at a reasonable price. 2. Fast and comfortable delivery.	1. Market place. 2. Digitalization of business and processes. 3. Optimization of marketing.	1. SberMarket. 2. Samokat. 3. 2GIS. 4. Farmacy. 5. 3p marketplace. 6. SberConnect. 7. SberLogistics

© The Authors, published by EDP Sciences. This is an open access article distributed under the terms of the Creative Commons Attribution License 4.0 (http://creativecommons.org/licenses/by/4.0/).

Table 3. Continued

2. Health service:	1. The right doctor is always available. 2. Medical content. 3. Preventive measures and healthy lifestyle.	1. Utilization of infrastructure and optimization of the doctor's time. 2. Access of clinics to the client base of PAO Sberbank. 3. Digitalization of back-office.	1. SberHealth 2. Telemed. 3. Medcard.
3. Education:	1. Personalized content. 2. Lifelong learning.	1. Digitalization of processes and learning programs. 2. Educational content.	1. SberClass. 2. SberEducation. 3. Personnel. 4. School. 5. Skills. 6. Information. 7. Platform for universities.

In contrast to the manufacturing sector, not only product innovations are customer-oriented, but also, not least, technological banking innovations. Reproduction of the emphasized functionality of any bank, turning it into a full-scale fintech center. In this context, an important factor of innovative development is the ability to competently cooperate with technology companies, both in terms of joint development and implementation of innovative solutions, and in outsourcing innovations. The list of strategic partners of banks includes aggregators of information of an extensive user orientation (for example, social networks and telecom operators) that have and provide access to external customer data [7].

For example, VTB Group is a member of the corporate accelerator GenerationS, which integrated institutions and startups aimed at finding a customer for project development, ready to work on the principles of "open innovation". The key initiatives and directions in the field of technological development and digitalization of the Group in 2020 are presented in Figure 5.

In 2019, VTB launched pilot projects with the following companies participating in the accelerator: Data Fabric; FreshDoc, Synpatic, VOCA-TECH, WantResult, Ziax, AIST.

The format of foresight projects is the basis of VTB's digital transformation methodology. Each project is focused on the effective promotion of digitalization products based on advanced development technologies: IoT (Internet of Things), Big Data, artificial intelligence, blockchain, biometrics. 2017 — the beginning of the launch of 5 foresight projects, 2018 — the beginning of the work of the Bank's Digital Transformation Management team on 17 new products, including developing the previous ones.

In 2020, Ak Bars Bank announced the search for startups in the field of SoftPOS-solutions for small and micro businesses to accept payments via SBP and card payments using smartphones. Initiated in 2020, the URALSIB Tech Corporate Accelerator is aimed at finding, piloting and implementing innovative technologies and solutions based on fintech startups in the Bank's business. The process within the program is based on direct interaction between the cross-functional team on the part of the Bank and the startup. Among the key areas of transformation identified by the bank: services to increase loyalty and optimize the customer path; products in the field of digitalization of the customer experience of small and medium-sized businesses; services for speech recognition and customer identification by voice; scoring of customers-individuals and legal entities; platforms and solutions for Machine Learning models; optimization of internal processes, automation and robotics (document recognition, archival digitalization) [15].

© The Authors, published by EDP Sciences. This is an open access article distributed under the terms of the Creative Commons Attribution License 4.0 (http://creativecommons.org/licenses/by/4.0/).

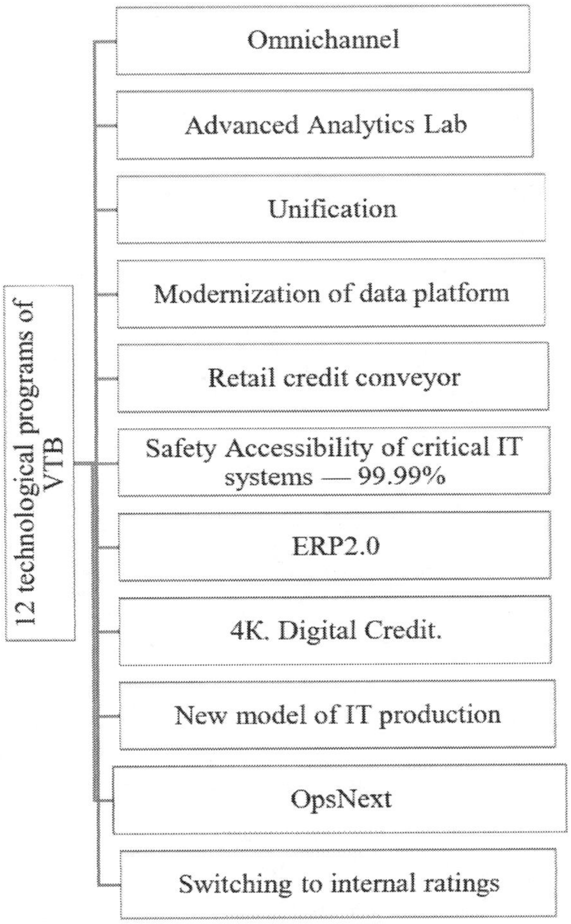

Fig. 5. The key initiatives and directions in the field of technological development and digitalization of the VTB Group in 2020 [4]

The bank's customers will be its specialized subdivisions, which form mini-product teams, for which an educational program is planned for the development of soft-skills for the quick launch of pilots and the formation of concepts (product vision).

Within the framework of the interbank collaborative accelerator, the partnership strategy is implemented by Absolut Bank, Gazprombank, and Credit Bank of Moscow (MKB). Banks selected 23 startups with new and promising services and products out of 90 teams that entered the team, aimed at improving the efficiency of the banking sector [15].

Based on the scale of its own financial resources. the path of creating its own "point" innovation laboratories was chosen by PAO Sberbank in 2017. 8 laboratories of the bank are working to increase expertise in fundamental technologies, to introduce new and new innovations in the bank's products, conduct scientific research in conjunction with scientific centers and national institutions (Figure 6).

© The Authors, published by EDP Sciences. This is an open access article distributed under the terms of the Creative Commons Attribution License 4.0 (http://creativecommons.org/licenses/by/4.0/).

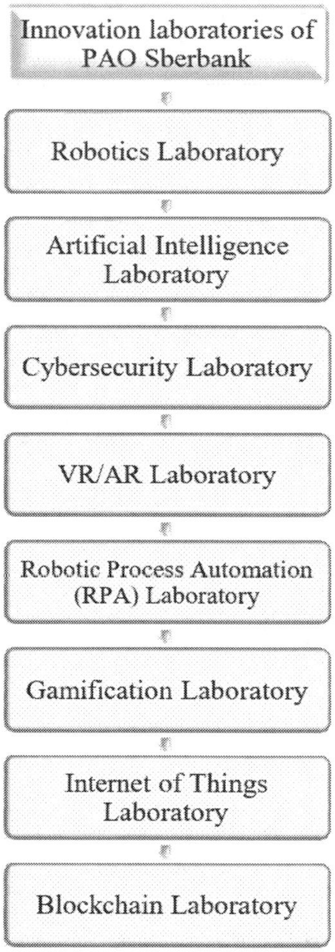

Fig. 6. Technological system of innovation laboratories of PAO Sberbank [8]

In 2019, the Robotics Laboratory implemented a pilot project for the delivery of correspondence by an autonomous office robot-courier "Gleb". The Navigation-As-A-Service and Robot-As-A-Service platforms are focused on promoting robotic solutions for indoor logistics. Testing of the ExoChair Mk3 exoskeleton was completed as part of a pilot implementation in the logistics and document accounting sector of the archive center in Tomilino.

The laboratory develops solutions for use in various processes of the bank. Thus, the "Digital Avatar" project includes a prototype of a solution for generating a video image of a speaker in real time based on artificial neural networks. Avatar allows to reduce the cost and time period of video content production. Since 2019, the robotics laboratory of PAO Sberbank and Microsoft Research has been conducting joint applied research on the use of AI in industrial robotics.

The dynamics and forecast of sales of logistics robots in monetary terms in the dynamics of 2016–2021 are shown in Figure 7.

© The Authors, published by EDP Sciences. This is an open access article distributed under the terms of the Creative Commons Attribution License 4.0 (http://creativecommons.org/licenses/by/4.0/).

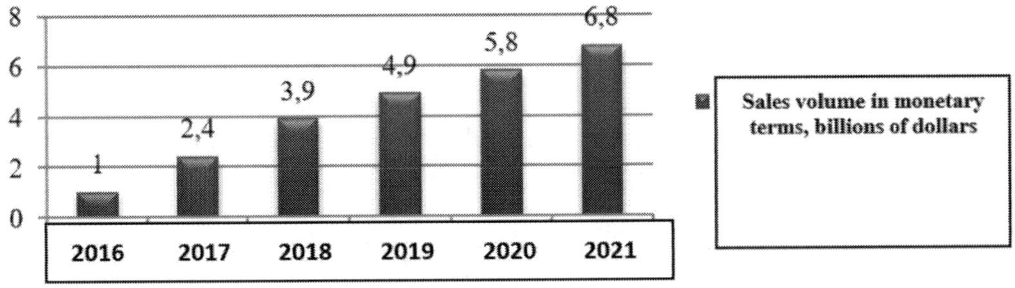

Fig. 7. Dynamics and forecast of sales of logistics robots in monetary terms in 2016–2021, billions of dollars [12]

The product of the research of the artificial intelligence laboratory is the AutoML library, which automates typical decision-making models. The Laboratory also created a minimum viable product (MVP) that allows to quickly prototype recommendation systems for various scenarios specific to the bank and the ecosystem. The importance of applying new approaches to solving NLP problems is emphasized. 4 pilots were recognized as successful:

- information "smart search" based on the data set of the crowdsourcing platform "SberbankIdea";
- identification of suspected duplicate ideas on the platform;
- prediction of expert opinion in the part of the proposal to unload the internal moderators of the platform;
- recommendations of ideas to the users based on their previous activity on the platform [19].

As a result of the activities of the cybersecurity laboratory in 2019, 4 patents for the invention and 3 patents for the design were obtained. In 2019, the VR/AR Laboratory completed the development of the basic functionality of a tactical VR simulator, which allows remote cash collection training. Blockchain Lab specialists developed and presented to the customer an updated version of the blockchain platform for the structure "Sberbank Factoring". Together with Rostech, the Government of the Russian Federation developed and approved the roadmap of the "Distributed Registry System", which is recognized as a strategic tool that outlines the priorities and prospects for the development of end-to-end digital technologies in the Russian Federation [14].

The Robotic Process Automation (RPA) laboratory of Sberbank automated 103 processes and received 2.8 million robot hours, which means there is no need to hire 930 people. In 2019, the following achievements of the Laboratory were recognised (Figure 8).

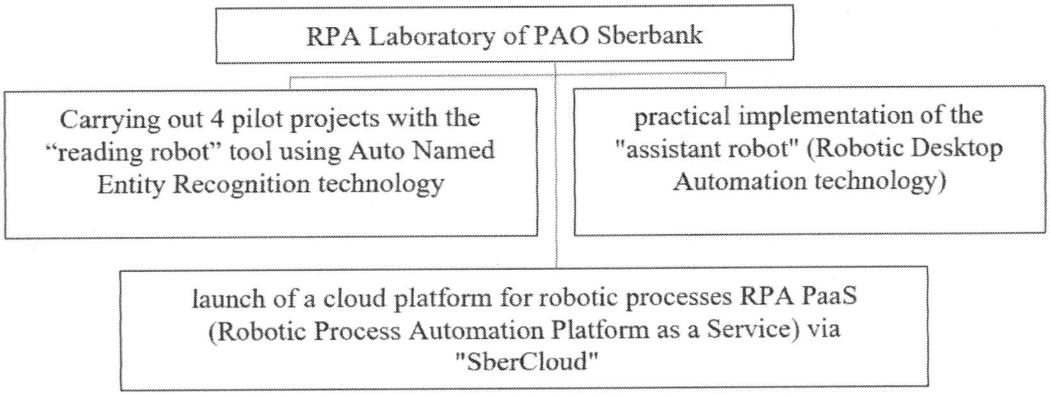

© The Authors, published by EDP Sciences. This is an open access article distributed under the terms of the Creative Commons Attribution License 4.0 (http://creativecommons.org/licenses/by/4.0/).

Fig. 8. Key developments of the Robotic Process Automation Laboratory of PAO Sberbank [8]

3 Conclusion

Thus, innovative strategies of modern commercial banks, openly focused on the partner, the client, maximizing the financial result in the conditions of growing competition of traditional banking institutions and non-banking sector entities that aggressively penetrate the financial markets, require the reproduction of banking innovations, further continuous digitalization of the banking system. It is noted, on the one hand, the development of the properties of consistency, scale, universality, multichannel provision of banking and non-banking services, on the other hand, specialization, maximum personalization, efficiency of processes, technologies and services. The paper emphasizes the equal importance in the content of banking innovation of reproducing the novelty of the product and the technology of its provision, which mediates new organizational forms of banking activity. Promising areas of work on digitalization of various segments of banking services are recognized as: the introduction of intelligent management systems, the creation of a system of innovation laboratories in the banking ecosystem, startup accelerators, the implementation of foresight projects. In this conditions, the Bank of Russia is playing a more significant role, its functional content is significantly expanding: in addition to the financial regulator, it is becoming a direct participant in the digital space, a consumer of innovative products, and an initiator of technological solutions that cannot be abandoned. The potential of new technologies in regulation and supervision – RegTech, SupTech [5].

References

1. O. I. Lavrushin, *Banking management*, 347 (2019)
2. The Bank of Russia saw risks in creating banking ecosystems, https://www.rbc.ru/
3. Innovations in banking sphere, http://edoc.bseu.by:8080/
4. Information technologies in VTB Group, Information portal Tadviser. State. Business. IT, https://www.tadviser.ru/
5. I. A. Ezangina, A .S. Kolchugina, V. A. Bondarenko, Financial business, **3**, 22 (2021)
6. E. A. Kazanskaya, Young scientists, **15**, 297 (2016)
7. How innovations will change the banking industry of Russia, https://trends.rbc.ru/
8. Laboratories of PAO Sberbank, https://sbergraduate.ru/laboratories/
9. O. I. Lavrushin, I. I. Vasilyeva, A. E. Ushanov, *Models and technologies of banking activity*, 112 (2020)
10. N. N. Martynenko, I. D. Machikhin, Innovations and investments, **3**, 184 (2020)
11. The main development directions for financial technologies for the period 2018–2020, https://cbr.ru/
12. Official web-site of PAO Sberbank Russia, https://www.sberbank.com/ru
13. PAO Sberbank launched a digital medical service, https://www.rbc.ru/
14. PAO Sberbank prepared a roadmap for the development of AI in Russia, https://www.cnews.ru/
15. S. P. Sazonov, I. A. Ezangina, E. E. Kharlamova, I. A. Chekhovskaya, Fundamental research, **2**, 50 (2021)
16. Russian banks will start testing digital mortgage in May 2021, https://www.finanz.ru/

© The Authors, published by EDP Sciences. This is an open access article distributed under the terms of the Creative Commons Attribution License 4.0 (http://creativecommons.org/licenses/by/4.0/).

17. N. E. Sokolinskaya, *Optimizing Market Instruments and Banking Management: A Compendium*, 152 (2017)
18. I. S. Lola, M. Bakeev, Management and Production Engineering Review, **11(3)**, 26 (2020)
19. I. Krasonikolakis, M. Tsarbopoulos, T.-Y. Eng, Journal of General Management, **46(1)**, 60 (2020)
20. L. B. Gouveia, M. Perun, Y. I. Daradkeh, International Journal of Open Information Technologies, **8(7)**, 124 (2020)
21. F. J. Forcadell, E. Aracil, F. Ubeda, Business Strategy and the Environment, **29(6)**, 2181 (2020)

© The Authors, published by EDP Sciences. This is an open access article distributed under the terms of the Creative Commons Attribution License 4.0 (http://creativecommons.org/licenses/by/4.0/).

Digitalization and Indian economy: patterns and questions

Inna Lazanyuk[1,*], and *Dr. Swati* Modi[2]

[1]RUDN University, Miklukho-Maklaya str.6, 117198 Moscow, Russia
[2]GLS Institute of Business Administration, GLS University, Opp. Law Garden, Ellisbridge, 380006 Ahmedabad, India

Abstract. Digital technologies play a crucial role in enhancing the competitiveness of the economy and promoting economic growth in many countries. The paper is dedicated to the development of digitalization in the modern economy of India and its impact on economic processes that determine the country's economic growth. We analyzed changes in unemployment, debit and credit card transactions, indirect tax revenues, exports, and trade volumes. In this work, we used analysis and methods of correlation-regression analysis. We found a positive relationship between the number of debit and credit card transactions for indirect tax collection. We confirmed that digitalization had influenced the growth of exports and trade volumes by increasing the role of e-commerce while not increasing the unemployment rate.

1 Introduction

Digitalization, being an objective process, affects all spheres of the socio-economic system. Modern information technologies have a significant impact on the economy and society, changing the prevailing economic structure and life of people.A study by the McKinsey Global Institute (MGI) notes that in 2017 the digital revolution entered a decisive phase - every second inhabitant of the Earth connected to the Internet.

UNCTAD's 2017 Report on the New Digital Economy and Development notes that it is challenging to measure the development impact of digital growth due to its rapidly changing nature and complex network of areas affected by the growth of the ICT sector. However, the study looks at a wide range of impact areas such as economic performance, employment, innovation (including research and development), privacy and security, education, health, citizen, individual and community participation, and the environment.

Since July 2015, the Make India Digital project launched in India. This project aims to increase the digital literacy of rural India by connecting them to the Internet. India is one of the fastest-growing countries with a very high youth population (~ 65%), less than 35 years old.The impact of digitalization can only clearly be observed when people can understand and accept them. Every person must be financially literate, and this makes it possible to make intelligent decisions. After introducing digitalization, financial transactions began to be

*Corresponding author: lazanyuk-iv@rudn.ru

© The Authors, published by EDP Sciences. This is an open access article distributed under the terms of the Creative Commons Attribution License 4.0 (http://creativecommons.org/licenses/by/4.0/).

carried out over the Internet, and digital financial literacy became critical. Digitalization can transform the business model and provide new opportunities for generating revenue and creating value.

Despite the active study of various economic processes by scientists from different countries, many unsolved problems exist. As a rule, research in this area involves studying the decline in employment due to the introduction of new technologies. At the same time, this issue looks pretty relevant given the high rate of destructive processes. American scientists have found that almost half of the workers are engaged in operations that new technologies and machines can potentially replace within one to two decades [1]. A similar study in Germany estimated the share of such workers at 42% [2]. The work of scientists from India, the digital divide issues, and the impact of digitization on the distribution of income and wage inequality [3, 4]. However, one should not consider only the negative aspects of this process. The advantages include the increase in labor mobility [5] due to the "blurring" the boundaries, opening up new opportunities and contributes to competitiveness.

For the first time, the term digital economy was used in 1995 by Canadian professor D. Tapscott in his book "Digital Economy: Promise and Danger in the Age of Network Intelligence" [6]. American scientist J. Negroponte drew attention to the role of information and digital technologies in the economy and society and their possible increasing importance in the future [7]. Since its first appearance in the mid-1990s, the definition of the digital economy has evolved to reflect the rapidly changing nature of technology and its use by businesses and consumers [8].The scientific literature focuses primarily on the current state and prospects for the development of digitalization of the economy [9]. Some authors emphasize the role and significance of transaction costs for digital business [10], others cover the issues of risk and the limitations that result from innovations of the digital economy, leading to social problems [11]. Some authors explore the development and analysis of business models in today's economy [12-14]. Currently, the research quite widely presents the general problems of business development and management in the conditions of a sharing economy [15, 16]. The sharing economy is based on information technology to provide people with information to optimize resources by moving surplus goods and services [17, 18]. However, research is under-represented on the contribution of the digitization of the economy on economic growth and the level of social welfare, primarily since the acceleration of these processes has not led to the rapid development of individual countries and the world economy a whole.

This paper focuses on the impact of accelerated digitalization on the Indian economy. The presented research is based on analyzing two main hypotheses: does digitalization reduce the level of employment in the Indian economy; whether debit and credit card transactions affect indirect tax revenue collection, export growth, and trade volumes.We are interested in why digitalization is essential and which sectors have been affected and how India could become the third-largest economy shortly.The work aims to identify the contribution of digitalization to the Indian economy in sustainable development.

2 Methods

The research methodology is based on secondary data analysis. Secondary data we collected from reliable government sources, research papers from prominent researchers, websites, journals, etc. The study also involves recognizing the challenges and opportunities of digital India in all aspects of excellence and improving the personal satisfaction of residents. The study provides an optimistic methodology for providing broader job prospects for youth to help the country's economy achieve positive growth.

© The Authors, published by EDP Sciences. This is an open access article distributed under the terms of the Creative Commons Attribution License 4.0 (http://creativecommons.org/licenses/by/4.0/).

3 Results and Discussion

The research literature describes the significant contribution that digitalization has made to economic growth and productivity growth and efficiency at the macroeconomic and microeconomic levels. While the overall conclusion is clear, the results vary widely, and the precise definition or quantification of effects is complex for many reasons. First, digitalization is one of the many drivers of productive activity, so it isn't easy to measure its impact on outcomes such as GDP or jobs. Moreover, digitalization contributes to production in many ways - by increasing the efficiency of other resources, such as the workforce (for example, by allowing skilled workers to shift their attention from routine tasks to more complex tasks, thereby increasing labor productivity, capital (for example, allowing more flexibility redirecting excess capacity to produce goods for other markets); and changing fundamental ways of using resources (for example, by enabling businesses to reorganize).

3.1 New digital realities

The Government's efforts facilitated shaping new digital realities in India, active actions of Indian business, and joint steps of Indian government agencies and private entrepreneurs. In 2015, India launched the campaign "Digital India» (Digital India), which included some critical government initiatives to enhance digital literacy, infrastructure development, and the creation of e-Government. Among the most significant successes achieved along this path is the completion of the design and implementation of the state system of digital identification of citizens "Adhar". An important step was creating a single digital tax system, extended to all states that previously had individual tax rules.

The spread of digital technologies in agriculture, education, energy efficiency, employment and labor market regulation, etc. is essential for the country and has significant potential. There is a widespread belief that the introduction of new technologies leads to a decrease in employment.

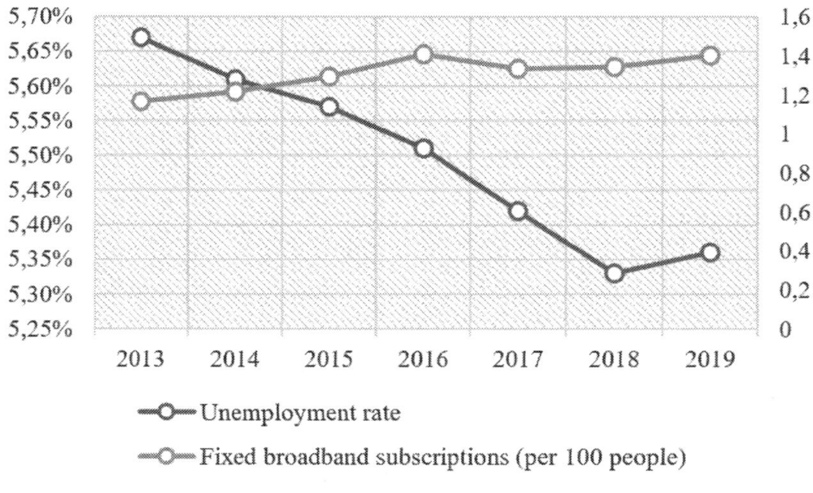

Fig. 1. Unemployment rate (left axis), Fixed broadband subscriptions (right axis) [19].

During the study period from 2013 to 2020, the number of broadband subscribers in India grew by an average of 20%. Fig. 1, you can see that the unemployment rate since 2013 declines and the number of subscribers continues to grow. It indicates that with an increase in the number of broadband subscribers, unemployment decreases, positively impacting digitalization. It does not support the hypothesis that digitalization leads to a decrease in

© The Authors, published by EDP Sciences. This is an open access article distributed under the terms of the Creative Commons Attribution License 4.0 (http://creativecommons.org/licenses/by/4.0/).

employment. Instead, it leads to a redistribution of work due to, for example, an increase in employment in large infrastructure projects, etc.

The development and implementation of the fast payment system by the Reserve Bank of India and the Association of Indian Banks led in the shortest possible time (only 5-7 years) to revolutionary changes in such industries as electronic payments and Fintech, e-commerce, telemedicine, entertainment. Debit and credit card transactions are sky rocketing (Fig.2)

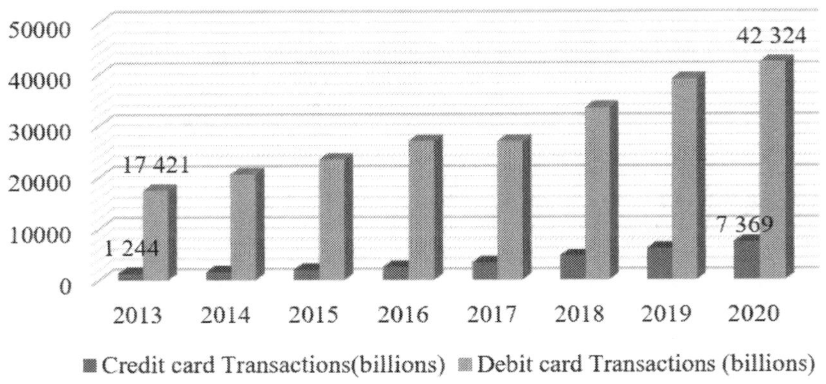

Fig. 2. The number of Card Transactions Source: Prowess Database

As seen in Fig. 2 for the period 2013-2020, the number of debit card transactions increased by an average of 2.5 times. It is also expanding the number of credit card transactions, the annual growth during the same period amounted to 28.9%.

The e-commerce market volume has grown almost 5 times compared to 2015 and in 2020 amounted to 64 billion US dollars. Due to the growth of Internet users, which in 2020 reached 697 million people and the number of mobile phones sold (3.6 billion units). The Indian e-commerce market is overgrowing and is projected to exceed $ 100 billion by 2023.

In just a few years, India has made a quantum leap in many areas related to digitalization, avoiding the intermediate stages that other countries have spent years on, and second only to Indonesia in the speed of this process. Data consumption per user in India is growing twice as fast as in the United States and China, increasing 152% annually. Internet users in India are expected to increase by about 40% to 750-800 million by 2023, and the number of smartphones will double to 650-700 million. With a high degree of confidence, we can assume that the new conditions associated with a pandemic significantly speed up these processes. Over the past few years, India has gone through a rapid digitalization journey. The current crisis has dramatically accelerated this process and also more noticeably indicated specific trends.

3.2 Correlation and regression analysis of the impact of digitalization in the country's economy

More than 560million internet users in 2020, India has become the second largest online market globally. The accessibility to internet facilities is increasing day by day. The government efforts towards digitalization have resulted in the increase in country's digital population amounting to approximately 688 million active users as of January 2020. The effect of digitalization in India has been pronounced and positive. In this study, the impact of digitalization has been studied on the unemployment rate, debit and credit card transactions, Indirect tax revenue collection, level of exports and trading volumes.

© The Authors, published by EDP Sciences. This is an open access article distributed under the terms of the Creative Commons Attribution License 4.0 (http://creativecommons.org/licenses/by/4.0/).

Table 1. Correlation Matrix between factors.

Corelation Matrix table No 1	*internet usage (millions)*	*Trading volumes*	*Indirect tax revenue (Millions)*
internet usage (millions)	1		
Trading volumes	0.218386547	1	
Indirect tax revenue (Million	0.981960691	0.363486324	1

We can infer that internet usage is highly significantly and positively related with indirect tax revenue (0.98) from the table 1. The trading volumes have increased due to digitalization and technology advancement. The E-trading helps investor to access the market easily and also increases their awareness. Due to the modernization of Indian stock market, especially online trading, the trading volumes have increased. There is a significant positive relationship between internet usage and trading volumes, indicating the up surge of activity in the stock market due to digitalization.

Table 2. Regression statistics results.

Regression Statistics							
Multiple R	0.981961						
R Square	0.964247						
Adjusted R Square	0.958288						
Standard Error	1015463						
Observations	8						
ANOVA							
	df	*SS*	*MS*	*F*	*Significance F*		
Regression	1	1.7E+14	1.7E+14	161.817	**1.4E-05**		
Residual	6	6.2E+12	1E+12				
Total	7	1.7E+14					
	Coefficients	*Standard Error*	*t Stat*	*P-value*	*Lower 95%*	*Upper 95%*	*Lower 95.0%*
Intercept	3183935	1085234	2.93387	0.02616	528463	5839407	528463
internet usage (millions)	30081.15	2364.73	12.7207	1.4E-05	24294.9	35867.4	24294.9

The model is highly significant (Tab. 2) The Null hypothesis that the tax revenue is not depending on the internet usage is rejected. There has been a tremendous and significant relationship between the two variables. The collection of Indirect Tax revenue has increased considerably due to the better internet facilities and other IT development initiatives by the Indian Government.

Table 3. Correlation Matrix between factors.

Correlation Matrix	Fixed broadband subscriptions (per 100 people) India	Unemployment rate	Credit card transactions	Debit card transactions	Transactions at POS	Exports (Millions)
Fixed broadband subscriptions (per 100 people) India	1					
Unemployment rate	-0.784	1				
Credit card transactions	0.721	-0.921	1			

© The Authors, published by EDP Sciences. This is an open access article distributed under the terms of the Creative Commons Attribution License 4.0 (http://creativecommons.org/licenses/by/4.0/).

Table 3. Continued.

Debit card transactions	0.826	-0.928	0.979	1		
Transactions at POS	0.666	-0.927	0.992	0.954	1	
Exports (Millions)	0.702	-0.790	0.855	0.884	0.833	1

The Unemployment rate has a negative correlation which clearly states with the increase in broadband subscriptions, the unemployment is reducing which is a positive impact of digitalization (Tab. 3). There is a significant direct positive relationship between fixed broadband subscriptions per 100 people with credit card transactions, debit card transactions, transactions at POS and exports. There has been a significant increase in credit card transactions by 388.35%, in debit card transactions by 124.14%, transactions at POS by 699.42% and in level of exports by 68.88%.

Table 4. Regression statistics results.

Regression Statistics						
Multiple R	0.784981					
R Square	0.616196					
Adjusted R Square	0.539435					
Standard Error	0.00088					
Observations	7					
ANOVA						
	df	*SS*	*MS*	*F*	*Significance F*	
Regression	1	6.21E-06	6.21E-06	8.02748	0.036534	
Residual	5	3.87E-06	7.74E-07			
Total	6	1.01E-05				
	Coefficients	*Standard Error*	*t Stat*	*P-value*	*Lower 95%*	*Upper 95%*
Intercept	0.069	0.0051	13.46799	4.04E-05	0.056284	0.082837
Fixed broadband subscriptions (per 100 people) India	-0.01116	0.003938	-2.833	0.03653	-0.02128	-0.00103

Regression model of fixed broad band subscription and unemployment rate where unemployment is a dependent variable on fixed broad band subscription (Tab. 4). The model is significant and the unemployment rate is decreasing with the increase in fixed broad band subscription. Individuals using the Internet (% of population) in India have increased from 15.1% to 20.08%

India is going for mass vaccination where lakhs of people are vaccinated per day. As soon as the vaccination is done, message is sent that you are partially or fully vaccinated. Maintaining of such huge database with all the details like registration, date of first dose, 2nd dose is all made possible through digitalization.

Indian E-commerce is overgrowing and is expected to go beyond the US economy by 2034. India e-commerce sector will reach US$99 billion by 2024 from US$30 billion in 2019,

© The Authors, published by EDP Sciences. This is an open access article distributed under the terms of the Creative Commons Attribution License 4.0 (http://creativecommons.org/licenses/by/4.0/).

expanding at a 27% CAGR, with grocery and fashion/apparel likely to be the key drivers of incremental growth. There has been increase in Internet and smart phone penetration. Due to these various developments in November 2020, Amazon India announced collaboration with Hindustan Petroleum Corporation Limited. Under this partnership, customers will be able to book and pay for their LPG cylinders until the delivery. Amazon India also collaborated with the Indian Railway Catering and Tourism Corporation (IRCTC) to enable users to book and reserve train tickets on Amazon. Amazon India invested over Rs. 700 crores (US$ 95.40 million) into its payment unit, Amazon Pay. Market size of e-commerce industry (Billion) has grown from 14% in 2014 to 64% in 2020.

Table 5. Correlation Matrix between factors

Correlation Matrix	*Secure Internet servers - India*	*GVA at basic prices*
Secure Internet servers - India	1	
GVA at basic prices	0.897972408	1

From the table 5 we can infer that Gross Value added at basic prices is highly correlated with the secure internet servers which are a clear indication that the increasing use of Internet is having a positive association with GVA [20].

4 Conclusion

The study showed evidence of the contribution of digitalization to the Indian economy. The hypothesis that connecting users to broadband increases the unemployment rate has not been confirmed. Use of smart phones affected digital transactions and increased transparency and accountability of the financial system. We reaffirm the importance of financial literacy, which can change the development phase of emerging economies such as India. Internet use is highly and positively associated with indirect tax revenue. Due to the modernization of the Indian stock market, especially online trading, trading volumes have increased. There is a significant positive relationship between Internet usage and trading volumes, indicating a surge in stock market activity due to digitalization.

The current crisis associated with the pandemic is shaping new consumer habits and increasing the demand for qualitative changes in approaches to economic development problems for many countries of the world in the future. Correct implementation of digitalization could change the development phase of the Indian economy and enable it to become the third-largest economy in the future.

References

1. C. Frey, M. Osborne, Technological Forecasting and Social Change, **114**, 254 (2017)
2. M. Arntz, T. Gregory, U. Zierahn, SSRN Electronic Journal (2019)
3. R. Kapoor, Technology, Accelerators of India's Growth—Industry, Trade and Employment, 301 (Springer, Singapore, 2020)
4. D. Maiti, F. Castellacci, A. Melchior, Digitalisation and Development, 3 (Springer, Singapore, 2020)
5. H. Matlay, Journal of Small Business and Enterprise Development, **11(3),** 419 (2004)
6. R. Yakob, International Journal of Advertising, **28(1),** 182 (2009)
7. N. Negroponte, R. Harrington, S. McKay, W Christian, *Computers in Physics*, **11(3),** 261 (1997)

© The Authors, published by EDP Sciences. This is an open access article distributed under the terms of the Creative Commons Attribution License 4.0 (http://creativecommons.org/licenses/by/4.0/).

8. K. Barefoot, D. Curtis, W. Jolliff, J. R. Nicholson, R. Omohundro, US Department of Commerce Bureau of Economic Analysis, Washington, DC, **15** (2018)
9. K. Schwab, *The Fourth Industrial Revolution: What It Means and How to Respond* (2015) https://www.foreignaffairs.com
10. M. Munger, Journal of Economic Sociology, **20(5),** 74 (2019)
11. D. Curran, *European journal of social theory*, **21(2),** 207 (2018)
12. B. W. Wirtz, Digital business models (Springer International Publishing, 2019)
13. D. Stojanovic, Economics: theory and practice, **10(1),** 80 (2017)
14. H. Ke, S. Chai, R. Cheng, Journal of Intelligent & Fuzzy Systems, **36(1),** 609 (2019)
15. S. Revinova, S. Ratner, I. Lazanyuk, K. Gomonov, Sustainability, **12(12),** 4855 (2020)
16. W., Nadeem, M. Juntunen, F. Shirazi, N. Hajli, Technological Forecasting and Social Change, **151,** 119786 (2020)
17. B. Fang, Q. Ye, R. Law, Annals of Tourism Research, **57,** 264 (2016)
18. S. Hofmann, O. Sæbø, A.M. Braccini, S. Za, Government Information Quarterly, **36(4),** 101399 (2019)
19. India: Internet subscribers, per 100 people, https://www.theglobaleconomy.com
20. Indian E-commerce Industry Analysis (2021) https://www.ibef.org

© The Authors, published by EDP Sciences. This is an open access article distributed under the terms of the Creative Commons Attribution License 4.0 (http://creativecommons.org/licenses/by/4.0/).

Eco-efficiency of regional development: strategical planning with DEA

Svetlana Ratner[1,2,*], and *Andrey* Kovalev[2]

[1]RUDN University, Miklukho-Maklaya str.6, 117198 Moscow, Russia
[2]V.A. Trapeznikov Institute of Control Sciences of Russian Academy of Sciences, Profsoyuznaya str., 65, 117997 Moscow, Russia

Abstract. Despite the implementation of the national "Ecology" project, attempts to develop low-carbon energy and transport, the implementation of "garbage reform" and other measures aimed at reducing the negative impact of the economy on the environment, the eco-efficiency of production systems in Russia remains low. In this paper, we propose a new approach to solving the problems of comparative assessment of the effectiveness of regional environmental management systems (R-EMS), taking into account the structural heterogeneity of the regional economies. The developed approach includes the simultaneous solution of two DEA tasks, one of which is formulated to assess the eco-efficiency of the region's economy, and the other to assess the efficiency of spending funds aimed at protecting the environment in the region. As a result of solving two problems, the researcher receives a large amount of information that can be used in developing strategies for regional development. In particular, the information obtained can be used to develop measures to stimulate the activities of regional authorities in the field of environmental management and increase their responsibility for the effectiveness of environmental protection measures. The proposed approach was tested on the example of solving the problem of comparative assessment of the effectiveness of the R-EMSs in Russia in the period 2015-2019. However, the method developed by us is not tied to specific territories and can be used to conduct similar studies in other countries and regions.

1 Introduction

Nowadays most emerging economies consider economic growth, increasing competitiveness in world markets, mastering advanced production technologies, and solving social problems of the population as the main strategic goals of their development. At the same time, environmental problems, although they are recognized by the authorities and the business community, do not receive enough attention due to limited resources, especially finances. This situation is typical for Russia as well as for many other countries. Funding of environmental protection in Russia is significantly lower than, for example, in the developed countries of the European Union (Fig. 1).

* Corresponding author: ratner_sv@pfour.ru

© The Authors, published by EDP Sciences. This is an open access article distributed under the terms of the Creative Commons Attribution License 4.0 (http://creativecommons.org/licenses/by/4.0/).

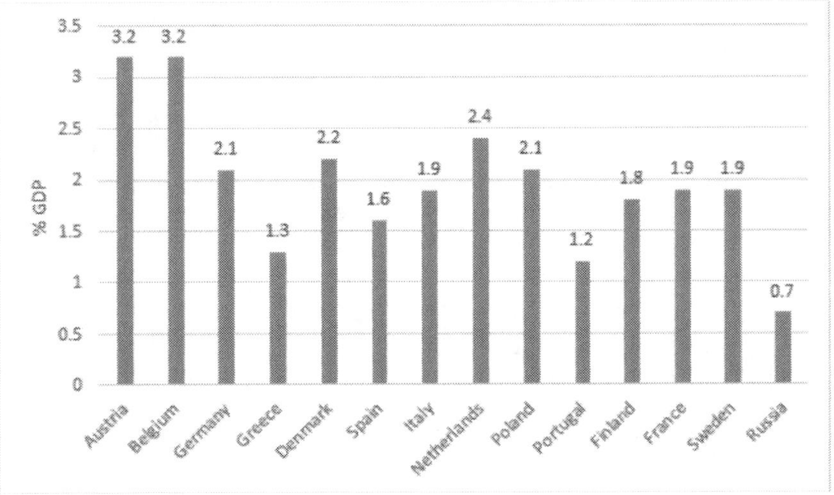

Fig. 1 Financing environmental protection (Source: authoring, based on data from Russian Federal State Statistic Service https://gks.ru/)

Despite the implementation of the national "Ecology" project, attempts to develop low-carbon energy and transport, the implementation of "garbage reform" and other measures aimed at reducing the negative impact of the economy on the environment, the eco-efficiency of production systems in Russia remains low. This statement can be easily confirmed by cross-country comparisons of statistical indicators of resource intensity, energy intensity and carbon intensity of the Russian economy.

The main environmental problems are solved in Russia at the federal level through the implementation of federal programs and projects. At the same time, everyday monitoring of compliance with environmental legislation, collection of data on the state of the environment, educational work with the population and business, distribution of funds received in the form of payments for the negative impact of enterprises on the environment, occur at the regional level. Therefore, in the scientific literature devoted to the problems of environmental management in Russia, it is customary to talk about territorial/regional environmental management systems, although this term is not officially used either in the standards for environmental management systems of the ISO 14000 series, or in legislation. The center of decision making and the center of responsibility in the regional environmental management system is the regional ministry of natural resources and ecology (a subdivision of the regional administration). However, so far the indicators of the eco-efficiency of the regional economy are not included in the system of indicators, according to which the activities of the regional authorities are assessed (for example, the KPI indicators of the heads of the regions). In our opinion, the introduction of such indicators into the system for assessing the effectiveness of regional administrations could significantly improve the quality of environmental management at the regional level.

It should be noted that there are no generally accepted approaches to assessing the effectiveness of regional environmental management systems in the Russian literature. In studies [1-2] authors propose an approach to assess the effectiveness of a regional environmental management system based on Data Envelopment Analysis (DEA) models, in which the costs of environmental protection in the region were considered as inputs, and positive changes in the eco-efficiency of the regional economy were considered as outputs. The advantage of this approach is the comprehensive accounting of a wide range of different categories of economic impact on the environment, as well as the calculation of target indicators for each region, the achievement of which allows it to become efficient in the ecological and economic sense. In addition, the advantage of this approach is its consistency

© The Authors, published by EDP Sciences. This is an open access article distributed under the terms of the Creative Commons Attribution License 4.0 (http://creativecommons.org/licenses/by/4.0/).

and naturalness, since the effectiveness of the regional environmental management system is understood in a purely economic sense, as the ratio of the positive results obtained to the costs incurred to achieve these results.

At the same time, this approach does not take into account the degree of eco-efficiency of the economies of different regions. It is well known, that the spatial development of Russia is highly heterogeneous. The regions of Russia differ significantly among themselves not only in natural and climatic conditions, area, population, level of economic development, but also in the structures of economies. Therefore, regional environmental management systems (R-EMS), as subjects of management, are initially not in equal conditions from the point of view of how complicated the object of the management is.

The aim of this study is to expand the approach originally proposed in [1-2], and adapting it to solve the problems of assessing the comparative effectiveness of regional environmental management systems, taking into account the significant heterogeneity of regions in terms of eco-efficiency.

2 Literature review

Eco-efficiency is a concept often used in environmental management literature. This concept was first explained in the works [3-6], who presented eco-efficiency as the ability of a production system to simultaneously achieve positive economic and environmental effects, i.e. increase profits while simultaneously reducing the negative impact on the environment. In the studies of these authors, the concept of eco-efficiency was put in contrast to the traditional idea that increasing the environmental efficiency of the production system inevitably leads to costs and, thereby, reduces economic efficiency. Thus, eco-efficiency was seen more as a process of introducing eco-innovation. This understanding is enshrined in the documents of the World Business Council for sustainable development [7], where eco-efficiency is considered as a process of constant reduction of negative environmental impacts.

Later in the literature, eco-efficiency began to be understood more "statically". For example, in [8], eco-efficiency is understood as the efficiency of using natural resources in the production of products. In [9], the eco-efficiency of a manufacturing system is understood as its ability to produce the lowest possible amount of emissions with the best available technologies. This approach to determining eco-efficiency has become widespread due to the active use of the DEA methodology in solving the problems of assessing the comparative efficiency of production systems, including regional production systems. For example, in [10-12] environmental efficiency is considered as the ability of a production system to minimize undesirable outputs (negative externalities of production activity) at fixed values of inputs (material and human resources) and desired outputs (GRP or other economic indicator of the results of the production system). In [13-14] eco-efficiency is considered to be the ability of a production system to produce the maximum desired outputs with minimum negative environmental impacts. This implies that the reduction of negative environmental effects is achieved through a more rational use of all types of input resources, i.e. eco-efficiency includes the efficient use of not only natural resources, but also human and financial resources. In addition, the desired output is not only the GRP as a useful result of the regional economy, but also the size of the population. That is, the eco-efficiency of the household sphere is taken into account (the housing and communal services of the region, the system of urban mobility, etc.). We consider this idea of eco-efficiency is the most holistic and will use it in our study.

With regard to understanding the effectiveness of the environmental management system in the region, as noted above, we rely on the approach proposed in [1-2]. In these studies, the environmental management system is considered effective if, at the lowest cost for

© The Authors, published by EDP Sciences. This is an open access article distributed under the terms of the Creative Commons Attribution License 4.0 (http://creativecommons.org/licenses/by/4.0/).

environmental protection, it allows achieving the greatest positive changes in the eco-efficiency of the region.

3 Methodology

The inputs and outputs of the two DEA models, solved in parallel, are presented in Table 1. The choice of specific statistical indicators for assessing the impact of the regional economy on the environment is due to the peculiarities of statistical accounting of the state of the environment in the Russian Federation. We took into account all the indicators published in the public domain on a regular basis.

The first problem is the problem of determining the comparative eco-efficiency of the region and can be solved both for one period (as a rule, one year), and for a time interval (for example, for 3 - 5 years). In the second case, the given problem is dynamic, and to solve it we can use the "window" method or other methods for solving dynamic problems. We solve Problem 1 over a period of 5 years (from 2015 to 2019) and use the "window" method. As shown in [13], this time period is optimal for solving dynamic problems of assessing the comparative effectiveness of the development of regional economic systems, since it fully fits into the time frame of cardinal technological changes.

Table 1. Inputs and outputs of DEA models

Problem	Inputs	Outputs
1. Determination of the comparative eco-efficiency of the region	1. Air emissions from stationary sources (thousand tons); 2. Air emissions from mobile sources (thousand tons); 3. Wastewaters (million cubic meters); 4. Fresh water use (million cubic meters); 5. Wastes (tons)	1. GRP (million rubles); 2. Population.
2. Determination of the comparative effectiveness of the regional environmental management system	*In the field of economy*	
	1. Current expenses for environmental protection (million rubles); 2. Investments in fixed assets aimed at environmental protection (million rubles)	1. Change in specific emissions into the atmosphere from stationary sources (thousand tons / million rubles); 2. Change in specific emissions into the atmosphere from mobile sources (thousand tons / million rubles); 3. Change in the specific collection of wastewater (million cubic meters / million rubles) 4. Change in specific consumption of fresh water (million cubic meters / million rubles); 5. Change in specific waste generation (tons / million rubles)
	In the field of housing and communal services and personal mobility	
	1. Current expenses for environmental protection (million rubles); 2. Investments in fixed assets aimed at environmental protection (million rubles)	1. Change in specific emissions into the atmosphere from stationary sources (thousand tons / person); 2. Change in specific emissions into the atmosphere from mobile sources (thousand tons / person); 3. Change in the specific collection of wastewater (million cubic meters / person) 4. Change in specific consumption of fresh water (million cubic meters / person); 5. Change in specific waste generation (tons / person)

© The Authors, published by EDP Sciences. This is an open access article distributed under the terms of the Creative Commons Attribution License 4.0 (http://creativecommons.org/licenses/by/4.0/).

Problem 2 is also solved over a period of 5 years, but with the usual "static method". Changes in specific negative environmental effects are calculated as the difference between the specific effect at time *t1* (beginning of the period) and the specific effect at time *t2* (end of the period). The specific negative environmental effect is understood as the volume of negative impact on the environment (for example, the volume of emissions into the atmosphere) per unit of GRP (or per person). This indicator is a lot like the indicators of resource intensity (energy intensity, carbon intensity, etc.). Current costs are summed up over the entire period. Investments in fixed assets are summed up in a similar way. With this formulation of the problem, we manage to avoid the need to assess the return on investment or take into account the corresponding time lags between the investment and the receipt of a positive result from them. Note that in both problems, all indicators measured in monetary units must be deflated, i.e. take into account the change in the value of money over time, since otherwise comparisons of indicators in different periods will be inappropriate.

Another feature of our approach is the complex formulation of problem 2. This complexity lies in the fact that the task of assessing the comparative efficiency of the regional environmental management system is divided into two in accordance with two areas of regional development: economics and the residential sector. Further, the overall efficiency is calculated as the arithmetic average of the two efficiency ratios.

The mathematical formulation of the DEA problem is traditional for BCC problems [15-16]. It is well known that, BCC problems are more realistic, since they take into account the basic laws of economic development, and their solution provides more information that can be used to make management decisions. [17]. In particular, this is information about the scale effect of each of the objects under study, about the ratio of the pure technical efficiency of the object and its scale efficiency, etc. In addition, BCC tasks allow you to deal with negative outputs (if any of the objects under study) by performing a scale shift procedure [18].

The solution of two DEA problems provides a large amount of information that can be used to develop a strategy for regional development (Table 2).

Table 2. Indicators obtained because of solving DEA problems

Problem	**Indicators**
1. Determination of the comparative eco-efficiency of the region	1. Eco-efficiency scores of the region for each year of the study period; 2. Benchmarks (reference objects) for each region for each year of the study period; 3. Projections (targets) of inputs (or outputs) that a region needs to achieve in order to become eco-efficient for each region in each year of the study period; 4. Direction of economies of scale (increasing, decreasing, constant) for each region in each year of the study period; 5. Scores of pure technical eco-efficiency and eco-efficiency by scale for each region in each year of the study period;
2. Determination of the comparative effectiveness of the regional environmental management system	1. Efficiency scores of regional RSEM for the study period; 2. Benchmarks (reference objects) for each RSEM for the period under study; 3. Projections (targets) of inputs (or outputs) that the R-EMS needs to achieve in order to become effective for each region over the period; 4. Direction of economies of scale (increasing, decreasing, constant) for each R-EMS for the period under study 5. Scores of pure technical efficiency and efficiency by scale for each R-EMS for the period.

© The Authors, published by EDP Sciences. This is an open access article distributed under the terms of the Creative Commons Attribution License 4.0 (http://creativecommons.org/licenses/by/4.0/).

The most interesting for practical applications are cases when a region with high eco-efficiency has an environmental management system with low efficiency, and vice versa, when a region with low eco-efficiency has R-EMS with high efficiency. The environmental management practices used in such regions can be considered the worst and the best, respectively, with all the ensuing administrative consequences (punishment / reward).

4 Results and Discussion

Our calculations showed that the regions with the highest eco-efficiency in the period from 2015 to 2019 are:

1) in the Central Federal District: Bryansk, Vladimir, Ivanovsk, Kaluga regions, and city of Moscow;

2) in the North-West Federal District: the Kaliningrad region;

3) in the Southern Federal District: Sevastopol and the Republic of Crimea;

4) in the North Caucasian Federal District: the Republic of Dagestan, Ingushetia, the Kabardino-Balkarian Republic and the Chechen Republic;

5) in the Volga Federal District: the Republic of Mordovia, the Chuvash Republic, the Saratov and Ulyanovsk regions;

6) in the Ural Federal District: Kurgan and Tyumen regions;

7) in the Siberian Federal District: Altai and Omsk regions;

8) in the Far Eastern Federal District: Sakhalin Region and Chukotka Autonomous District.

Eco-efficiency in 67 regions during 2016-2019 is growing (although not always monotonously), in 13 regions it is falling, in 2 regions it remains stable. Regions with diminishing eco-efficiency are the Oryol and Tambov regions, the Stavropol Territory, the city of Sevastopol, the Republic of Crimea, the Chuvash Republic, the Saratov Region, the Chelyabinsk Region, the Altai Territory, the Omsk and Tomsk Regions, and the Chukotka Autonomous District.

Regions with the highest efficiency of the R-EMS during 2015-2019 are:

1) in the Central Federal District: Bryansk, Ivanovskaya, Kostroma, Moscow and Tver regions;

2) in the Northwestern Federal District: the Komi Republic, Vologda, Leningrad, Murmansk, Novgorod and Pskov regions,

3) in the Southern Federal District: the Republic of Adygea, the Republic of Kalmykia and the city of Sevastopol;

4) in the North Caucasian Federal District: the Republic of Dagestan, Ingushetia, the Kabardino-Balkarian Republic, the Karachay-Cherkess Republic, the Republic of North Ossetia - Alania, the Chechen Republic;

5) in the Volga Federal District: no R-EMS with high efficiency;

6) in the Ural Federal District: Chelyabinsk Region;

7) in the Siberian Federal District: Altai Territory, Republic of Tyva, Republic of Buryatia, Transbaikal Territory;

8) in the Far Eastern Federal District: the Jewish Autonomous Region and the Chukotka Autonomous District.

Almost all regional environmental management systems have diminishing economies of scale, both in the economic sphere and in the sphere of housing and communal services and personal mobility. This means that an increase in the financing of the R-EMS (current costs and investments in fixed assets) leads to a disproportionately small increase in the eco-efficiency of the regions. The only exceptions are the R-EMS of the Bryansk Region, the Republic of Ingushetia, the Republic of Kalmykia, the Chechen Republic and the Republic of Tyva.

© The Authors, published by EDP Sciences. This is an open access article distributed under the terms of the Creative Commons Attribution License 4.0 (http://creativecommons.org/licenses/by/4.0/).

The largest number of questions arise about the activities of the R-EMS of the Kursk region, Moscow city, the Kaliningrad region, the Republic of Crimea, the Republic of Mordovia, the Chuvash Republic, the Saratov region, the Kurgan region, the Altai Territory, the Omsk region and the Sakhalin region, where, with high eco-efficiency of the regions themselves, the efficiency of the RSEM is low. Based on the study, it can be concluded that in these regions funds for environmental protection are spent ineffectively, although the structure of the economy is quite favorable.

Regional environmental management systems with the most effective approaches and practices are R-EMSs of the Kostroma region, the Komi Republic, the Vologda region, the Leningrad, Murmansk and Novgorod regions, the Republic of Adygea, the Republic of Karachay-Cherkessia, the Republic of Buryatia and the Zabaikalsky region, where, with an unfavorable economic structure and the low eco-efficiency of the regions themselves, the costs of environmental protection bring significant changes in eco-efficiency.

5 Conclusion

In this paper, we propose a new approach to solving the problems of comparative assessment of the effectiveness of regional environmental management systems, taking into account the structural heterogeneity of the regional economies. The developed approach includes the simultaneous solution of two DEA tasks, one of which is formulated to assess the eco-efficiency of the region's economy, and the other to assess the efficiency of spending funds aimed at protecting the environment in the region. As a result of solving two problems, the researcher receives a large amount of information that can be used in developing strategies for regional development. In particular, the information obtained can be used to develop measures to stimulate the activities of regional authorities in the field of environmental management and increase their responsibility for the effectiveness of environmental protection measures. The proposed approach was tested on the example of solving the problem of comparative assessment of the effectiveness of the R-EMSs in Russia in the period 2015-2019. However, the method developed by us is not tied to specific territories and can be used to conduct similar studies in other countries and regions. In addition, the developed method is not rigidly linked to a set of statistical indicators used to assess the environmental situation in any territory. The statistical indicators used in our calculations can be replaced with others if they reflect the negative impact of the economy and the housing and utilities sector of the region on the environment.

The limitations of the proposed approach include the fact that, due to the peculiarities of constructing efficiency boundaries in the DEA methodology, objects with zero inputs will always have the highest efficiency. In real practice of environmental management, such situations are possible. They occur when there are no environmental costs in the region at all. In our practical example, this is precisely the situation that is observed in the Bryansk region, where throughout the entire study period, investments in fixed assets aimed at protecting the environment are zero. This practical case requires additional research to understand why this is happening: the region does not need this kind of investment, or the regional authorities cannot create the necessary incentive system for enterprises to make such investments. Elimination of these limitations is the subject of further research of the authors.

References

1. S. Ratner, P. Ratner, Applied Computer Science, **13(2)**, 48 (2017)
2. S. Ratner, M. Zaretskaya, Quality: Access to Success, **21(175)**, 120 (2020)
3. A. King, M. Lenox, Management Science, **48**, 289 (2002)

© The Authors, published by EDP Sciences. This is an open access article distributed under the terms of the Creative Commons Attribution License 4.0 (http://creativecommons.org/licenses/by/4.0/).

4. M. Porter. America's green strategy. Scientific American, **168** (1991)
5. M. Porter, C. Van der Linde, Journal of Economic Perspectives, **9**, 97 (1995)
6. M. Porter, C. Van der Linde, Harvard Business Review, 120 (1995)
7. WBCSD. Eco-Efficiency: Creating More Value with Less Impact. World Business Council for Sustainable Development, Geneva. (2000)
8. S.-O. Holm, G. Englund, Ecological Economics, **68(3)**, 879 (2009)
9. J.-X. Chen, Y. Zhang, S. Zheng, Journal of Cleaner Production, **233**, 611 (2019)
10. N. Zhang, Y. Choi, The Social Science Journal, **50 (2)**, 225 (2013)
11. J. Zhang, W. Zeng, H. Shi, Ecological Indicators, **71**, 218 (2016)
12. J. Huang, J. Xia, Y. Yu, N. Zhang, Ecological Indicators, **85**, 674 (2018)
13. S. Ratner, P. Ratner, Measuring Efficiencies of Regional Eco-Management Systems with DEA. Proceedings of 2017 IEEE 11th International Conference on Application of Information and Communication Technologies (AICT), 2, 256 (2017)
14. S. Ratner, Automation and Remote Control, **81**, 1330 (2020)
15. R.D. Burnett, D.R. Hansen, Accounting, Organizations and Society, **33(6)**, 551 (2008)
16. R.D. Banker, A. Charnes, W.W. Cooper, Management Sciences, **30**, 1078 (1984)
17. A. Emrouznejad, G. Yang, Socio-Economic Planning Sciences, **61**, 4 (2018)
18. R. Lin, W. Yang, H. Huang, Computers & Industrial Engineering, **135**, 39 (2019)

© The Authors, published by EDP Sciences. This is an open access article distributed under the terms of the Creative Commons Attribution License 4.0 (http://creativecommons.org/licenses/by/4.0/).

Eco-innovations for a circular economy: the EU countries rankings

Konstantin G. Gomonov*

Peoples' Friendship University of Russia (RUDN University), 6 Miklukho-Maklaya Street, 117198 Moscow, Russian Federation

Abstract. This article reveals the importance of implementing eco-innovation for the circular economy and sustainable development. The author provides different approaches to measuring the level of development of eco-innovation in the countries of the European Union. The two main indices for measuring eco-innovation are revealed in more detail. Described the pillars of each of the indices. Clustering of EU countries was carried out using the k-means method and groups of countries were identified by the level of implementation of eco-innovation. The clear leaders are Denmark, Sweden. Bulgaria, Cyprus, Poland are catching-up countries.

1 Introduction

Currently, in most developed countries, the government direction of development is associated with the formation of a "green" economy and the concept of sustainable development. This concept is based on a harmonious combination of environmental, economic and social factors. At the same time, the priorities of society's development are built from environmental restrictions through a fair distribution of public resources - to the efficiency of economic development. Thus, almost the entire world community is currently confronted with the challenge associated with the threat of environmental safety. The renewal of natural resources is hindered by a significant increase in production and consumption of products. This leads to tangible signs of an environmental crisis, such as environmental pollution, climate change and an increase in the frequency of natural and man-made disasters [1].

Eco-innovations are necessary factor in preventing a global environmental disaster. The disruption of the natural balance and the deterioration of the interaction between society and the environment are a consequence of the development of the technogenic economy in the world. The key direction in solving the problems described above is the use of the concept of "green growth", which contributes to sustainable economic development and allows funds to be redirected to meet the needs of society. The concept of sustainable development combines three main aspects: economic, social and environmental [2]. In a market economy, sustainable development is achieved by increasing the efficiency of

*Corresponding author: gomonov-kg@rudn.ru

© The Authors, published by EDP Sciences. This is an open access article distributed under the terms of the Creative Commons Attribution License 4.0 (http://creativecommons.org/licenses/by/4.0/).

 https://doi.org/10.1051/shsconf/202111401012

resource use through the introduction of more advanced and environmentally friendly technologies, recycling production waste and their further consumption.

Eco-innovations are considered a key element in making the transition from a linear economy to a circular system of production and consumption. The transition to a circular economy depends on systematic eco-innovation, which includes not only technology intensive but also dynamic and holistic combinations of innovative services and new efficient organizational structures and management systems.

The concept of eco-innovation has emerged as an approach that contributes to the sustainable development of all societies around the world. Eco-innovation can be defined as "all efforts from relevant actors that introduce, develop, and apply new ideas, behaviors, products and processes and contribute to reducing environmental burdens or ecologically specified sustainability targets" [3]. Eco-innovation plays a key role in promoting and implementing green growth because it promotes all forms of innovation that reduce environmental impacts and strengthen resilience to environmental pressures [4].

2 Materials and methods

There are various approaches to measuring eco-innovation, for example, eco-innovation can be measured directly or indirectly [5]. Another approach [6] proposes to classify measures into four types for quantifying technological change considering the eco-innovation process: input measures, output measures, direct outcome measures and indirect impact measures. A review of the various approaches shows that in order to create a measurable index, it is important to understand the details of eco-innovation concepts. Based on the research done to date, the types of innovation that need to be considered when measuring eco-innovation are as follows: Product Innovation, Process Innovation, Marketing Innovation, Organizational Innovation, Material Flow Innovation, and Social Innovation [7,8]. According to the OECD methodology, eco-innovation considers only product innovation, technological innovation, marketing innovation and organizational innovation as the main types of eco-innovation. The EIO approach [8] highlighted the importance of innovation in material flows and social innovation. As the scope expanded, eco-innovation began to embrace different environments around companies and interactions between stakeholders. This means that material flows are also considered together with economic activities, both in their physical and socio-economic dimensions.

The purpose of the study is to consider two approaches to measuring eco-innovation in the European Union, on the basis of statistical data, to determine the leading countries in the development of this direction, that is, to carry out the clustering of countries.

In order to evaluate the eco-innovation performance across all EU Member States, a composite index has been developed by the Eco-Innovation Observatory. [9]. The Eco-Innovation Index (Eco-IS) shows the effectiveness of eco-innovation in the EU member states. The index covers various areas of eco-innovation by contributing 16 indicators grouped into five areas: eco-innovation inputs, eco-innovation activities, eco-innovation outputs, resource efficiency and socio-economic outcomes. The Eco-Innovation Index measures how well individual member states perform in various aspects of eco-innovation. The Eco-Innovation Index complements other approaches to measuring innovation in EU countries and aims to promote a holistic view of economic, environmental and social performance. Table 1 illustrates the relationship between the components of the Eco-Innovation Index and the sustainable development goals.

© The Authors, published by EDP Sciences. This is an open access article distributed under the terms of the Creative Commons Attribution License 4.0 (http://creativecommons.org/licenses/by/4.0/).

Table 1. The Eco-Innovation Index and the sustainable development goals [10].

Component	Indicator	SDGs
ECO-INNOVATION INPUTS	Governments environmental and energy R&D appropriations and outlays (% of GDP)	Goal 8 – Decent work and economic growth Goal 9 – Industry, innovation and infrastructure
	Total R&D personnel and researchers (% of total employment)	Goal 8 – Decent work and economic growth
	Total value of green early stage investments (USD/capita)	Goal 9 – Industry, innovation and infrastructure
ECO-INNOVATION ACTIVITIES	Implementation of resource efficiency actions among SMEs (Score)	Goal 9 – Industry, innovation and infrastructure Goal 12 – Responsible consumption and production
	Implementation of sustainable products among SMEs (% of surveyed firms)	Goal 7 – Affordable and clean energy Goal 9 – Industry, innovation and infrastructure Goal 12 – Responsible consumption and production
	Number of ISO 14001 certificates (per mln population)	Goal 9 – Industry, innovation and infrastructure Goal 12 – Responsible consumption and production
ECO-INNOVATION OUTPUTS	- Eco-innovation related patents (per mln population)	Goal 9 – Industry, innovation and infrastructure Goal 12 – Responsible consumption and production
	- Eco-innovation related academic publications (per mln population)	Goal 9 – Industry, innovation and infrastructure
	- Eco-innovation related media coverage (per mln population)	Goal 9 – Industry, innovation and infrastructure
ECO-INNOVATION SOCIO-ECONOMIC OUTCOMES	- Exports of products from eco-industries (% of total exports)	Goal 9 – Industry, innovation and infrastructure
	- Employment in environmental protection and resource management activities (% of workforce)	Goal 8 – Decent work and economic growth Goal 9 – Industry, innovation and infrastructure
	- Value added in environmental protection and resource management activities (% of GDP)	Goal 9 – Industry, innovation and infrastructure
ECO-INNOVATION RESOURCE EFFICIENCY OUTCOMES	- Material productivity (GDP/Domestic Material Consumption)	Goal 9 – Industry, innovation and infrastructure Goal 12 – Responsible consumption and production
	- Water productivity (GDP/total fresh water abstraction)	Goal 6 – Clean water and sanitation
	- Energy productivity (GDP/gross inland energy consumption)	Goal 7 – Affordable and clean energy Goal 9 – Industry, innovation and infrastructure
	- GHG emissions intensity (CO2e/GDP)	Goal 13 – Climate action

© The Authors, published by EDP Sciences. This is an open access article distributed under the terms of the Creative Commons Attribution License 4.0 (http://creativecommons.org/licenses/by/4.0/).

Eco-innovation inputs comprise investments (financial or human resources) aiming to trigger eco-innovation activities. Eco-innovation activities includes indicators to monitor the scope and scale of eco-innovation activities undertaken by companies. The component focuses on efforts and activities rather than on actual results of innovation activity. Eco-innovation outputs describe the immediate results of eco-innovation activities. Indicators in this component are used to monitor the extent to which knowledge outputs generated by businesses and researchers relate to eco-innovation. Socio-economic outcomes of eco-innovation depict wider effects of eco-innovation activities for society and the economy. This includes changes in employment, turnover or exports that can be related to broadly understood eco-innovation activities. Resource efficiency outcomes relate to wider effects of eco-innovation on improved resource productivity. Eco-innovation can have a twofold positive impact on resource efficiency: it can increase the generated economic value, while at the same time decrease pressures on the natural environment.

The second significant index for assessing eco-innovation – ASEI. It attempts to measure actual eco-innovation both quantitatively and qualitatively using data sources compiled and calculated from various international organizations. This index consists of the following main pillars 1) Eco-Innovation Capacity, 2) Eco-Innovation Supporting Environment, 3) Eco-Innovation Activity and 4) Eco-Innovation Performance. Eco-Innovation Capacity measures competitive potential of nation's eco-innovation level in terms of both social, economic and human resources. Eco-Innovation Supporting Environment measures nation's political and financial support on eco-innovation based on government spending, political priorities and etc. Eco-Innovation Activity reflects actual level of actions taken by government, academy and companies with respect to enhancing level of nation's eco-innovational growth. Eco-Innovation performance demonstrates what nation has accomplished in order to tackle environmental, economic and social standpoint of eco-innovation [11]. Table 2 shows the relationship between the ASEI and the Sustainable Development Goals.

Table 2. The ASEI and the sustainable development goals [11].

Component	Indicator	SDGs
ECO-INNOVATION CAPACITY	1.1. Nation's Economic Competitiveness	Goal 8 – Decent work and economic growth Goal 12 – Responsible consumption and production
	1.2. Nation's General Innovation Capacity	Goal 8 – Decent work and economic growth Goal 12 – Responsible consumption and production
	1.3. Green Technology R&D Institution Capacity	Goal 9 – Industry, innovation and infrastructure
	1.4. Green Technology possessed/acquired Enterprises	Goal 9 – Industry, innovation and infrastructure
	1.5. Awareness of Sustainability Management	Goal 12 – Responsible consumption and production
ECO-INNOVATION SUPPORTING ENVIRONMENT	2.1. Government's R&D expenditure in Green Industry	Goal 9 – Industry, innovation and infrastructure Goal 12 – Responsible consumption and production
	2.2. Implementation of Environmental Regulations	Goal 16 – Peace, justice and strong institutions
	2.3. Maturity of Investment Setting for Green Technology Industry	Goal 9 – Industry, innovation and infrastructure
	2.4. Investment Scale of Green Technology SMEs	Goal 9 – Industry, innovation and infrastructure

© The Authors, published by EDP Sciences. This is an open access article distributed under the terms of the Creative Commons Attribution License 4.0 (http://creativecommons.org/licenses/by/4.0/).

Table 2. Continued

ECO-INNOVATION ACTIVITIES	3.1. Commercialization Level of Green Technology	Goal 12 – Responsible consumption and production
	3.2. Enterprises' Participation on Environmental Management System	Goal 12 – Responsible consumption and production
	3.3. Economic Influence of Leading Environmentally Responsive Enterprises	Goal 9 – Industry, innovation and infrastructure
	3.4. Green Patents	Goal 12 – Responsible consumption and production
	3.5. Activeness of Renewable Energy Utilization	Goal 7 – Affordable and clean energy
ECO-INNOVATION PERFORMANCES	4.1. Level of Environmental Impact on Society	Goal 3 – Good health and well-being Goal 12 – Responsible consumption and production
	4.2. CO2 Emission Intensity	Goal 13 – Climate action
	4.3. Nation's Energy sustainability level	Goal 7 – Affordable and clean energy
	4.4. Water Consumption Intensity	Goal 6 – Clean water and sanitation
	4.5. Jobs in Green Technology Industry	Goal 8 – Decent work and economic growth Goal 9 – Industry, innovation and infrastructure
	4.6. Green Industry Market Size	Goal 8 – Decent work and economic growth Goal 9 – Industry, innovation and infrastructure

The k-means method was chosen as the method of clustering. Data processing was carried out in the object-oriented Python environment. KMeans algorithm clusters data by trying to separate samples in n groups of equal variances, minimizing a criterion known as the inertia or within-cluster sum-of-squares. This algorithm requires the number of clusters to be specified. It scales well to a large number of samples and has been used across a large range of application areas in many different fields.

The k-means algorithm divides a set of N samples X into K disjoint clusters C, each described by the mean μj of the samples in the cluster. The means are commonly called the cluster "centroids"; note that they are not, in general, points from X, although they live in the same space. The K-means algorithm aims to choose centroids that minimize the inertia, or within-cluster sum-of-squares criterion [12]:

$$\sum_{i=0}^{n} \min_{\mu_j \in C}(||x_i - \mu_j||^2) \qquad (1)$$

Inertia can be recognized as a measure of how internally coherent clusters are [12].

3 Results

The analyzed period is explained by the presence of calculated indices: for Eco-IS index is 2010-2019 and for ASEI index 2015-2017. We used the Elbow method to determine the optimal number of clusters in both indices. The basic idea behind partitioning methods, such as k-means clustering, is to define clusters such that the total intra-cluster variation [or total within-cluster sum of square (WSS)] is minimized. The complete WSS measures the

© The Authors, published by EDP Sciences. This is an open access article distributed under the terms of the Creative Commons Attribution License 4.0 (http://creativecommons.org/licenses/by/4.0/).

compactness of the clustering and we want it to be as small as possible [13]. The optimal number of clusters in both cases is four clusters. The results of the Elbow method are shown in Figure 1.

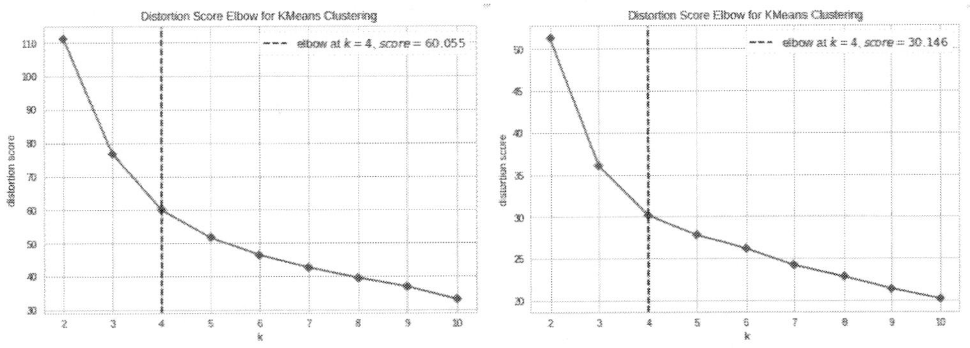

Fig. 1. The optimal number of clusters according to the Elbow method. (left) Eco-IS index; (right) ASEI index.
Source: Compiled by the authors

Silhouette analysis shows the distance between the resulting clusters. The silhouette plot (Figure 2) displays a measure of how close each point in one cluster is to points in neighboring clusters. Thus, it allows for a visual assessment of parameters such as the number of clusters.

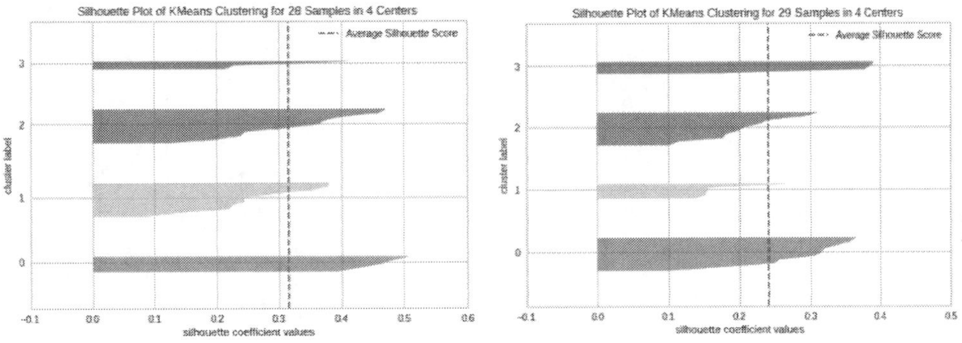

Fig. 2. Silhouette analysis. (left) Eco-IS index; (right) ASEI index.
Source: Compiled by the authors

We see that in the case of Eco-IS index clusters 1 and 2 are more uniform. And in the case of ASEI index clusters 0 and 2 are more uniform. Note that the distribution by clusters primarily considers the dynamics of cluster changes in the period under consideration. The distribution of countries within each cluster for each of the indices is shown in Figure 3.

© The Authors, published by EDP Sciences. This is an open access article distributed under the terms of the Creative Commons Attribution License 4.0 (http://creativecommons.org/licenses/by/4.0/).

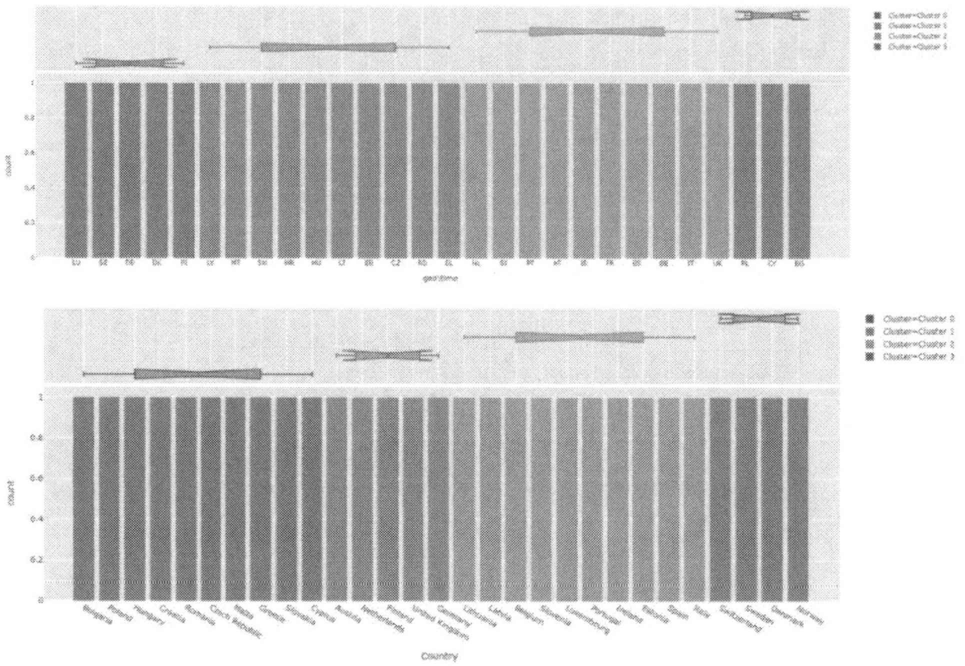

Fig. 3. Distribution of countries within each cluster. (top) Eco-IS index; (bottom) ASEI index.
Source: Compiled by the authors

Conducting a comparative analysis of the clustering of the European Union countries by two indices, one can notice that the leaders in this direction have been identified. The characteristics of each cluster are presented in Table 3. The clear leaders in promoting eco-innovation policy are Denmark, Sweden. These countries are among the leaders in two indices. Bulgaria, Cyprus, Poland are catching-up countries and are included in the corresponding clusters of two indices. The rest of the countries have values around the average. However, it is worth noting that a number of countries: the Netherlands, Germany, Finland, Great Britain are also pursuing an active policy of introducing eco-innovations, which is confirmed by their high values of the indices.

Table 3. Clustering results

Index	Cluster	Min	Max	Mean	Country	Type
Eco-IS	Cluster 0	110	165	132,68	Germany, Denmark, Finland, Luxemburg, Sweden	Liders
	Cluster 1	42	100	67,91	Czech Republic, Estonia, Greece, Croatia, Hungary, Lithuania, Latvia, Malta, Romania, Slovakia	Average
	Cluster 2	71	134	103,43	Austria, Belgium, Spain, France, Ireland, Italy, Netherlands, Portugal, Slovenia, UK	Developers
	Cluster 3	20	66	45,13	Bulgaria, Cyprus, Poland	Catching-up
ASEI	Cluster 0	0,23	0,37	0,30	Czech Republic, Malta, Greece, Slovakia, Cyprus, Croatia, Hungary, Poland, Romania, Bulgaria	Catching-up
	Cluster 1	0,46	0,56	0,51	Germany, Great Britain, Finland, Netherlands, Austria	Developers
	Cluster 2	0,35	0,45	0,40	Spain, Estonia, Ireland, Portugal, Luxemburg, Belgium, Slovenia, Italy, Latvia, Lithuania	Average
	Cluster 3	0,57	0,66	0,62	Norway, Denmark, Sweden, Switzerland	Liders

© The Authors, published by EDP Sciences. This is an open access article distributed under the terms of the Creative Commons Attribution License 4.0 (http://creativecommons.org/licenses/by/4.0/).

4 Conclusions

The study showed that the k-means algorithm allows to divide all countries of the training sample (EU countries) into several clusters, differing in the level of development of eco-innovations according to two complex indices. And the object-oriented approach allows you to quickly process an unlimited amount of data, including statistical evaluations and building multiple models. We were able to find that the two indices have approximately the same distribution of countries by number among the clusters. The statistic estimate of the silhouette differs slightly. Leaders and catching-up countries stand out clearly.

The results of the study demonstrate the fundamental applicability of classical statistical approaches to solving the problems of strategic management of the development of a circular economy, in particular, eco-innovations. And its' potential effectiveness from the point of view of visibility and information content of reflecting the development of circular processes.

Acknowledgements

Funding: The publication has been prepared with the support of the grant of the President of the Russian Federation for young scientists candidates of sciences MK-5006.2021.2.

References

1. P. Mealy, A. Teytelboym, Research Policy (2020)
2. A. Frigon, D. Doloreux, R. Shearmur. J. Clean. Prod., **275** (2020)
3. E. K. Jang, M. S. Park, T. W. Roh, K. J. Han. Sustainability (2015)
4. M. Geissdoerfer, S. N. Morioka, M. M. de Carvalho, S. Evans J. Clean. Prod., 190 (2018)
5. G. Huppes, R. Kleijn, R. Huele, P. Ekins, B. Shaw, M. Esders, S. Schaltegger, *Measuring Eco-Innovation: Framework and Typology of Indicators Based on Causal Chains; Final Report of the ECODRIVE Project* (CML, University of Leiden: Leiden, The Netherlands, 2008)
6. R. Kemp, P. Pearson, Maastricht, 10 (2007)
7. OECD Oslo Manual: Guidelines for Collecting and Interpreting Innovation Data. Oslo Manual (2005)
8. Eco-Innovation Observatory, Methodological report (2012) https://www.eco-innovation.eu
9. The Eco-Innovation Scoreboard (Eco-IS) and the Eco-Innovation Index, https://www.eco-innovation.eu
10. M. S. Park, R. Bleischwitz, K. J. Han, E. K. Jang, Sustainability (2017)
11. Comparing Eco-innovation Indices: ASEM Eco-Innovation Index (ASEI) & Eco-Innovation Scoreboard (Eco-IS) (2016) http://www.aseic.org
12. Clustering — scikit-learn 0.22.1 documentation, https://scikit-learn.org
13. Determining The Optimal Number Of Clusters: 3 Must Know Methods, https://www.datanovia.com

© The Authors, published by EDP Sciences. This is an open access article distributed under the terms of the Creative Commons Attribution License 4.0 (http://creativecommons.org/licenses/by/4.0/).

E-commerce effects for the sustainable development goals

Svetlana Revinova*

RUDN University, Miklukho-Maklaya str.6, 117198 Moscow, Russia

Abstract. Achieving sustainable development goals is the task of all countries, both developed and developing. At the same time, almost all countries are moving to a digital economy, with the core of e-commerce. The study aims to identify the connections between these two areas and analyze the effects that may arise as a result. The analysis showed that e-commerce is linked to 10 of the 17 UN Sustainable Development Goals. For the most part, these effects are positive. However, to achieve the most positive impact, the interest of both Internet companies and consumers is required.

1 Introduction

In September 2015, during the United Nations Summit on Sustainable Development in New York, 193 UN member states officially adopted a new sustainable development agenda "Transforming our world: the 2030 Agenda for Sustainable Development". This plan includes 17 goals and 169 tasks [1]. The goal of this program is to improve the quality of life and prospects for humanity.

One of the goals of sustainable development is to ensure the transition to sustainable consumption and production patterns. Sustainable consumption and production involve promoting the efficiency of resource and energy use, constructing sustainable infrastructure, providing access to essential social services, ensuring green and decent jobs and a better quality of life for all. It involves engaging consumers through education and training initiatives on sustainable consumption and lifestyles. The UN plan assumes that the implementation of the action strategy for the transition to sustainable consumption and production patterns should be carried out with the participation of all countries, and developed countries should be the first, taking into account the development and potential of these countries.

At the same time, the government of all countries aims to develop the digital economy, which has a significant impact on all spheres of life. The digital economy is becoming the basis for the implementation of sustainable consumption. One of the critical elements of the digital economy today is e-commerce. Over the past few years, e-commerce has become an integral part of the global retail system. As Internet access overgrows worldwide, the number of online shoppers continues to grow every year. In 2020, more than two billion

* Corresponding author: revinova-syu@rudn.ru

© The Authors, published by EDP Sciences. This is an open access article distributed under the terms of the Creative Commons Attribution License 4.0 (http://creativecommons.org/licenses/by/4.0/).

people purchased goods or services online, and during the same year, Internet retail sales worldwide exceeded $ 4.2 trillion [2].

E-commerce is the marketing, purchase, and sale of goods and services on the Internet. E-commerce growth is observed in all countries without exception, but it is growing especially rapidly in developing countries. In recent years, China has been the leader in e-commerce turnover in the world [3]. E-commerce includes the distribution and sale of goods and services over the Internet. The backbone of e-commerce is distributing goods, but today an increasing number of services are also provided online. The Covid-19 pandemic has given a new impulse to the growth of e-commerce, especially in areas such as education, medicine, etc.

As more and more shoppers move to the Internet, there are more and more opinions about the negative consequences of this transition. Still, at the same time, the positive aspects of this transition often go unnoticed [4, 5]. While research results show that, for example, e-groceries are potentially more sustainable than traditional shopping, with 10-30% lower emissions [6]. Most researchers talk about the adverse effects of e-commerce in connection with product packaging. Packaging functions must progress with the new requirements of e-commerce (for example, an increase in the number of packaging materials for each product, an increased need for product protection, life management, environmental sustainability, etc.) [7]. There is quite a lot of research related to the environmental impact of the last mile, the final stage of delivery to the buyer. Research shows that clicking and collecting for those customers who live within 5 kilometers of a store is a much more sustainable way to shop. However, if that distance increases, the cars' emissions they are likely to use to travel to the store will exceed the emissions produced by the carrier, making online shopping more sustainable. Last mile logistics are becoming increasingly important with the growth of e-commerce, thus leading to inevitable changes in the functioning of modern cities and the potential consequences cannot be ignored [8, 9].

E-commerce can be influenced not only by Internet companies but by consumers themselves [10]. The carbon footprint that remains after the purchase depends on which choice of delivery, packaging, and fitting of goods they choose.

The purpose of this article is to determine the possible contribution of e-commerce to the achievement of sustainable development goals and to determine the positive or negative impact of e-commerce.

2 Methods

We used expert, statistical and comparative analysis; data analysis using methods of grouping, generalization, methods of classification, induction, deduction, analogy, comparison, formalization. The data for the study were open sources of information on the Internet and research presented in the scientific literature.

3 Results and Discussion

There are various approaches to analyzing and classifying the 17 UN Sustainable Development Goals (SDG) in modern literature. For example, the classification of the Goals by the "dimensions" of sustainable development [11], classification of SDG based on their significance for business [12], etc. After analyzing the connection between e-commerce and the achievement of sustainable development goals, we propose the following classification (Table 1).

© The Authors, published by EDP Sciences. This is an open access article distributed under the terms of the Creative Commons Attribution License 4.0 (http://creativecommons.org/licenses/by/4.0/).

Table 1. The connection between e-commerce and achieving development goals

E-commerce can impact the achievement of the SDGs	The SDGs can affect the development of e-commerce	SDGs and e-commerce are not linked
Goal 1. End poverty in all its forms everywhere Goal 2. End hunger, achieve food security and improved nutrition and promote sustainable agriculture Goal 3. Ensure healthy lives and promote well-being for all at all ages Goal 4. Ensure inclusive and equitable quality education and promote lifelong learning opportunities for all Goal 5. Achieve gender equality and empower all women and girls Goal 8. Promote sustained, inclusive and sustainable economic growth, full and productive employment and decent work for all Goal 10. Reduce inequality within and among countries Goal 12. Ensure sustainable consumption and production patterns Goal 13. Take urgent action to combat climate change and its impacts Goal 17. Strengthen the means of implementation and revitalize the global partnership for sustainable development	Goal 9. Build resilient infrastructure, promote inclusive and sustainable industrialization and foster innovation	Goal 6. Ensure availability and sustainable management of water and sanitation for all Goal 7. Ensure access to affordable, reliable, sustainable and modern energy for all Goal 11. Make cities and human settlements inclusive, safe, resilient and sustainable Goal 14. Conserve and sustainably use the oceans, seas and marine resources for sustainable development Goal 15. Protect, restore and promote sustainable use of terrestrial ecosystems, sustainably manage forests, combat desertification, and halt and reverse land degradation and halt biodiversity loss Goal 16. Promote peaceful and inclusive societies for sustainable development, provide access to justice for all and build effective, accountable and inclusive institutions at all levels

We have identified one goal, the achievement of which is necessary for the further development of e-commerce and the growth of the e-market, Goal 9. Innovation and technological development have an impact on all areas of life. The Internet and mobile communications have become a necessity for almost the entire population of the planet. These two types of communication are at the core of e-commerce technologies. In 2019, 96.5% of the world's population was covered by one or another network of at least 2G standard, and more than 60% were Internet users [13]. The further growth of e-commerce is inextricably linked with the increase in the number of Internet users, the acquisition of digital skills by the population not yet involved in the Internet, and the introduction of new technologies into the business processes of internet companies. Without sustainable infrastructure and innovation, the future development of e-commerce will be weak. Today we are faced with apparent inequalities in infrastructure development. So the penetration of the Internet differs by region of the world (Fig. 1).

© The Authors, published by EDP Sciences. This is an open access article distributed under the terms of the Creative Commons Attribution License 4.0 (http://creativecommons.org/licenses/by/4.0/).

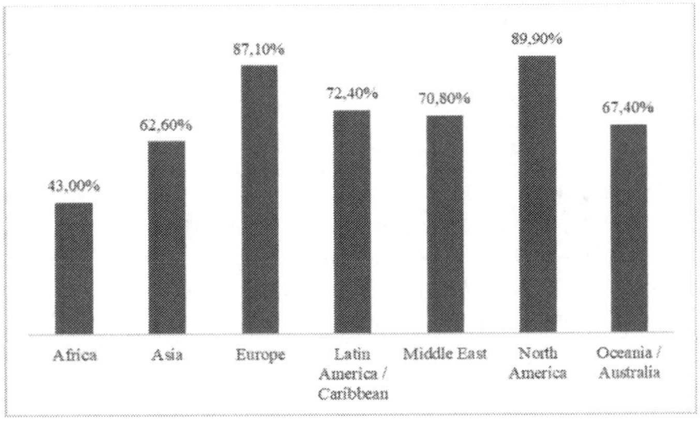

Fig. 1. Penetration of the Internet into the world's regions [14]

Let's take a look at 10 goals that e-commerce can impact. These impacts can be both positive and negative. Goal 1, on the one hand, can be partially achieved through the provision of new jobs by Internet companies. At the same time, participation in e-commerce requires costs, for example, on the Internet and communication devices. Also, poor citizens are more likely to lack the education and digital skills to participate equally in e-commerce.

Goal 2 : End hunger, achieve food security and improved nutrition and promote sustainable agriculture. The e-commerce marketplace provides equal access to both businesses and consumers. Information on the Internet on food markets and food prices is becoming publicly available and timely.

One of the tasks of Goal 3 is universal health coverage, access to quality essential health services and access to safe, adequate, quality, and affordable essential drugs and vaccines for all. Today we see how the Covid-19 pandemic has changed consumer preferences in the electronic market. More and more internet users are turning to e-commerce to buy medicines. E-Health issues: the use of electronic communication technologies for patient care, health worker education, disease detection, and patient monitoring has become even more pressing. Medical and health centers have begun to provide electronic services more and more often.

Many educational Internet projects are now presented on the electronic market, starting with such well-known projects in mass online education as Coursera and ending with small projects of a narrow focus, for example, teaching children to draw. The Covid-19 pandemic has given a new impulse to the development of e-commerce projects in this area. Educational organizations that previously did not provide online services were forced to either introduce modern teaching methods and enter the electronic market or to close. Even the end of the pandemic will not stop the rapid growth of this segment of the electronic market. Today's schoolchildren and students have adapted to receiving education through the Internet and will undoubtedly use it throughout their lives. Simultaneously, educational organizations have realized the advantages of conducting educational activities online. For example, the absence of additional classrooms and handouts, the lack of restrictions on the size of groups, etc., will also use e-commerce projects more and more. It will contribute to the achievement of Goal 4: Ensure inclusive and equitable quality education and promote lifelong learning opportunities for all.

The contribution of e-commerce to Goal 8 and, in part, Goal 5 is linked to the labor market. E-commerce can influence these goals, as in goal 1, by providing equal access for all labor markets. The e-commerce market is a fast-growing market, and the need for new employees, from programmers to couriers, is very high. On popular job search sites, vacancies are created every day to find professionals selling goods and services on the

© The Authors, published by EDP Sciences. This is an open access article distributed under the terms of the Creative Commons Attribution License 4.0 (http://creativecommons.org/licenses/by/4.0/).

Internet. The demand for employees exceeds the supply since the new field has not yet been fully mastered by young specialists. E-commerce projects have generated many unique specialties: web designers, web analysts, SMM marketers, etc. The most popular professional fields are internet marketing, performance advertising, social media marketing management, loyalty programs, and purchasing.

Goal 10: reduce inequality within and among countries. Inequality between countries is determined by many factors and often has historical roots. Developing countries are trying to overcome the lag behind developed countries through intensive development based on institutions, technologies, and the population's quality of life. Inequality within and between countries is a constant cause of concern for national leaders. Such inequality leads to social tensions, and vulnerable segments of the population become increasingly vulnerable. E-commerce provides an opportunity to provide access to essential services for all parts of the people, both domestically and internationally. But achieving this e-commerce effect requires equal access to technology and infrastructure. So, in Russia, we note a different level of Internet penetration in the subjects of the Russian Federation. In other regions of Russia, we see different levels of digital skills. E-commerce is also developing unevenly.

As already mentioned, the e-commerce market has neither geographic nor time limitations. It operates 24 hours 7 days a week and can be accessed by any company or country resident with access to the Internet. Electronic commerce can significantly increase the exports of developing countries and ensure timely provision of all least developed countries with market access according to the decisions of the World Trade Organization. E-commerce can help strengthen the means of implementation and revitalize the Global Partnership for Sustainable Development, in line with Goal 17.

According to experts, the most significant contribution of e-commerce is achieving goals 12 and 13, related to each other. Sustainable consumption and production aim to do more and better with less costly resources. They also aim to overcome the direct relationship between economic growth and environmental degradation, improve resource efficiency and promote more sustainable lifestyles.

Distribution, packaging and return of online goods are considered to be the primary source of carbon emissions. Online shopping needs to be greener. It requires measures from the outside, both on Internet companies and the Internet buyers themselves.

Each product delivered from an online store has individual packaging, most often a plastic bag or even several bags. If customers order the same product in several sizes at once, they use their packaging. Every day, thousands of trucks and cars deliver these goods to customers and take back unclaimed but already unpacked goods. It creates tons of carbon emissions. Online stores need to take a responsible approach to the type of packaging used and its quantity, giving preference to reusable packaging. For example, Zara moved away from paper and plastic packaging a few years ago. Items are shipped in a recycled carton and are not individually packed, significantly reducing waste. These boxes can then be reused for storage or easily recycled. Also, the product must be packed in boxes of suitable size. New technologies such as additive manufacturing and 3D printing can help optimize the volume and shape of the packaging, thereby contributing to greener production, such as by reducing CO_2 emissions. Currently available technologies can help rethink the whole paradigm of e-commerce packaging, which has changed very little over the past few decades [15].

Amazon, which delivers more than 10 billion units a year, said it emitted 44.4 million metric tons of carbon dioxide last year, which is about 85% of Switzerland's or Denmark's emissions. Nearly a third of U.S. solid waste comes from e-commerce packaging. To measure its carbon footprint, Amazon looked at emissions from all of its businesses, including the planes it operates and the energy it uses to produce Echos, Kindles, and its other devices. Amazon even included customer trips to its grocery chain Whole Foods [16].

© The Authors, published by EDP Sciences. This is an open access article distributed under the terms of the Creative Commons Attribution License 4.0 (http://creativecommons.org/licenses/by/4.0/).

These are the results of only one company. The most successful Russian company Wildberries in 2019 delivered 1,1 billion goods to customers, 400000 km was the average mileage of one truck. During the day, the truck covered about 1070 km. Each car transported nearly 2 million pieces of goods per year [17]. At the same time, no one yet considers the carbon footprint left by online shoppers when searching for goods. One request for information on Google leaves the same carbon footprint as one boiled kettle of water or a 16-meter trip by car.

The mailing of letters by Internet companies also leaves no trace for the environment. Billions of e-mails consume the same amount of energy every day as heating 2 million American homes. The associated greenhouse gas emissions are comparable to the emissions of 3 million cars. When sending one letter, four grams of carbon are formed, and if a picture is attached to the message, then the carbon footprint rises to 50 grams. Emissions when sending 65 e-mails - comparable to 1 kilometer of driving car [18].

Large companies have already begun to solve the emissions problem. For example, Amazon plans to achieve zero carbon emissions by 2040 and wants other companies to join it. Google intends to use only carbon-free energy sources by 2030. Google said it has been eliminating the company's carbon footprint since 1998, becoming the first company to achieve zero carbon emissions in 2007 and entirely switching to renewable energy since 2017 [19]. Many well-known brands have switched to packaging products with more environmentally friendly materials and less harmful delivery methods, which often leads to higher prices for the product.

Nevertheless, a lot depends on Internet companies. For example, Green Courier and DPD use bicycles and electric vehicles for delivery. Electric cars reduce greenhouse gas emissions by 17% (in the case of 20 km /day) to 54% (in the case of 120 km/day), or even more if the daily mileage increases [20].

The more retailers educate their customers about the environmental impact of their purchases and possible ways to reduce damage, the more likely they are to choose these greener options. One of the options for more rational consumption is to refuse express delivery. Suppose the online store must deliver the package to the buyer as quickly as possible. In that case, he will not wait for the full load of the car, while with standard delivery, the courier can deliver several packages in one trip, which significantly reduces the carbon footprint of the last mile. Pick-up points are becoming more and more popular in Russia. Delivery to pick-up points can increase the time but has a positive effect on the environment. In this case, the online store can deliver multiple orders at one time. In addition, in courier delivery, there are situations where the courier is forced to come a second time if the buyer is not at home, which doubles the carbon emissions. In using pick-up points, the buyer picks up his order at a convenient time for him. Informing the environmental and social impacts of last mile delivery affects e-commerce customers and encourages them to choose more sustainable last-mile delivery [21].

The free return service is attractive to buyers. However, many returned goods go to landfills and pollute the environment. Returns are currently costing retailers £ 60 billion a year, with fashion retailers having the highest return rate at 56%. The damage to the environment is even more significant. Indeed, in the United States, recycling generates 5 billion pounds of landfill waste and 15 million tons of carbon emissions annually. It is equivalent to the amount of garbage generated by 5 million people a year [22]. Barclaycard study found that 30% of shoppers deliberately overfill and then return unwanted items [23]. Thus, if retailers refuse a free return, shoppers might reject to order additional products. E-commerce can become more environmentally sustainable by encouraging consumers to reduce purchases and increase the number of items in each delivery [24]. Content on e-commerce websites must contain detailed product descriptions and product illustrations. Websites should provide objective, complete and valuable information about products to

© The Authors, published by EDP Sciences. This is an open access article distributed under the terms of the Creative Commons Attribution License 4.0 (http://creativecommons.org/licenses/by/4.0/).

reduce the number of searches and create customer confidence in the correct choice [25]. Companies must do their utmost to ensure that the buyer does not need to order a size range of the same product or the need for a free return. New technologies, such as artificial intelligence or augmented reality, may gradually replace the need for face-to-face product fitting.

4 Conclusion

The analysis showed that the contribution of e-commerce to achieving sustainable development goals is relatively high. E-commerce can help achieve 10 of the 17 SDG in one way or another. E-commerce has a positive impact on the sustainable development of both individual countries and the world. This impact is especially evident in the labor market, as the number of Internet companies are overgrowing, and with them, the number of jobs increases. Most vacancies require some qualifications, but in 2020 we saw a rise in areas such as courier delivery, where additional education is not required. E-commerce offers opportunities for access from anywhere globally to the same market for education and health services. Often, these services obtained over the Internet are cheaper. It includes primary and additional education, advanced training, getting medical consultations via the Internet from specialists from other countries, etc. At the same time, e-commerce can have adverse effects on the environment. Searching, packaging, shipping and returning items purchased through online stores leave their carbon footprint. E-companies can affect their carbon footprint by using rational packaging and shipping methods, although this can come with additional costs. They can influence sustainable consumption by educating their customers about sustainable behavior and opportunities to reduce environmental pollution. E-companies can significantly reduce their negative contribution to sustainable development goals.

Acknowledgements

This paper has been supported by the RUDN University Strategic Academic Leadership Program.

References

1. Transforming our world: the 2030 Agenda for Sustainable Development, Department of Economic and Social Affairs, https://sdgs.un.org/2030agenda
2. Topic: E-commerce worldwide, Statista, https://www.statista.com/
3. S. Revinova. International Journal of Economic Policy in Emerging Economies, **12(4)**, 377 (2019)
4. T. Tokar, R. Jensen and B. Williams. Business Horizons, **64(3)**, 323 (2021)
5. G. H. Popescu. Economics, Management and Financial Markets, **10,** 80 (2015)
6. C. Siragusa and A. Tumino. International Journal of Logistics Research and Applications (2021)
7. R. Alberto, S. Giulia, G. Mauro and M. Cristina. International Journal of Engineering Business Management, **6**, 14 (2014)
8. M. Viu-Roig and E. Alvarez-Palau. Sustainability, **12(16)**, 6492 (2020)
9. D. Lazarevic, L. Švadlenka, V. Radojicic, M. Dobrodolac. Sustainability, **12** (2020)

© The Authors, published by EDP Sciences. This is an open access article distributed under the terms of the Creative Commons Attribution License 4.0 (http://creativecommons.org/licenses/by/4.0/).

10. J. Oláh, N. Kitukutha, H. Haddad, M. Pakurár, D. Máté and J. Popp. Sustainability, **11(1)**, 89 (2018)
11. E. Zavyalova, E. Starikova. Journal of Law and Administration, **3**, 107 (2019)
12. Novak K., Zajceva N. Rethinking Sustainability, BRICS Business Magazine, http://bricsmagazine.com
13. E. Tairova, Sustainable development (2021)
14. World Internet Users Statistics and 2021 World Population Stats, https://www.internetworldstats.com/
15. S. Escursell, P. Llorach-Massana, M. B. Roncero. Journal of Cleaner Production, **280** (2021)
16. A. News, Middle of the herd' no more: Amazon tackles climate change", 2021, https://abcnews.go.com/
17. Wildberries presented the results of 2019 in numbers, 2021, https://www.retail.ru/
18. How does one Google search affect the environment? - Digitark , https://digitark.ee/
19. Contribution to the future, Google Sustainability, https://sustainability.google/
20. C. Siragusa, A. Tumino, R. Mangiaracina, A. Perego, International Journal of Sustainable Transportation (2020)
21. B. Ignat, S. Chankov, International Journal of Logistics Management **31,** 521 (2020)
22. J. Schiffer, Vogue Business (2021)
23. Press releases, Barclaycard, *Home.barclaycard*, https://home.barclaycard/
24. P. van Loon, L. Deketele, J. Dewaele, A. McKinnon, C. Rutherford, Journal of Cleaner Production, **106**, 478 (2015)
25. L.Xiao, F. Guo, F. Yu, S. Liu. Sustainability **11,** (2019)

© The Authors, published by EDP Sciences. This is an open access article distributed under the terms of the Creative Commons Attribution License 4.0 (http://creativecommons.org/licenses/by/4.0/).

Estimation of Global Public IaaS market concentration by Linda index

Timur Musin*

RUDN University, Miklukho-Maklaya str.6, 117198 Moscow, Russia

Abstract. Infrastructure as a Service (IaaS) is one of three key cloud market segments, that include also Platform as a Service (PaaS) and Software as a Service (SaaS). Representing less than a third of the entire market, this segment is projected to keep the higher pace of growth over the next few years. But this growth is mainly driven by market leaders - technology companies with a diversified international business, thereby increasing the concentration of the largest players in the market. This article defines the degree of global IaaS market concentration and whether an oligopoly is being formed. The Linda index is utilized for determination of "oligopoly boundary"- number of key market players.

1 Introduction

Infrastructure as a Service is a cloud-based service with which an organization can outsource its data center capabilities to an external cloud service provider receiving from the latter basic infrastructural services like computing capacities, storage capacities and network resources. IaaS providers on their side take advantage of scale economies via technologies like virtualization and multi-tenancy. [1]

As economy becomes more cloud-based, IaaS along with other cloud segments become more popular. [2] Small and mid-size companies use it to save on facility and personnel costs [3], whilst large enterprises leverage cloud benefits to implement large scale digital transformation. [4]

However, despite such heterogeneous demand that allows providers to locate in different parts of the price-quality space [5], the IaaS market is growing mainly due to certain group of providers, presented solely by cloud divisions of technology giants like Amazon, Microsoft, Google and Chinese Alibaba and Tencent. In this situation, the probability of an oligopoly increases.

That oligopoly of several companies creates strong incentives for these providers to leverage their IaaS market power to distort competition in the diverse markets that depend on access to IaaS. [6] This could firstly affect PaaS and SaaS segments, as these solutions are vertically integrated (from IaaS to PaaS to SaaS) and each layer includes the services of underlying layers. [1]

*Corresponding author: 1042200217@rudn.university

© The Authors, published by EDP Sciences. This is an open access article distributed under the terms of the Creative Commons Attribution License 4.0 (http://creativecommons.org/licenses/by/4.0/).

High concentration gives prospects to act anticompetitively to market leaders. Moreover, that could lead to structural problems like ineffective competition, which will stimulate anticompetitive conduct.

The article examines the reasons of domination of several players on the global IaaS market and defines whether some type of oligopoly (or duopoly) is being formed on IaaS market. It is only Public IaaS market being considered, as Private IaaS market size is hard to be measured.

Although IaaS represents a segment of entire cloud market, business models differ among the market segments, especially between IaaS and SaaS. This difference concerns product offering, target audience, profitability and entry barriers, that is reflected more clearly in Table 1.

PaaS segment is historically combined into the one market segment with IaaS, as the PaaS services do not have separate providers. Basically, PaaS products were offered by IaaS providers in order to increase the product portfolio and get higher margins. Despite the fact, in this study IaaS and PaaS segments are considered separately.

Table 1. Comparison of Public Cloud Market segments

Market	IaaS	PaaS	SaaS
Key players	Amazon, Microsoft, Alibaba, Google, Tencent	Amazon Microsoft, Alibaba, Google, IBM	Microsoft, Salesforce, Oracle, SAP, Google
Market concentration. Top-5 players market share	High. 80%	High. 70%	Low. 35%
Product portfolio	Limited	Medium	Wide
Target audience	IT administrators	Software developers	End-users
Gross Margin	50-60%	Average of IaaS & SaaS	80-90%
Entry barriers	High	Medium	Low

As demonstrated in Table 1, SaaS market is a highly competitive market, despite the presence of such giants like Microsoft and Salesforce. Its top-5 players market share is 35% [16]. IaaS and PaaS markets are much more consolidated. Major 5 IaaS players have almost 80% of market with Amazon Web Services (Amazon Cloud subsidiary) market share of more than a third. [17]

To better explain this difference, below is a comparison with SaaS and PaaS segments are provided.

Key factors that affect the difference in market concentration are provided below.

1. Product portfolio

- SaaS companies provides to their customers a wide range of products: CRM, ERP, HRM, content management tools, communication platforms, messengers, etc. Moreover, SaaS services are divided in two streams – horizontal (targeted to a wide audience of business users, regardless of their industry) and vertical (targeted to particular industry). This variety allows new companies to take place on the market, despite the presence of major players. [9]
- PaaS' main product offering is a platform: virtual environment for software development. Platforms provide everything needed to develop new software and include debugging, testing, and deployment tools that allow to distribute software to end users. Key features include software for application development, data management, analytics and artificial intelligence, and middleware for integration and orchestration. [10]

© The Authors, published by EDP Sciences. This is an open access article distributed under the terms of the Creative Commons Attribution License 4.0 (http://creativecommons.org/licenses/by/4.0/).

- IaaS product offerings represent basic infrastructural services: virtual servers, storages and network. With the development of cloud economy, these services become a commodity. [11]

2. Target audience

- SaaS products are designed for end-users and do not require IT skills, which obviously affects to their higher adoption.
- PaaS services are generally in demand by software developers, who design applications for end-users. Developers need to code, test and deploy their applications by using platforms with full set of tools needed. As development languages along with developers' goals and objectives differ, platforms vary with their functionalities and toolsets.
- IaaS target audience is IT administrators, that control, manage and deliver pool of IT resources to developers. As IaaS requires extensive expertise on the part of the customer to manage the computing infrastructure, thus stimulating small to medium size enterprises to use infrastructure-ready PaaS and SaaS products.

3. Gross Margins

- Typically, SaaS business model should have a gross margin of about 80-90%, [13, 14] which means low cost of sales even for new companies that have recently entered the market.
- IaaS products have average gross margin in a range between 50-60%. [13, 15]
- PaaS gross margin are estimated as an average between IaaS and SaaS business models.

4. Entry barriers

IaaS market has the highest entry barriers for new players. As being mentioned, IaaS providers are generally large technology companies that benefit economies of scale, making enormous investments in facilities and the equipment, as well as in the technology to increase infrastructure utilization and provide essential cloud characteristics such as high elasticity and scalability. [6]

In addition to datacenter investments, the leveraging of the economies of scale requires innovative virtualization, automation, dynamic scaling, and metering technology, that enable IaaS providers to pool resources to dynamically serve multiple consumers with varying demands for physical and virtual resources. Market leaders also invest their money into research and patent licensing, that let them improve the efficiency.

Moreover, if the new player enters the market, it faces the reputational & security barriers. [6] Market leaders are well-known brands with high brand recognition. These companies have proven that they provide reliability for both their own services and their customer data. And it's not only the brand power that challenges new enterer's expansion: migrating data may be the crucial reason to not change the provider, as most customers face data loss issues and other cyber risks. [18]

All that factors above (commoditized services, low margins and high barriers to enter the market) significantly affect the level of concentration on Public IaaS market, thus incentivizing an oligopoly being formed. In next chapter the level of IaaS market concentration over last years will be estimated.

2 Methods of market concentration measurement

To determine the level of market concentration for major companies, Concentration ratio (CR) can be used. This ratio gained the greatest popularity due to relative simplicity of calculations. In the most common practice this ratio is calculated by summarizing shares of each particular market companies in the total production of the product in the relevant market. For the largest n manufacturers, concentration ratio is calculating as in formula below

© The Authors, published by EDP Sciences. This is an open access article distributed under the terms of the Creative Commons Attribution License 4.0 (http://creativecommons.org/licenses/by/4.0/).

$$CR_n = s_1 + s_2 + \cdots + s_n = \sum_{i=1}^{n} s_i \quad (1)$$

where: si – is share of the producer (company) *i* in the production of a given product, Qi – total production of the producer *i* for that product, and Q = ΣQi – total production of that product in the relevant market.

$$s_i = Q_I / Q \quad (2)$$

Another metric that measures a level of market concentration is a Herfindahl-Hirschman index, which is defined as a sum of squares of the share of all manufacturers.

$$HHI = \sum_{i=1}^{m} (s_i)^2 \quad (3)$$

Herfindahl-Hirschman index varies from 10000 (in the case of monopoly with one player holds 100% market share) to 10000/n, where n is a number of companies in industry. [12]

The usage of Herfindahl-Hirschman index increased in relation with its application in practice of the anti-monopoly legislation, especially in the USA, where it has been legally made an important indicator when assessing the admissibility of mergers and acquisitions:

- at HHI<=1000 (low concentrated markets), mergers and acquisitions are freely allowed
- at 1000<HHI<=1800 (moderately concentrated market), a Department of Justice audit is required;
- at 1800<HHI (a high concentrated market), mergers and acquisitions are not allowed if ΔHHI<=50, and a Department of Justice audit is required if ΔHHI>100

Being a convenient tool for determining the level of market concentration, the Herfindahl-Hirschman index cannot determine the balance of power and the number of leading companies in the market. The Linda index can be utilized for this purpose. It was proposed in 1976 by Remo Linda, Head of the Market Structure Division of EU Commission. [8] This index, as well as the concentration ratio, is calculated only for a few largest companies, so it also does not take into account the situation for the "rest of the market". However, unlike the considered indices of CRn concentrations, it seeks to describe a criterion for evaluating oligopolistic structures that may be present.

$$LI = \frac{1}{n(n-1)} \sum_{i=1}^{n-1} \left(\frac{n-i}{i} \frac{CR_i}{CR_n - CR_i}\right)$$

Where n – number of companies for which index is calculated, CRi – is a cumulative share of n market players, CRi – cumulative share of first i players.

The key feature of Linda index is that it allows to determine a "boundary of oligopoly". Representing initially declining broken curve, it sets the boundary of oligopoly (duopoly or monopoly) for the first i companies, if LI_{i+1} exceeds the value of LI_i.

3 Measurement of market concentration on Global Public IaaS market

As being said before, IaaS market is a highly consolidated market, represented generally by technology giants – Amazon, Microsoft, Alibaba, Google, Tencent, IBM. These companies

© The Authors, published by EDP Sciences. This is an open access article distributed under the terms of the Creative Commons Attribution License 4.0 (http://creativecommons.org/licenses/by/4.0/).

leverage enormous resources investing in capital expenditures and research and development required for continuous growth.

In this chapter of the study the level of IaaS market concentration will be measured. In addition, the type of market –duopoly/oligopoly will be determined.

Concentration ratio requires information about number of goods produced by each company and the whole market. However, it seems difficult to calculate total production or consumption of such services as neither cloud vendors nor their customers report such information. Thus, this study uses vendors' revenue instead of production in order to estimate the degree of market concentration.

For this goal Gartner's annual IaaS market size estimations are used. In Gartner's Worldwide IaaS Public Cloud Services Market [17], the company provides information about both the market size (Table 2) and market shares of top-5 vendors (Table 3).

Table 2. Worldwide IaaS Public Cloud Services Market by Gartner

$bn	**2015**	**2016**	**2017**	**2018**	**2019**
Market size	16,8	22,2	24,7	32,4	44,5

Table 3. Market share of top-5 IaaS companies by Gartner

2015		**2016**		**2017**		**2018**		**2019**	
Top-5 companies	**Market share, %**	**Top-5 companies**	**Market share, %**	**Top-5 companies**	**Market share, %**	**Top-5 companies**	**Market share, %**	**Top-5 companies**	**Market share, %**
Amazon	39,8	Amazon	44,20	Amazon	49,40	Amazon	47,90	Amazon	45,00
Microsoft	5,8	Microsoft	7,10	Microsoft	12,70	Microsoft	15,60	Microsoft	17,90
Rackspace	2,7	Alibaba	3,00	Alibaba	5,30	Alibaba	7,70	Alibaba	9,10
Alibaba	1,8	Google	2,30	Google	3,30	Google	4,10	Google	5,30
Google	1,5	Rackspace	2,20	IBM	1,90	Tencent	1,90	Tencent	2,80
Top-5 players Market Share	51,6	Total share	58,8	Total share	72,6	Total share	77,2	Total share	80,1

As can be seen, in the period from 2015 to 2019 a cumulative share of major 5 IaaS providers grew from 51,6% in 2015 to 80% in 2019. In addition, in 2019 Amazon and Microsoft represent more almost two thirds of the market. Taking into account that in this period the IaaS market grew 165,2% [16], the rest of the market grew slower than its leaders, thereby underlying the reasons for further concentration described in previous chapters. This tendency is clearly demonstrated on Fig. 1.

© The Authors, published by EDP Sciences. This is an open access article distributed under the terms of the Creative Commons Attribution License 4.0 (http://creativecommons.org/licenses/by/4.0/).

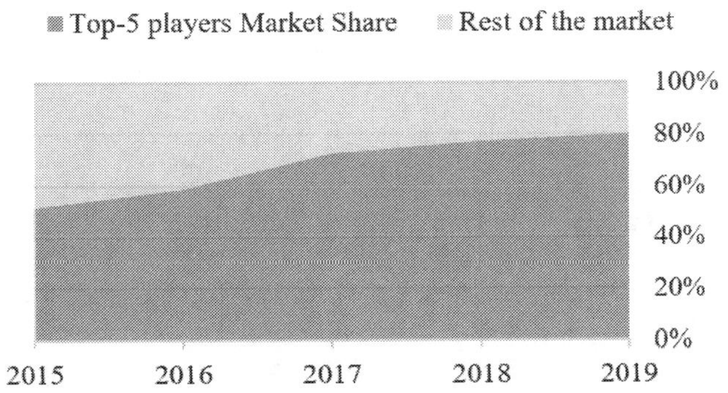

Fig. 1. Market share of top-5 players on IaaS Market, 2015-2019

Application of Herfindahl-Hirschman index to the Gartner's data requires some limitations and assumptions. It applies to the rest of the market – the lack of information about number of all IaaS providers along with their market shares. However, as these shares are less than 1,5-3% in any considered year, the study assumes that they don't add substantial values to the index. The top-5 players HHI on the Fig. 2 shows evolved from moderately concentrated in 2015 to highly concentrated in 2016 – 2019.

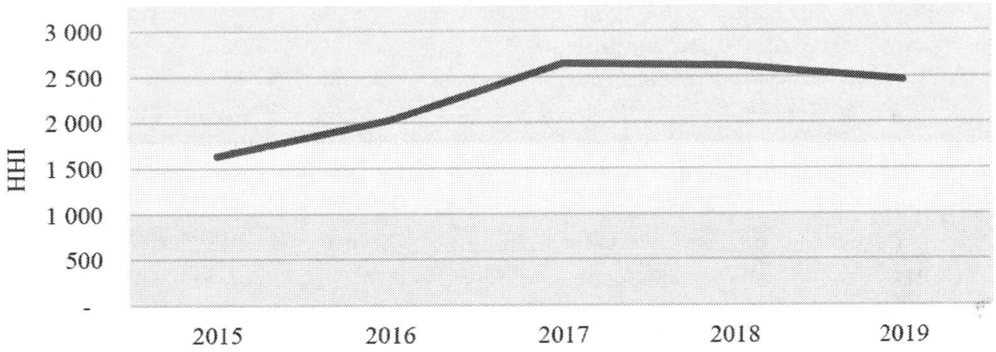

Fig. 2. Top-5 IaaS players Herfindahl-Hirschman Index

The obtained results also support the line that IaaS is a high-concentrated market, even considering the applied assumption of smaller players' insignificant contribution to index values.

However, the results do not give an indication of the balance of power between key players in the market. Thus, the Linda index should be applied to determine the number of major IaaS market providers. Based on the data from Table 3, the Linda indexes for each year were calculated. These results are presented in Figure 3.

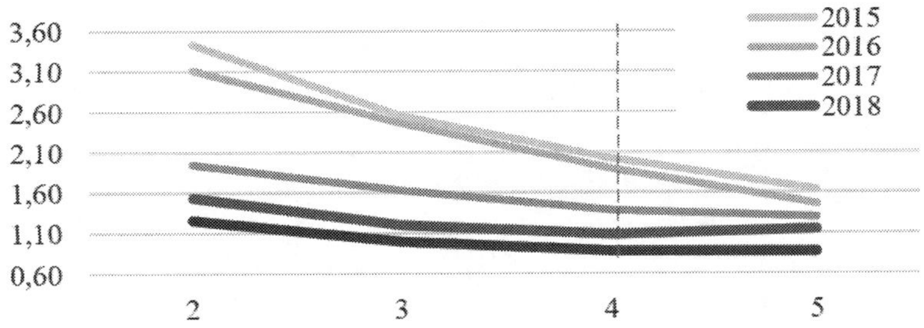

Fig. 3. Linda Ratio for top-5 IaaS providers, 2015-2019

© The Authors, published by EDP Sciences. This is an open access article distributed under the terms of the Creative Commons Attribution License 4.0 (http://creativecommons.org/licenses/by/4.0/).

Figure 3 shows that since 2018 there is a noticeable tendency to form an oligopoly of 4 companies (LI_5 tends to be greater than LI_4). That is clearly expressed in 2018, where $LI_5 > LI_4$. Value i=4 means classic or "hard" oligopoly in a competitive environment. That means Global Public IaaS market is forming or has already formed an oligopoly of the 4 largest players: Amazon, Microsoft, Alibaba and Google. In addition, the decline of Linda index for the 1st market player means that it loses its monopolistic position.

Despite the appropriateness of Linda index to the study, there are assumptions that were used to made it. The data used in this study are based on Gartner's estimation of IaaS market and revenue of IaaS providers. As there are no IaaS-only companies in Gartner's researches, IaaS represents a certain portion of their revenue.

Gartner calculations are based on its methodology of IaaS market determination. The company defines IaaS as "standardized, highly automated offering in which computing resources owned by a service provider, complemented by storage and networking capabilities, are offered to customers on demand. Resources are scalable and elastic in near real time and metered by use. Self-service interfaces, including an API and a graphical user interface (GUI), are exposed directly to customers. Resources may be single-tenant or multitenant and are hosted by the service provider or on-premises in a customer's data center." The calculation of the market size and revenue of the 5 largest companies are both based on this methodology.

Furthermore, Gartner data provides information for only first 5 companies, thus the study can't discover an oligopoly of first 5 companies, as Linda index requires data of 6th company's market share for this purpose.

Therefore, there is a need to analyze the sustainability of results obtained with Linda index if the data used deviate. IaaS market size and percentage of IaaS revenue for major players are considered as potentially deviating from their original estimations.

- **Market size estimation.** To assess the impact of IaaS market size on Linda index, the deviation at the level of 15% from the Gartner's estimation is considered.
- **Percentage of IaaS revenue estimation.** As all of major 5 IaaS providers are public companies, we use financial statements, that these companies disclose to calculate a calendar year revenue. For each company we use a revenue of its cloud business department or subsidiary. As fiscal periods vary among these companies (e.g. Alibaba fiscal year ends on March 31 and Microsoft – June 30) we calculate revenues for calendar 2018 and 2019, by summarizing their quarterly earnings ending March 31, June 30, September 30 and December 31. Then obtained revenues are compared with IaaS revenue of these companies, calculated by Gartner. As revenues of cloud business do not consist of only IaaS, IaaS revenue represents certain share of the entire cloud business revenue.

Table 4. AWS share of IaaS Revenue

$ mln	2018	2019
AWS Revenue for	25 655	35 026
Gartner IaaS AWS Revenue	15 495	19 990
as % of AWS Revenue	*60%*	*57,1%*

For example, AWS is historically known as IaaS pioneer, that were specialized on basic cloud infrastructural services and then expanded its cloud portfolio to PaaS and SaaS services. So, the calculated shares of its IaaS business look credible. In further research we will evaluate the sensitivity of "oligopoly boundary" of 4 companies, by changing companies' shares of IaaS revenue at the deviation level of 10%.

- **Revenue of the 6th company.** As Linda index requires to calculate LI_{i+1} to determine the oligopoly boundary of the first i companies, we need to take into account a

© The Authors, published by EDP Sciences. This is an open access article distributed under the terms of the Creative Commons Attribution License 4.0 (http://creativecommons.org/licenses/by/4.0/).

market share of the 6th company on the market to determine whether the IaaS oligopoly consists of 5 companies. Thus, we assume the range of revenue starting by revenue of fifth company and twice less. This measure is required only for modeling scenarios of Linda index for 5th company in the rating.

The results of implementing these different scenarios of Linda index to assumptions we described before are as follows.

In 2018, where Linda index curve trend was clearly reversed on 5th company thus determining "oligopoly boundary" of first 4 companies, no significant impact of the above factors on the Linda index was found. That means that the result of 4 companies form an oligopoly is sustainable.

In 2019, where Linda index curve was not reversed explicitly on 5th company and thus meaning absence of the 4 firms oligopoly, there are scenarios when such oligopoly exists. Table 5 shows whether an oligopoly boundary would be found if market size, share of IaaS in Amazon Web Services revenue changed: 4 means there is an oligopoly of 4 firms, 0 means there is no oligopoly. Linda index would determine an oligopoly of 4 companies in scenarios where IaaS market size was bigger for more than 5% and (or) AWS IaaS revenue share was more than 2% lower, than estimated by Gartner.

Table 6 also demonstrates a possibility of an oligopoly of 5 firms would be formed if market share of 6th company would lower than 1,83%.

Table 5. Linda index values by Market size and % of IaaS Revenue of AWS

Amazon	Market Size, $mln (% of deviation)						
% of IaaS in AWS Revenue	37 788 (-15%)	40 010 (-10%)	42 233 (-5%)	**44 456**	46 679 (+5%)	48 902 (+10%)	51 124 (+15%)
49%	4	4	4	4	4	4	4
50%	4	4	4	4	4	4	4
52%	4	4	4	4	4	4	4
54%	0	4	4	4	4	4	4
55%	0	0	0	4	4	4	4
57%	0	0	0	0	4	4	4
59%	0	0	0	0	0	4	4
60%	0	0	0	0	0	0	4
62%	0	0	0	0	0	0	0
64%	0	0	0	0	0	0	0
66%	0	0	0	0	0	0	0

Table 6. Linda index values by Market size and Market share of 6th company

Amazon	Market Size, $mln						
Market share of 6th company	37788	40010	42233	**44456**	46679	48902	51124
2,8%	0	0	0	0	4	4	4
2,6%	0	0	0	0	4	4	4
2,5%	0	0	0	0	4	4	4
2,3%	0	0	0	0	4	4	4
2,2%	0	0	0	0	4	4	4
2,0%	0	0	0	0	4	4	4
1,9%	0	0	0	0	4	4	4
1,7%	5	5	5	5	4	4	4
1,6%	5	5	5	5	4	4	4
1,4%	5	5	5	5	4	4	4
1,3%	5	5	5	5	4	4	4

© The Authors, published by EDP Sciences. This is an open access article distributed under the terms of the Creative Commons Attribution License 4.0 (http://creativecommons.org/licenses/by/4.0/).

For the rest major players (Microsoft, Alibaba, Google, Tencent) there are opposite results: for each company Linda index would determine an oligopoly of 4 companies in scenarios where IaaS market size was smaller for more than 5% and (or) their IaaS revenue share was more than 2% higher, than estimated by Gartner. (Table 7)

Table 8 shows a possibility of an oligopoly of 5 firms would be formed if market share of 6th company would lower than 1,83%, as it was already demonstrated in Table 6.

Table 7. Linda index values by Market size and % of IaaS Revenue of Microsoft

Microsoft	Market Size, $mln						
% of IaaS in Revenue	37788	40010	42233	**44456**	46679	48902	51124
17%	0	0	0	0	0	0	0
18%	4	0	0	0	0	0	0
19%	4	0	0	0	0	0	0
19%	4	4	0	0	0	0	0
20%	4	4	4	0	0	0	0
20%	4	4	4	0	0	0	0
21%	4	4	4	4	0	0	0
22%	4	4	4	4	4	0	0
22%	4	4	4	4	4	4	0
23%	2	4	4	4	4	4	0
23%	2	4	4	4	4	4	4

Table 8. Linda index values by Market size and Market share of 6th company

Microsoft	Market Size, $mln						
Market share of 6th company	37788	40010	42233	**44456**	46679	48902	51124
2,8%	4	4	4	0	0	0	0
2,6%	4	4	4	0	0	0	0
2,5%	4	4	4	0	0	0	0
2,3%	4	4	4	0	0	0	0
2,2%	4	4	4	0	0	0	0
2,0%	4	4	4	0	0	0	0
1,9%	4	4	4	0	0	0	0
1,7%	4	4	4	5	5	5	5
1,6%	4	4	4	5	5	5	5
1,4%	4	4	4	5	5	5	5
1,3%	4	4	4	5	5	5	5

Consequently, although the Linda Index does not clearly demonstrate an oligopoly of 4 firms, as it does in 2018, there are many scenarios where small deviations from Gartner's estimates show the existence of such oligopoly. Therefore, global IaaS market can be considered as an oligopoly of 4 firms.

4 Conclusion

The growth of Global Public IaaS market is mostly driven by its leaders – cloud departments of global technology companies, thereby the concentration of the largest players in the market is increasing. The last 5 years the share of 5 major players increased by almost 30 percentage points - from 51,6% in 2015 to 80,1% in 2019. Based on Linda Index, a noticeable tendency to form an oligopoly of 4 companies was defined since 2018. In 2018 Linda Index demonstrated a clear "oligopoly boundary" of 4 companies ($LI_5 > LI_4$). In 2019, despite the absence of an explicit boundary, values of LI_4 and LI_5 were close, which may indicate the stability of that tendency. The sensitivity analysis of the results obtained showed that with

© The Authors, published by EDP Sciences. This is an open access article distributed under the terms of the Creative Commons Attribution License 4.0 (http://creativecommons.org/licenses/by/4.0/).

minor changes in the estimates used in the study, the Linda Index would also set an oligopoly boundary of 4 companies in 2019.

5 Prospects for further research

The study conducted above was the first step for further research development. Based on results of concentration degree on Global market, the further works could be focused on regional IaaS markets, that could include countries like USA, China and Russia. Their Cloud markets are generally represented by local players, that could be explained by the requirements of regulators, market attractiveness for foreign providers and other factors. These countries have different level of Cloud adoption thus providing to determine is it possible to prevent an oligopoly or its negative consequences.

References

1. D.G. Arce, Computers & Security, 93 (2020)
2. K. Katsantonis, P. Mitropoulou, Filiopoulou E., C. Michalakelis, M. Nikolaidou, 19th Panhellenic Conference on Informatics (2015)
3. S. Balashova, I. Mikhaylov, M. Lazyrin, SHS Web of Conferences (2021)
4. V. Matyushok, V. Krasavina, A. Berezin, J. S. García. Econ, Res. Istraživanja (2021)
5. C. Kilcioglu, J.M. Rao, 25th International Conference on World Wide Web (2016)
6. K. Benzina, Berkeley Technology Law Journal, **34,** 1 (2019)
7. R. Bukvić, R. Pavlović, I.M. Gajić, MRPA Paper, 81707 (2014)
8. R. Linda, Methodology of Concentration Analysis Applied to the Study of Industries and Markets (1976)
9. D. Ma, A. Seidmann, Information Systems Research, **26,** 2 (2015)
10. A. J. Ferrer, D.G. Perez, R.S. Gonzalez, Procedia Computer Science, 97 (2016)
11. R. Harmon, H. Demirkan, B. Hefley, N. Auseklis, IEEE (2009)
12. M. Naldi, M. Flamini. Dynamics of the Hirschman-Herfindahl Index under new market entries (2018)
13. ARK-Invest. PaaS: The Goldilocks of IaaS and SaaS (2015) https://ark-invest.com
14. Statista, Private SaaS companies' gross margin from only subscription / SaaS revenue worldwide (2016) https://www.statista.com
15. Microsoft. Q4 2018 investor call (2018) https://www.microsoft.com
16. Statista. Public Cloud market outlook (2021) https://www.statista.com
17. Gartner. Worldwide IaaS Public Cloud Services Market (2016-2019) https://www.gartner.com
18. Oracle, KPMG. Cloud Threat Report (2020) https://home.kpmg

© The Authors, published by EDP Sciences. This is an open access article distributed under the terms of the Creative Commons Attribution License 4.0 (http://creativecommons.org/licenses/by/4.0/).

Evaluation of the Digitalization Development level in the Latin-American Economy

Chavarry Galvez Diana Pamela*, and *Chavarry* Galvez Wilmer Paul

Peoples' Friendship University of Russia, Peru

Abstract. Latin American and the Caribbean (LAC) countries have experienced an economic slowdown in the last five years, to accelerate economic growth, LAC countries need to find new instruments to transform the economy. digitalization in all sectors of the economy gives a boost to the country's economic growth. Digital tools also promote access to better services, such as education, government management and the foreign trade sector. The article analyzes the current state of the digitalization of the economy in LAC. To determine the digitalization of the economy, only indicators found in official sources were included.

1 Introduction

Digital transformation has become an important global trend that creates challenges and opportunities and that, if accompanied by effective policies, can help overcome development challenges in LAC countries. While some steps have already been taken, new approaches and policies will be needed to seize the opportunities, in addition to additional investments from the public and private sectors. The 2019 Latin American Economic Outlook: Development in Transition report identifies four main development problems: low productivity, social vulnerability, institutional weakness, and environmental sustainability [1]. LAC governments cannot find a way out of the crisis that has been occurring in recent years, the LAC region has a history of facing developmental challenges: due to digital inequality, environmental degradation, the erosion of democracy, and financial debt, which have caused deep discontent among the people of the region, and this dissatisfaction has been intensified by the crises derived from the COVID-19 Pandemic [2]. Also, the sharp decline in global demand for commodities, the significant decline in commodity prices, financial instability, and other effects related to declining investment, reduced tourism, and possible reduced remittances pose a complex scenario. Overall, the region's economic growth will decline by more than 7.7% by 2021[3]. The Internet penetration rate in the LAC countries is 69.5%, 10% higher than the world rate at 59.5%. The great challenge and challenge of the LAC countries, in the context of falling prices of raw materials, minerals, one of the most important levers to stimulate economic growth is productivity growth, for this to increase, the digital ecosystem must grow. An idea of the scale of the challenge is given by the contribution of the knowledge economy to the growth of labor productivity: in an advanced digitalization country like South Korea, it reached 3.1 between 1995 and 2012; the average

* Corresponding author: diana_p48@hotmail.com

© The Authors, published by EDP Sciences. This is an open access article distributed under the terms of the Creative Commons Attribution License 4.0 (http://creativecommons.org/licenses/by/4.0/).

of the five largest countries in LAC is 0.3 [4]. The political and economic situation is held back by inequality in access to political goods and also economic goods. In total, 68% of the region is located in the lower class, 30% in the growing middle class, and 2% in the upper class. That 68% of the lower class have half of their members in a precarious situation, some are not enough to buy food. With this, we have two groups in LAC, the one that enjoys the benefits of growth and the one that observes this growth but is not benefited [5]. The digital economy could allow governments to advance rapidly through programs that drive digitalization, innovation, and the application of these resources to the well-being of their populations. The virtual world has grown at a very high rate in the last decade, but Latin American countries for the adoption of digital products are still underdeveloped. With each passing day, more digital companies are trying to break into this digital world, as the pandemic works as a catalyst for this digital transformation. The starting point in this transformation is the development of infrastructures, such as Internet access and broadband for the public, followed by the introduction of online government services that simplify paperwork procedures for the public and foreign trade in particular. On the other hand, the digital divide between companies and the government can create serious problems for the future development of this area.

The objective of the article is to evaluate the level of development of the digital economy in Latin American countries, identifying the leading and lagging countries.

2 Material and methods

Methods of comparative, statistical, and logical analysis were used. For comparative analysis, socio-economic indicators were used from the databases of the World Bank; Inter-American Development Bank; the Statista internet portal; Global Innovation Index; and other open sources. A generalization method was used to formulate the conclusions. To analyze the current situation of the formation of the digital economy in Latin America and the Caribbean, the main economic sectors of each country were analyzed.

3 Results and Discussion

The digital economy is the economic activity that results from billions of online connections between people, businesses, institutions, and devices [6]. In today's world, technology plays a crucial role in transforming societies and economies by improving efficiency, connectivity, and access to resources and services [7]. In addition, there are several factors that influence the development of the use of digital technologies. These include: government regulation, human capital, IT research and development, innovation, business environment, information security, digital sector and digital infrastructure [8]. There are three main groups of prerequisites for the formation of the digital economy:

1. Infrastructure (computers, Internet, etc.);
2. Technologies (artificial intelligence, software);
3. Human capital (human resources, programmers, education);
4. Introduction of modern information technologies to expand foreign trade opportunities;

Each of the prerequisite groups above is related to each other, as each group is a component of the other. Today, digital technology creates another reality for governments, individuals, companies, organizations, global markets and provides a promising new growth path for any country [9]. Currently, LAC is going through a transition period towards a complete transition of the "information society". As in any transit, and the transition to digitalization, there are contradictions. On the one hand, the virtual space develops dynamically, the number of users is constantly growing and, on the other hand, the

© The Authors, published by EDP Sciences. This is an open access article distributed under the terms of the Creative Commons Attribution License 4.0 (http://creativecommons.org/licenses/by/4.0/).

infrastructure remains undeveloped. At present, LAC countries are far from leading countries. To date, practically all LAC countries have adopted plans for the transition to the digital economy and have included measures such as:

I. Improvement of communication between universities and labor groups;

II. Improve the quality and use of data;

III. Increased Internet access and use of technology

IV Using Internet technologies to expand foreign trade opportunities on B2B and B2C platforms, etc.

One of the main factors that allow the development of the digital economy is the population's access to the Internet, then it must be the high speed of the Internet, that is, the development of the ICT infrastructure. If the LAC countries do not implement digitalization plans, they will be left out of the benefits of digitalization and technology, which will lead to a loss of competitiveness on the world stage and thus lag behind the developed countries and the difference gap will increase even more [10].

Table 1 analyzes the current situation of the formation of the digital economy in such sectors as public administration and education. For the analysis in the public administration sector, the following indicators were considered: Infrastructure for Innovation 2020; the ICT Development Indicators Index (maximum1.0); and the E-Government Development Index, 2018. For the analysis in the Education sector, the following indicators were considered: Human capital and research, place in the world 2020; Institutions and/or organizations, innovation creation 2020; and Public spending on education (% of GDP-2019).

Table 1. Analysis of the current situation of the formation of the digital economy in the sectors of public administration and education in Latin America*.

Country	PUBLIC ADMINISTRATION			EDUCATION		
	Infrastructure for Innovation 2020	Index development ICT (maximum1.0)	E-Government Development Index, 2018	Human capital and research, a place in the world 2020	Institutions and / or organizations, creating innovation 2020	Government spending on education (%of GDP) - 2019
Panama	47	0,454	85	101	67	н/д
Colombia	50	0,441	61	82	57	4.5
Chile	51	0,537	42	55	38	4.9
Uruguay	52	0,696	34	71	46	н/д
Mexico	59	0,441	64	58	74	5.3
Brazil	61	0,522	44	49	82	5.9
Costa Rica	62	0,634	56	66	66	7.1
Peru	68	0,391	77	57	72	3.8
Argentina	70	0,592	43	48	97	5.9
Dominican Republic	77	0,365	95	100	98	н/д
Ecuador	82	0.369	84	91	126	5
Paraguay	89	0.35	108	98	109	5
El Salvador	101	0.381	100	105	100	3.5
Bolivia	104	0,314	103	56	129	7.3
Honduras	109	0,226	123	99	125	5.9
Guatemala	113	0,294	113	123	117	2.8

*Source: https://www.globalinnovationindex.org/; https://www.statista.com/; https://internetworldstats.com/, https://publicadministration.un.org/; https://www.weforum.org/; http://hdr.undp.org/ & InterAmerican Development bank.

© The Authors, published by EDP Sciences. This is an open access article distributed under the terms of the Creative Commons Attribution License 4.0 (http://creativecommons.org/licenses/by/4.0/).

The analysis revealed (tab.1) that LAC countries are divided into 3 groups into developed, medium, and underdeveloped. The developed countries of the LAC region in public administration with intervention and use of digital transformation are Chile, Uruguay, Brazil, Colombia, and Argentina. The medium-developed countries in public administration are Panama, Mexico, Peru, the Dominican Republic, and Ecuador. With much difference and much less development, occupying the last places in the region and the world, countries such as Honduras, Guatemala, Bolivia, Paraguay, and El Salvador, have little development of infrastructure for innovation and the development of ICT is very low. The role of government management is vital for the success of digital development, but it must fulfill 5 synergistic steps to achieve this successful transformation which is: leadership, action plan, resources, legislative framework, digital architecture [11]. Data from LAC show that a higher rate of development of electronic government and public bodies directly affects organizations and institutions that in turn directly affect the formation of the population, that is, human capital. On the other hand, it is directly related to institutions and organizations dedicated to training, learning, and innovation. The electronic government uses the ICT infrastructure, and especially the Internet, as a tool to improve public administration. The countries with the highest development of infrastructure for innovation and the ICT development index are leaders in the development of public administration, so it should be emphasized that investment in ICT infrastructure is the starting point for the development of the digital economy and also public administration and the need to think about how digitalization will be used, so the government must create programs that are attractive to the population and increase trust in public administration. In this way, it is possible to reduce costs and streamline in time and efficiency the procedures of the population that make them more reliable, efficient, inclusive, and innovative. On the other hand, the creation of central citizen service agencies with online procurement procedures such as specialized procurement and contracting centers, and the development of electronic procurement and contracting solutions (for example, "Chilecompra" and "Colombiacompraefectivo») transform traditional practices. Managing digital transformation is a fundamental issue of public policy. Changes in institutions, regulations, and markets are necessary to ensure that the digital transformation proceeds fairly and equitably. Governments face new regulatory challenges not only in managing the resulting aspects of digital transformation but also in ensuring that its benefits reach the entire population [12].

Developed countries have invested in human talent in less common fields, such as systems engineering, data science, or artificial intelligence. The countries with the highest investments in their GDP are at the forefront in indicators such as human capital and the desire for innovation. The accumulated human capital realized through labor potential becomes in turn a key factor in the economic development [13]. The developed countries of LAC in education with intervention and use of digital transformation are Brazil, Costa Rica, Chile, Uruguay, and Argentina. These countries have higher spending on education, compared to their share of GDP, and they also have the highest human capital and research level in the Latin American region.

The countries of medium development in the education sector are Panama, Colombia, Peru, Mexico, the Dominican Republic, and Bolivia, although the latter lacks the creation of innovative institutions and organizations, it is noteworthy that it invests the highest percentage of GDP in education in the LAC region 7.3%.

The underdeveloped countries in education are Guatemala, El Salvador, Honduras, Paraguay, Ecuador. These countries lack innovative institutions and organizations and have a low economic investment in education, for example, Guatemala uses only 2.8% of its GDP. Increasing the human capital of countries allows them to be at the forefront and to be competitive in the globalization process. LAC needs new strategies, new leaders, new markets to improve its economy, and it urgently needs innovative ideas that create an

© The Authors, published by EDP Sciences. This is an open access article distributed under the terms of the Creative Commons Attribution License 4.0 (http://creativecommons.org/licenses/by/4.0/).

environment conductive to economic growth in the region. One of the uses of digitalization is to be able to educate the population, investment in education has a direct return on human capital, which allows socio-economic progress. One of the alternatives is through online schools, it will allow educating the population and will be a great step to begin to eradicate the deficit in the education of the LAC population, more than 5% of the students had access to the Internet and other digital technologies exclusively through your educational institution. Therefore, it is also necessary to expand the Internet connection in educational institutions. For example, around 95% of computers in OECD educational institutions had an Internet connection, compared to 74% in LAC [14]. In general, the region's economy is not technological and has low labor productivity. Therefore, the innovation is weak and the results are very low. Regional leaders in innovation 2020 include Chile, Panama, and Colombia. The middle group of countries includes Uruguay, Mexico, Brazil, and Costa Rica. The lagging countries are Bolivia, Honduras, and Guatemala.

Table 2 analyzes the current situation of the formation of the digital economy the LA countries were analyzed in the following sectors: business sector, population and households. For the analysis in the business sector, the following indicators were considered: e-commerce turnover (millions of dollars in 2020); and Business Development in 2020. For the analysis in the population and household sector, the following indicators were considered: Technology development and the knowledge economy 2020; Creative activity results -2020; Human component capital 2018; and Internet penetration rate in %.

Table 2. Analysis of the current situation of the formation of the digital economy in the business, population and household sectors in Latin America*.

Country	BUSINESS SECTOR		POPULATION AND HOUSEHOLDS			
	E-commerce turnover, millions of dollars, 2020	Business Development 2020	Results obtained from the implementation of innovations-2020		Human Component Capital-2018	The level of Internet penetration, in %
			Development of technologies and the knowledge economy-2020	Results of creative activity -2020		
Chile	5269	49	64	61	0.8339	92.4
Mexico	18764	59	55	54	0.7044	69.5
Costa Rica	1083	48	53	53	0.7933	72.1
Brazil	21173	35	56	77	0.7525	70.1
Colombia	5386	52	72	80	0.7382	70.6
Uruguay	529	85	63	62	0.7719	88.1
Panama	457	123	91	55	0.7137	67.2
Peru	2943	43	112	87	0.7276	66.8
Argentina	5526	61	75	71	0.8579	92
Dominican Republic	950	83	99	82	0.6927	75.8
El Salvador	379	76	110	74	0.6348	58.6
Paraguay	387	84	115	78	0.6701	88.5
Ecuador	904	97	105	92	0.7395	76.4
Honduras	117	74	97	104	0.6015	41.7
Bolivia	279	90	114	109	0.7148	75.5
Guatemala	330	82	116	81	0.5524	44.2

*Source: https://www.globalinnovationindex.org/; https://www.statista.com; https://internetworldstats.com/. https://publicadministration.un.org/; https://www.weforum.org/; http://hdr.undp.org/ and InterAmerican Development bank.

In the analysis in Table 2, the developed countries in the LAC region in the business sector with intervention and use of digital transformation are Brazil, Costa Rica, Chile,

© The Authors, published by EDP Sciences. This is an open access article distributed under the terms of the Creative Commons Attribution License 4.0 (http://creativecommons.org/licenses/by/4.0/).

Colombia, and Mexico. These countries have the highest volume of electronic commerce, and this is the engine of business development; But these countries have achieved quite high e-commerce growth rates, but even so, the gap and lag with world leaders is significant. The countries with medium development in the business sector are Uruguay, Ecuador, Peru, Argentina, the Dominican Republic, and El Salvador. The underdeveloped countries in the business sector are Panama, Guatemala, Bolivia, Honduras, and Paraguay.

In the business sector in many countries of the world, the penetration of digital technologies into the lives of the population began with the emergence and spread of electronic commerce, electronic commerce is becoming the core of the digital economy. In LAC countries, the only Internet company is "Mercado Libre". With digital platforms, e-commerce can expand markets and achieve improved efficiency. Firms have been developing innovative marketing methods to learn about consumer preferences. One of the requirements to increase the use of electronic commerce is the transport infrastructure and logistics that should promote competitiveness and electronic commerce. On the other hand, the diffusion and impact of digital technologies on productivity and entrepreneurship in LAC are not automatic but depend on necessary elements such as adequate access and diffusion of digital technologies, business dynamics, participation of small and medium-sized enterprises in digital transformation, and sufficient competition in the digital economy.

The developed countries of LAC in the population and household sector with the intervention and use of digital transformation are Chile, Costa Rica, Brazil, Argentina and Mexico, although the latter country has a low percentage of Internet penetration. These countries are leaders due to their development in innovation; human capital and a large percentage of its population have access to the Internet. The medium-developed countries in the Latin American region in terms of population and households are Colombia, Uruguay, Panama, Peru, the Dominican Republic and Paraguay. The underdeveloped countries are El Salvador, Ecuador, Honduras, Bolivia and Guatemala. These countries have a very low Internet penetration rate of less than 50% in the case of Guatemala and Honduras. Besides not developing and promoting innovations; they occupy the last places in the region and the world; they have a large deficit in their human capital indicators.

According to comparative analysis in different sectors of public administration, the education, business sector, and population, and households for this digital transition vary greatly between countries in the Latin American region. After analyzing the current situation of the formation of the digital economy, the countries of Latin America were grouped into the most developed, medium-developed, and underdeveloped.

- The developed countries are Chile, Mexico, Argentina, Brazil, and Uruguay.
- Middle-developed countries are Costa Rica, Colombia, Peru, Panama, and the Dominican Republic.
- Underdeveloped countries are Ecuador, Paraguay, El Salvador, Bolivia, Honduras, and Guatemala.

4 Conclusions

The data shows that the first group of the developed LAC countries, in the population sector also have a high level of development with the use of digital instruments in the sectors of public administration, education, business sector, and have implemented digital plans in their governments, have all the requirements to achieve greater socioeconomic development and be globally competitive. In many indicators and indices, these countries occupy the Top 50 in the world. In other words, they are in the first third of all the countries in the world. The governments of these countries are adopting public policies to guarantee a rapid response and effectively articulate the digital transformation process as a key factor in social security.

© The Authors, published by EDP Sciences. This is an open access article distributed under the terms of the Creative Commons Attribution License 4.0 (http://creativecommons.org/licenses/by/4.0/).

The difference between developed and middle-developed countries in the LAC region in some indicators is quite large. This second group of countries are competing for development and have a quality ICT infrastructure and also the number of their population with Internet access is greater than the LAC average. But this group has a deficit in search and innovation and a lack of e-commerce development that could be a catalyst for the development of the digital economy. To prevent existing inequalities in the region from deepening, and technology benefits everyone equally, countries must increase access to these technologies and ensure that the necessary skills reach educational institutions, students, households, and workers.

The third group of underdeveloped countries in LAC does not have a qualitative development of ICT, the percentage of the population that has access to the Internet is lower than the average for the LAC region. On the other hand, its human capital and its educational system have a shortage of research and innovation. These countries must carry out reforms and changes in their main sectors, otherwise, they will continue to lag behind global competitiveness and, in the future, they will not be able to take advantage of all the opportunities that digital transformation will allow them. The digital transformation affects the quality of life of households, employment, and education. New technologies offer ample opportunities to access better public services, improve education, as well as to better integrate sectors with fewer economic resources in public policies. In the near future, workers who do not have access to new technologies or the necessary skills will be vulnerable to economic losses, they will not be able to take advantage of all the benefits of digitalization such as working remotely, selling products online, and access to platforms educational.

References

1. OECD, et al., Latin American Economic Outlook 2019: Development in Transition (2019)
2. M. A. Gonzalez-Perez, M. Mohieldin, G. T. M. Hult, J. Velez-Ocampo, Management Research (2021)
3. CEPAL, *Enfrentar los efectos cada vez mayores del COVID-19 para una reactivación con igualdad: nuevas proyecciones* (2020)
4. R. Katz and F. Callorda, Telecommunications Policy, **42** (2018)
5. L.V. D'yakova, World Economy and International Relations, **64** (2020)
6. Technology and Innovation - A Catalyst for Development (2017)
7. Innovation and Technology for Sustainable Development (2017)
8. National program «Digital economy of the Russian Federation», State Corporation ROSATOM (2018)
9. A. Khalimon, M. N. Guseva, I. Z. Kogotkova, I. S. Brikoshina, The European Proceedings of Social & Behavioural Sciences EpSBS, **57** (2018)
10. S. Yu. Revinova, D. Chavarri, Proceedings of the 2nd International Scientific and Practical Conf. "Modern Management Trends and the Digital Economy: from Regional Development to Global Economic Growth" (2020)
11. M. Acosta, A. Y. Velastegui, Perez, 3C COMPANY, **7** (2018)
12. OECD, Profiling the digital transformation in Latin America: Higher productivity for a better life (2019)
13. O. G. Ogiy, V.I. Kuzin, A.G. Harin, IOP Conf. Series: Earth and Environmental Science, 689 (2020)
14. Latin American Economic Outlook, Digital transformation for better reconstruction (2020)
15. Oleg N. Zhilkin, Wilmer Paul Chavarry Galvez Analysis of current trends in assessing the country's potential in international trade (on example of Peru) (2020)

© The Authors, published by EDP Sciences. This is an open access article distributed under the terms of the Creative Commons Attribution License 4.0 (http://creativecommons.org/licenses/by/4.0/).

Foreign Direct Investments in Kosovo

Fatos Geci*

University of Prishtina, Agim Ramadani str., 10000 Prishtina, Kosovo

Abstract. Purpose: The purpose of studying this topic is because in Kosovo the foreign direct investment has been consistently considered as a key factor in the potential development of the country. Knowing that investments are one of the most important macroeconomic aggregates of a country. Undoubtedly, their study has great importance with a special emphasis on the development of countries such as Kosovo, where economic, social and political circumstances require the continued absorption of investments by other countries. In general, for the country's economy, it is important to know what impact foreign investments have on economic growth, the factors that influence the growth of these investments and the actions of policy makers to attract investors. Design/methodology/approach: The data mainly obtained from the World Bank have compared Foreign Investments with the countries of the region. We consider the information to be accurate. Findings: From the findings we have made we consider that the main obstacles to foreign direct investment in Kosovo are at a high level of corruption, and politics, where we have disclosed several years of small investments due to non-approval of laws in the Assembly of Kosovo. Originality/value: The findings from this study can contribute to the improvement of policies, so that Kosovo as a transition country needs a lot of foreign direct investment that will change the economic conditions of the citizens.

1 Introduction

The impact of foreign direct investments are a very important segment of the economic activity of a country, due to the effects it brings to the economy of a country. Foreign direct investment in Kosovo are often considered as the leading aspect of the country. In our country, FDI is one of the main mechanisms that move the economy, as their focus has been on important sectors of the economy. Foreign direct investment in our country continues to be a very important source of capital growth flows and the economies development. They are an important indicator of the confidence of foreign investors in a country, and they decrease in the conditions of economic and political crises, and increase in the conditions of improving the economic and political environment. Investments are a very strong basis for the economic development of each country, especially countries in transition, such as Kosovo. Foreign direct investment are the most effective way to create new jobs in Kosovo and to create the preconditions for sustainable economic growth in the cities future. Kosovo has great potential to become an attractive place for foreign direct investments as well as to attract capital

* Corresponding author: tosi_star@hotmail.com

© The Authors, published by EDP Sciences. This is an open access article distributed under the terms of the Creative Commons Attribution License 4.0 (http://creativecommons.org/licenses/by/4.0/).

information and advanced technology. The main policies of the country are directed towards the liberalization of the economic framework, the improvement of business conditions and the absorption of foreign direct investments, as an important objective. In this context, our country is also in a constant effort to create development policies and other important measures to create a more appropriate business environment and a more favorable investment climate. In all these efforts, there are many institutions working together with the international community and donor agencies both in human capacity building, as well as in financing and co-financing. Various projects related to the creation of the basis for advancing the production of Kosovo products based on the European standards and increasing competitiveness. Not only in the regional market but also abroad.

2 Literature review

A lot has been said about the investments in Kosovo, especially from foreign investments, knowing the impact it can have on the country's GDP growth. Kosovo has not been doing well in recent years in terms of foreign direct investments. Since 2008, these investments have been steadily declining. Recently, in addition to the political crisis, the instability of institutions at both the local and central level has affected the decline in foreign investment. This situation has created space for this issue to be written and discussed by various economists and institutes. Every investor who wants to invest in Kosovo starts with a basic goal - profit. He or she does not come to invest because it is just Kosovo. They come mainly for profit, so we have to create such opportunities that they do not have in the countries of the regions to stand apart from everyone else and become noticed. This should be Kosovo's differential advantage or competitive advantage over the countries in the region.

There are small differences in the perception and importance given to specific aspects of the business support environment. For current investors, the rule of law, economic stability and political stability are the most important factors in determining the location of investments with 70%, 63%, and 53% of respondents who consider these factors as very important for investment (number 4 on a scale of 1-5). Bilateral trade agreements, double taxation treaties and the quality of infrastructure seem to be the lowest on the list of priorities of current investors, followed by the experiences from other foreign investors and the support services of the government agencies which ranks lower. Private sector development is critical, as it creates enterprises that brings more jobs and empowerment. It's also adds new technologies and innovation. As a result, it increases the country's revenue. In many developing countries, this private sector is considered to be the main driver of economic growth. Despite its important role, the figures of this sector in Kosovo are not very encouraging. Especially when compared to neighboring countries. Since entering the path of transition to a market economy. Kosovo has constantly faced obstacles of various that in one way or another, have reflected the business environment and as a result have hindered the development of the private sector in the country. Despite some recent reforms, Kosovo still ranks 86th out of 189 countries according to the 2014 World Bank Business Report, worse than all other neighboring countries except Albania. Macedonia. For example, it ranks 25th, or 61 places higher than Kosovo. Kosovos businesses operate in an environment where unfair competition, corruption, a dysfunctional judicial system, and other harmful factors constitute the "dominant rules of the game." This unfavorable business climate has been a burden to the cities existing businesses (54 percent of them have seen their sales decline), while at the same time sending negative signals to potential foreign investors, reducing the flow of investments. Direct Loan (FDI). What is the performance of FDI in Kosovo in recent years? In recent years there has been a downward trend in investments and this situation can be considered a major setback in the economic development of the city of Kosovo. According to many economists, the factors for the non-arrival of foreign investorments are a lot, but the first and

© The Authors, published by EDP Sciences. This is an open access article distributed under the terms of the Creative Commons Attribution License 4.0 (http://creativecommons.org/licenses/by/4.0/).

most important is that Kosovo has never became an attractive place for foreign investors. There is a severe environment of bureaucracy, corruption, non-functioning of the courts. Not only do these problems pose major barriers against the city. But also to the arrival of new foreign investors. In recent years, Kosovo has created an inadequate environment for doing business. As a result of high corruption, lack of more favorable fiscal policies and tax evasion. The high cost of the start of FDI, makes investors very sensitive to uncertainties, including uncertainties arising from government institutions.

What are the factors that have influenced the reduction of FDI in Kosovo? Among the factors that are considered to have influenced the level of low FDI in recent years is the global financial crisis that has gripped Europe and the world in the last 2-3 years. Despite this difficult factor to avoid, there are a number of other factors whose responsibility lies with the functioning institutions in the country. Return of investors can be mainly due to reasons such as improper state of law and order, existence of bureaucracy, high level of corruption, high bank interest rates, improper level of public procurement procedures and others. The problems are starting with law and order, eliminating bureaucracy, fighting corruption and removing conditionalities between different licenses, whether it is between the local level the central one.

3 Determining factors of foreign direct investment

Today FDI is seen as one of the most potential beneficiaries of economic growth, especially for emerging countries, and is sometimes needed to boost a country's economic growth (Adewumi). FDI can assist the host country especially in technology transfers and it is used also for in-house investments, know-how transfers and managerial or professional skills. From a sectoral point of view, the flow of FDI in the private sector depends on macroeconomic factors, development factors, institutional and qualitative factors. Regarding the second and third sectors, in different ways they are influenced by the level of national income, by the change of the exchange rate as well as by development indicators such as financial depth or by institutional factors of such as the independence of the judiciary and labor market flexibility. The effect of each of these factors are different depending on the economies where they occur, developing economies or developed economies. There are a number of factors that depicts the inflow of foreign investment in a country, factors which are different in different countries. Different depending on economic, social or political circumstances.

3.1 General Factors

Foreign Direct Investments are a very important segment of a country's economic activity, due to the effects that it brings. In theory, in the literature and also in practice, the effects of foreign direct investment are recognized. Various economic factors support FDI which include low interest rate loans, tax relief, grants, subsidies, and removal of restrictions and limitations, etc. One of the most determinants important of FDI, is the size and economic growth prospects of the country where it is expected to flow. It is assumed that if the country has a large market, it can grow rapidly. It has also been concluded that investors will be able to increase their investments in that country. In certain cases where FDI is export-based, the dimensions of the host country are important because the opportunities for economies of scale are great. The population of a country plays an important role in attracting foreign investors for direct investment in a country. In such cases, the prospect of a large customer base is an even greater investment attraction. Free labor is also an important determinant of FDI attraction. Infrastructural factors such as telecommunications and railways play an important role in the development of FDI in a particular country. Foreign Direct Investmetst is of great

© The Authors, published by EDP Sciences. This is an open access article distributed under the terms of the Creative Commons Attribution License 4.0 (http://creativecommons.org/licenses/by/4.0/).

importance in developing countries. Their importance is constantly increasing with the growth of the global economy. Not only developing countries but also the developed world also has its benefits from investing abroad, but in a different way from that of host countries. These benefits mainly relate to the expansion and acquisition of mature companies.

3.2 Cost-related factors

Use of foreign factors of production Labor and land costs can vary drastically from one country to another. So often, FDI's tends to shift the production process to those countries where these factors are cheaper, namely labor and land. Use of foreign raw materials often foreign investors buy raw materials from different countries because their cost varies significantly from country to country.

4 Data Methodology

4.1 Organizations and ethical consideration of the study

Knowing that the data included in this study were not obtained from the field, from the survey then we relied on data evaluated from secondary sources. Knowing that the data were mainly obtained from the CBK and KAS, we consider that they are as accurate as they can be, for the fact that the illegal economy is still present, and the data can never be accurate. Data on foreign direct investment are for the last seven years, in order to better compare the change in investment in general. Normally, the paper does not lack the opinions of economics experts, on the state of foreign direct investment in Kosovo and how to improve this situation.

4.2 Description of the study methodology

As a method of data collection for this paper are used secondary data, research on the impact of foreign direct investment on economic growth in Kosovo, there is no way to research otherwise than through the authorities which deal with maintaining their statistics as they are Central Bank of Kosovo and Kosovo Agency of Statistics. In addition to the data provided by the above-mentioned institutions, the paper is intertwined with prior knowledge from studies and beyond, but also from the opinions of connoisseurs of economic issues expressed in books, newspapers or portals.

The data collection is done in such a way as to expose the actors that generally influence the attraction and creation of a suitable environment for foreign investors.

4.3 Data Source

The data used in this paper are provided by the Central Bank of Kosovo and the Kosovo Agency of Statistics. Regarding the Kosovo Agency of Statistics, the statistics in the interest for this paper have been mainly national accounts statistics, to better demonstrate the relationship between GDP and FDI over the years. The Kosovo Agency of Statistics covers almost completely the territory of the Republic of Kosovo, being organized in the structuring of statistical circles and reporting units as the basic and only unit in the country from which it receives first-hand information. All field surveys use the sample size in these statistical circles but also the methodology according to international recommendations. When collecting statistical data and reports from reporting units engage; professionals, technicians, administrators, civilians of the country offices, field registrars from the Regional Offices, etc. As for the Statistics received from the Central Bank of Kosovo, external sector statistics

© The Authors, published by EDP Sciences. This is an open access article distributed under the terms of the Creative Commons Attribution License 4.0 (http://creativecommons.org/licenses/by/4.0/).

include balance of payments statistics and statistics on the international investment position. Foreign Trade Statistics include data on imports and exports of goods, as a function of balance of payments statistics. These data relate to general goods (regular import and export), goods in processing, repair. The paper also contains statistics provided by the Bank of Albania and UNCTAD. Regarding this objective we have put forward two hypotheses Hypothesis 1: Foreign direct investment affects economic growth A country's economic growth is synonymous with real GDP growth. Investment growth has a positive impact on economic growth, especially in developing countries. Investments encourage the consumption of venous products, which in turn has the reduction of a country's imports. Developing countries, such as Kosovo, need to encourage FDI to achieve greater economic growth. If we refer to the latest statistics, the share of FDI in GDP in 2014 was 2.81%, compared to 2007 where it was 12.74%. This means a reduction of FDI for seven years of 78%, while the nominal value of GDP for this period has increased by 55%.

Hypothesis 2: Foreign direct investment increases the level of employment According to the latest statistics, the unemployment rate in Kosovo is 35.5% of the labor force. The increase of foreign direct investments, in addition to bringing new capital, would affect the increase of production and as a result of consumption, will have a positive effect on the increase of the number of employees. Kosovo has a high unemployment rate where about 30.2% of them belong to the age group 15-24; this labor force is an important factor of production and such potential should be exploited more through increased investment, because the private sector has greater opportunities for generating new jobs. • Fiscal policies should be oriented towards creating economic stability, this is the main factor that pushes investors to invest, because if there are large economic fluctuations in the economic cycle then it is not an attractive place and investors do not invest. • Functioning of legislation - judiciary - a state in which the law does not work, investors are not interested in pouring their capital due to facing various problems such as administrative work, bureaucracy, corruption, etc. • Stable currency - we have the advantage of using Euro currency, because it is an indicator of foreign exchange stability that is very important for investors. Policymakers need to create a conducive environment for doing business.

5 Concluding Results

Given that Kosovo is a developing country then we can conclude that the effects of FDI on our economy have a positive result. • Foreign direct investment as a very important indicator of a country's economic growth, in the last decade has been characterized by a continuous increase and decrease. The lowest decrease was from 2011 to 2012 of 155 million euros and in 2014 where compared to the previous year FDI decreased by 129 million euros.

Foreign direct investment has a positive effect on GDP growth, the increase in FDI causes an increase in employment, domestic consumption which as a result has the mitigation of the large gap between imports and exports, economic stability or the achievement of other macroeconomic goals.

Concerning is the fact that Kosovo has a low level of FDI, not because of their non-promotion but because of ineffective policies that encourage the attraction of foreign direct investment.

The sectors most interested in foreign investors are real estate, renting and other business activities, financial services and industry.

The low level of employment in the country is a result of small investments, investments give rise to the creation of new jobs.

The main reasons for this level and FDI situation are:

1. Kosovo has not yet managed to convince investors to invest their capital in this country

© The Authors, published by EDP Sciences. This is an open access article distributed under the terms of the Creative Commons Attribution License 4.0 (http://creativecommons.org/licenses/by/4.0/).

2. The political stalemate that lasted six months of 2014 has led investors to hesitate during this period to invest in Kosovo

3. Law and order, elimination of bureaucracy, fight against corruption and removal of conditions through different licenses, whether between the local or central level, reduction of bank interest rates, improvement of infrastructure, stabilization of energy supply, improvement of public procurement procedures.

5.1. Key recommendations Kosovo has attracted a high value of foreign direct investment in recent years. However, foreign direct investment remains low compared to other countries in the region. This suggests that there is room for increased foreign direct investment in the country. To maintain the level of foreign direct investment in the country, or to increase it in the coming years, one needs:

• Improving the institutional, administrative and technical framework for FDI, proper targeting and government support.

• Regulation of law and order, elimination of bureaucracy, fight against corruption.

• At the same time, it is important that the types of foreign direct investment that will come to the country be those from which the domestic economy could benefit. Fiscal policies need to be further reformed to serve the manufacturing sector, as this is one of the sectors in which there is potential for attracting foreign investment. What we need to do is follow the examples of the countries of the region and their offers to foreign investors. Given Kosovo's favorable geographical position, as well as given the low cost of labor, we can say that fiscal favors and efforts others have been low compared to Kosovo's needs as a developing country.

• Further reforms are also needed in the areas of tax and customs administration, land use and building permits, the administrative appeals system, the inspection system and e-government. Kosovo has a lot of potential to attract foreign direct investment, it has resources in energy, it has resources in mineral resources, in the sector of development of agriculture and food products, in the sector of tourism, in the good geographical position it has in the region. Then there are the human resources, the very young population and a very good mood for work, but adequate fiscal policies need to be applied in Kosovo to be very suitable for business development. This can be achieved using • Fiscal incentives, or various tax breaks, through which it could become more or similarly competitive as the countries of the region. Efforts should also be stepped up to strengthen the rule of law, increase the efficiency of the courts and root out corruption. • Fiscal policies need to be changed to enable the competitive advantage of private sector enterprises, especially small and medium-sized enterprises, youth-run enterprises and women. Based on the experiences of the economies of developed countries that have gone through the transition phase in recent times, we can say that FDI is the key and plays a key role in the development of countries in transition such as our country. Therefore, local institutions should make efforts to improve the economic and political situation in Kosovo in order to attract foreign investors and consequently the economic development of the country.

Table 1. Regression results.

	(1)
	FDICapita
o.disctanca	
lfba	-38.293
	(32.958)
lfae	49.709*
	(27.822)
lfie	25.918
	(40.429)
tlf	-0.001
	(0.002)

© The Authors, published by EDP Sciences. This is an open access article distributed under the terms of the Creative Commons Attribution License 4.0 (http://creativecommons.org/licenses/by/4.0/).

 https://doi.org/10.1051/shsconf/202111401016

Table 1. Continued.

gdp	0.000
	(0.000)
gdppcapita	-0.194
	(0.161)
trade	4.373
	(5.462)
lspi	-293.026
	(1004.598)
eri	-175.174
	(545.451)
cogdp	53.686*
	(31.911)
co	-0.000*
	(0.000)
untlf	-18.692
	(52.467)
papun	0.002
	(0.002)
popull	-0.001
	(0.001)
abe	10.645
	(293.477)
emploiag	-212.187
	(664.703)
emploiin	-70.853
	(673.485)
emplopopu	18.489
	(97.766)
emploiser	-175.970
	(668.796)
punesimi	0.002
	(0.002)
gdpgro	29.973***
	(11.310)
ggfc	0.000
	(0.000)
ggfcgdp	70.271
	(59.495)
totaxcp	-12.020
	(14.187)
vaaggdp	-36.902
	(102.097)
vaag	-0.000
	(0.000)
produktinindu	0.062
	(0.058)
vaindu	-0.000
	(0.000)
vasergdp	-18.454
	(74.193)
vaser	-0.000
	(0.000)
produktinsher~m	0.070
	(0.092)
produktinbujq~i	0.002
	(0.020)
_cons	9528.585
	(68092.455)

© The Authors, published by EDP Sciences. This is an open access article distributed under the terms of the Creative Commons Attribution License 4.0 (http://creativecommons.org/licenses/by/4.0/).

Table 1. Continued.

Obs.	122
R-squared	0.542
Standard errors are in parenthesis	
*** $p<0.01$, ** $p<0.05$, * $p<0.1$	

Table 2. Summary statistics.

	N	St.Dev	kurtosis	skewness	min	max	t-value
fdipcapita	388	410.157	9.39	2.141	-549.302	2557.689	16.336
disctanca	425	430.994	1.726	-.367	904	2261	82.75
lfba	239	8.164	6.119	1.307	5.2	55.2	42.404
lfae	239	4.739	7.303	-1.22	49.4	86.4	251.795
lfie	239	6.283	11.333	-1.825	22.8	70.3	145.05
tlf	400	4390000	7.234	2.241	243246	1.84e+07	16.338
gdp	389	9.07e+10	15.156	3.26	7.09e+08	5.45e+11	12.613
gdppcapita	389	5855.573	3.206	1.022	218.492	27501.81	25.298
trade	383	32.285	2.79	.478	23.216	185.747	60.415
lspi	362	.88	3.146	-.962	1	4	66.478
eri	362	.753	2.648	-.531	1	3.67	64.72
cogdp	375	12.592	9.082	1.565	46.061	152.142	100.783
co	369	7.16e+10	15.605	3.334	9.05e+08	4.29e+11	12.796
untlf	400	7.19	4.054	1.124	3.3	37.3	38.577
papun	400	528000	16.581	3.489	27808.39	3460000	15.569
popull	425	9170000	8.015	2.371	604950	3.87e+07	16.589
abe	425	.459	1.751	.867	0	1	13.518
emploiag	400	12.665	5.02	1.514	2.6	62.6	26.092
emploiin	400	6.248	3.335	-.336	13.4	46	97.368
emplopopu	400	6.813	2.659	-.524	30.6	63.9	141.597
emploiser	400	10.512	3.296	-.596	23.5	77	100.95
punesimi	400	4790000	7.012	2.198	242000	2.04e+07	16.065
gdpgro	382	6.407	86.263	6.321	-14.814	88.958	10.863
ggfc	381	1.68e+10	14.368	3.153	1.40e+08	9.93e+10	12.869
ggfcgdp	376	3.311	4.1	-.466	9.195	29.941	110.3

© The Authors, published by EDP Sciences. This is an open access article distributed under the terms of the Creative Commons Attribution License 4.0 (http://creativecommons.org/licenses/by/4.0/).

Table 2. Continued.

totaxcp	199	12.728	2.331	-.39	7.4	66.8	41.019
vaaggdp	373	7.888	13.569	2.682	1.68	55.812	21.91
vaag	373	2.77e+09	8.278	2.231	1.11e+08	1.51e+10	17.28
produktinindu	362	9571.272	2.7	.759	677.809	42377.35	27.555
vaindu	373	2.80e+10	13.345	3.049	1.65e+08	1.60e+11	11.948
vasergdp	373	7.786	8.736	-1.862	21.69	74.729	151.936
vaser	373	5.14e+10	15.36	3.302	1.78e+08	3.09e+11	12.136
produktinsherbim	362	10175.07	2.824	.734	406.307	48082.03	29.117
produktinbujqesi	362	7668.509	6.063	1.801	321.293	41723.73	20.14

Table 3. Matrix of correlations.

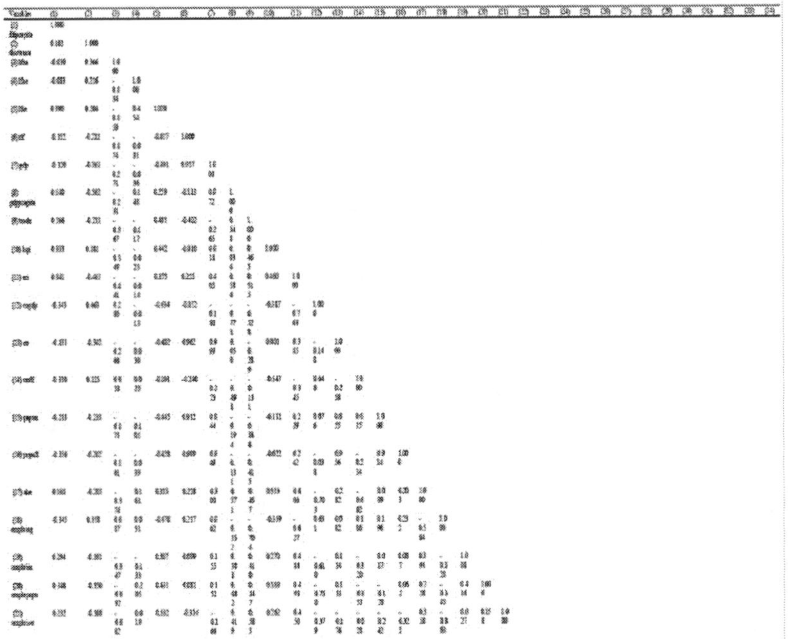

© The Authors, published by EDP Sciences. This is an open access article distributed under the terms of the Creative Commons Attribution License 4.0 (http://creativecommons.org/licenses/by/4.0/).

Table 3. Continued.

[illegible]	[illegible]
[illegible]	[illegible]
[illegible]	[illegible]
[illegible]	[illegible]
[illegible]	[illegible]
[illegible]	[illegible]
[illegible]	[illegible]
[illegible]	[illegible]
[illegible]	[illegible]
[illegible]	[illegible]
[illegible]	[illegible]
[illegible]	[illegible]
[illegible]	[illegible]
[illegible]	[illegible]

The annual turnover is taken as a dependent variable of this regression, while all other variables are taken as independent variables, from this we see that the level of significance for all independent variables is not very high, but the coefficients are appropriate.
This shows all the relationships between the variables that are not independent of each other, as well as the dependent variable.

References

1. Banka Qendrore e Kosovës- BQK, http://www.bqk-kos.org/
2. Statistic Agency of Kosovo- SAK, http://ask.rks-gov.net/
3. H. Xhafa, Foreing Direct Investments in Albania, Efective Analise, Politics and Economic Growth (2012)
4. United Nations Conference on Trade and Development, http://unctadstat.unctad.org/
5. Ministry of Trade and Industry, http://www.invest-ks.org
6. World Development Indicatores, https://datatopics.worldbank.org/
7. Central Bank of Kosovo, FDI based on sectors (2019)
8. Central Bank od Kosovo, "Report on Macroeconomic Growth", http://bqk-kos.org/repository/
9. Instituti Riinvest, "Klima e Biznesit ne Kosove" (2014)
10. Bizneses and Investments in Kosovo promoted in Washington, http://www.ekonomia-ks.com/
11. G. Rruci, IHD ne Kosove sipas vendit te origjines, Open Data Tirane (2014)
12. A. Zeka, Kosovo its in a bad position with foreign investments, Alb Info (2014)
13. A. Bevan, S. Estrin, The Determinants of Foreign Direct Investment in Transition Economies (2000)

© The Authors, published by EDP Sciences. This is an open access article distributed under the terms of the Creative Commons Attribution License 4.0 (http://creativecommons.org/licenses/by/4.0/).

High-Tech industry budget assessment and planning model

Evgenii Khrustalev[1], *Oleg* Khrustalev[1], *Nina* Baranova[2, *], and *Daria* Loginova[1,2]

[1]Laboratory of simulation modeling and the interaction of economic objects, Central Economics and Mathematics Institute RAS (CEMI RAS), Nakhimovsky prospect, 47, 117418 Moscow, Russia
[2]Peoples' Friendship University of Russia (RUDN University), Miklukho-Maklaya Str, 6, 117198 Moscow, Russia

Abstract. In the conditions of a constant budget deficit allocated to the research and manufacturing complex, the problem of value for money take on particular significance. The solution to this problem largely depends on the scientific validity of planning budget expenditures for innovative development and, first of all, on the optimality of the plans being developed in the context of crises, pandemics or sanctions. In this connection, the article analyzes modern research and manufacturing complex development processes under fiscal stress. As a result, economic and mathematical tools based on the analytic hierarchy process will be scientifically substantiated and developed. This method is designed to build a model that allows to quantify and efficiently distribute funds allocated by the State. This is necessary for the formation of that part of the budget item, which refers to the innovative development of knowledge-based industry.

1 Introduction

The efficiency disbursing public funds allocated to enterprises of the research and manufacturing complex (RMC), and the assessment of their contribution to the rate and quality of economic progress [1-4] cannot be assessed by a single indicator [5]. Even such an important indicator as the scientific and technological potential (scientific and technological capabilities) of individual enterprises and corporations of the RMC cannot serve as a sufficient measure for assessing the effectiveness. This is due to the fact that the pattern of this spending is characterized by interpenetration in various knowledge-intensive branch of the economy, as well as in the organization of the social sector, science and education [6]. In modern conditions of the rapid growth of knowledge-intensive industries and high-tech progress, in a context of uncertainty associated with a pandemic, crisis, etc., the efficient spending of funds is determined not so much by the value of the potential scientific and technical capabilities of the RMC, but in the indicator (indicators) characterizing its survival (structural, staff, etc.) and subsequent regeneration [7]. Thus, it is imperative to develop and apply multi-criteria optimization and tool method, which is

* Corresponding author: baranova_nm@pfur.ru

© The Authors, published by EDP Sciences. This is an open access article distributed under the terms of the Creative Commons Attribution License 4.0 (http://creativecommons.org/licenses/by/4.0/).

currently formalized and poorly and ill-defined, in order to formulate a strong mathematical problem to optimize a complex system of existing norms for disbursing public funds.

However, when identifying the dependences of the efficiency of spending the allocated resources, fundamental difficulties were identified. The fact of the matter is that these dependencies are so complex that their identification is possible in most cases only by parameterization based on extremely complex simulation models [8]. It is possible to apply the existing models to assess the impact on scientific and technical potential only for certain types of resources, but these models are not applicable for the majority, and special ones have not been developed. Moreover, there are no simulation models for other indicators of the resource use efficiency.

Thus, the objective function of the optimization problem of resource use in the form of an explicit, analytically specified function of the controlled parameters could not be represented. This circumstance determines the need to improve the processes of financial planning in the field of knowledge-intensive industries [9-10].

2 Materials and Methods

The proceedings of world scientists Astashova N. [1], Baranova N.M. [2, 10], Shevtsova N.A., Dmitrieva E.G. [2], Pilipenko A.I., Dikhtiar V.I., Pilipenko Z.A. [10], Sukharev O.S. [3], Sydor I.P., Koval S.L. [4], Tishutina O.I., Mikhailov A.V. [5], Tedeeva Z.B., Kulova A.Z. [6], Zimarin K.A., Zakirova Yu.A. [7], Canay I.A., Shaikh A.M. [8], Khrustalev E.Yu., Strelnikova I.A. [9], Saaty Thomas L. [11-14], Kearns K. [11], Blanchard O., Gopinath G., Rogoff K. [15], Zhang L. [16], Gómez M.A. [17], Crainich D., Eeckhoudt L., Le Courtois O. [18], Menegatti M., Peter Richard [19], Gollier C. [20] were studied to solve the problem. As a result, an optimization model was developed using the analytic hierarchy process. The proposed model will make it possible to effectively distribute allot funds for the development of science-intensive industries.

3 Results and Discussion

3.1 Building an objective function using the analytic hierarchy process

Despite the problems that arose in the formulation and solution of the optimization problem of the disbursing public funds for the development of science-intensive industries, it turned out to be possible to present the objective function in a different way. The analytic hierarchy process has already been sufficiently tested and can be proposed as one of the methods for the formalized construction of a linearized objective function [11-13]. This method involves the sequential decomposition of the global (generalized) objective function into its individual criterion components and the processing of expert judgements about them using the pairwise comparisons.

The essence of using the method of detailed analytic hierarchy process in the sequential calculation of the values of global and particular priorities, and then in checking the degree of consistency of the results obtained at each analyzed stage. The calculation of these vectors is based on the construction of matrices of paired comparisons of all criterial components of each hierarchical level and on the execution of a certain specific sequence of operations on them. These operations are performed for all existing hierarchical levels [14]. At the last level, the vectors of resource priorities are determined, and the degree of their impact on the criterial components of the penultimate (previous) level is also revealed.

The weighing coefficients K_i for each considered particular criterion are considered by an expert and established the preference pattern chosen by him. The coordinated (i.e.,

© The Authors, published by EDP Sciences. This is an open access article distributed under the terms of the Creative Commons Attribution License 4.0 (http://creativecommons.org/licenses/by/4.0/).

logically consistent) priorities of each selected particular criterion are established to determine K_i by pairwise comparison of them on a certain scale of relative importance.

In practice, it is the following scale of relative importance in pairwise comparison that has become widespread:

1 – equal importance of resources;
3 – weak importance for one resource over another;
5 – essential or strong importance of one resource over another;
7 – demonstrated importance of one resource over another;
9 – absolute importance of one resource over another;
2, 4, 6, 8 – intermediate values between the two adjacent judgments.

Using this scale, the inverse of symmetric square matrices of all paired comparisons made are constructed for each particular criterion at all levels and resources at the final (last) level (1) [13]:

$$\|a_{ij}\|,\ (a_{ij} = \frac{1}{a_{ji}}) \tag{1}$$

i=1...n; j=1...n – the number of criteria (or resources) compared at each level.

A set of local priorities is determined from the formed groups of matrices of pairwise comparisons, which express K_i. The determination of K_i is based on the calculation of the eigenvectors of the obtained matrices, their normalization and consistency check. So, K_i are calculated as follows for the criteria of the first level of the hierarchy.

The weights are assigned for each criterion of the first level (2), in accordance with the used preference scale, and a matrix of pairwise comparisons $\|a_{ij}\|$ is constructed.

$$\begin{pmatrix} 1 & a_{12} & \cdots & a_{1,n-1} & a_{1n} \\ a_{21} & 1 & & a_{2,n-1} & a_{2n} \\ \vdots & & \ddots & \vdots & \\ a_{i1} & a_{i2} & & 1 & a_{in} \\ a_{n1} & a_{n2} & \cdots & a_{n,n-1} & 1 \end{pmatrix} \tag{2}$$

Next, the components of the eigenvector of the matrix are calculated (3):

$$a_1 = \sqrt[n]{\prod_j^n a_{1j}}\,; \cdots;\ a_n = \sqrt[n]{\prod_j^n a_{nj}} \tag{3}$$

Then, the normalized estimates of the local priority vectors of the first level are determined:

$$K_1 = \frac{a_1}{\sum_i a_i}; \cdots;\ K_n = \frac{a_n}{\sum_i a_i} \tag{4}$$

Finally, the consistency of the obtained local priorities is checked by calculating the consistency index (CI) and comparing it with random consistency (5):

$$\mathrm{CI} = \frac{\sum_{i=1}^{n} K_i \sum_{j=1}^{n} a_{ij} - n}{n-1} \tag{5}$$

If the CI is no more than 10-15%, the obtained K_i values are considered to be sufficiently consistent [12], otherwise the expert should carefully check and clarify the preferences assigned to him. The similar operations are performed at all levels of the hierarchy. It provides to determine all the values of the weighing coefficients for each parameter of the last (lower) level. These values will be the final priority of each specific normalized resource for a linear objective function [11-14].

3.2 The principles for constructing a mathematical model

Let us consider the proposed method in the context of solving a task of reducing financial exposure [9-10, 18-19] and determining the rational volume of funding cuts for the main

© The Authors, published by EDP Sciences. This is an open access article distributed under the terms of the Creative Commons Attribution License 4.0 (http://creativecommons.org/licenses/by/4.0/).

object of expenditure on the development of high-tech industries while limiting the state budget of the RMC.

The objective function should be constructed to solve this problem and the degree of influence on it by each of the object of expenditure on the development of high-tech industries should be assessed. In this case, the main objects of expenditure are the types of normalized resources for which the unconditional specific consumption indicator is known (in this case, the annual consumed amount of financing). Not absolute values of these indicators will be used, but relative values (expressed as a percentage of needs). The task can be reformulated: to determine the most rational percentage of satisfaction of needs for each item of expenditure in the conditions of a given funding deficit.

As noted, there is no explicit objective function for determining the effective disbursing public funds. Therefore, we will make the assumption that the objective function will take the maximum value, i.e. equal to 1, with 100% satisfaction of needs for all objects of expenditure. Let us construct the given objective function by the analytic hierarchy process. In order to do this, build a three-level hierarchy:

the 0 level corresponds to making decisions across the entire vector of target priorities;

the 1st level corresponds to the components of the vector of target priorities;

the 2nd level corresponds to the components of the priority vector of expenditure items.

The main problem that arises when determining the objective function is the decomposition of ideas about the efficiency of disbursing public funds into simpler components. It means a sequence of judgments is formed by the person making a decision on the priority of standardized items of expenditure in terms of their impact on the total efficiency. Therefore, the main stage will be the building of a hierarchy of judgments between the 0 and last levels (in this case, this is only one, 1st level).

The task of presenting judgments of the 1st level is the formulation of priority goals to be achieved as a result of disbursing allocated public funds, their ranking according to the degree of importance by the decision maker. The task of formulating goals is solved on the basis of an assessment of the totality of political, economic, social and other factors in the current timeframe.

Let us define the priority vector of the 1st level of the hierarchy through a set of ranked goals [9].

The *number of the RMC employees* (the priority of $\boldsymbol{K_1}$, determines the importance of maintaining the existing staff number).

The *scientific and technical potential level* of the RMC (the priority of $\boldsymbol{K_2}$, determines the importance of the average level of innovativeness of the RMC enterprises).

The *qualification of the RMC key personnel* and the science intensity of the RMC production capacity (the priority of $\boldsymbol{K_3}$).

The *conversion of spending on the RMC development* in solving social and economic problems of the state (the priority of $\boldsymbol{K_4}$).

$\boldsymbol{K_i}$ is an alternative set of goals that a decision-maker wants to achieve in the process of spending the limited financial resources allocated to him.

The absolute priority of $\boldsymbol{K_1}$ means that the number of the RMC employees affects the ability of the state to create innovative science-intensive and high-tech products. The absolute priority of $\boldsymbol{K_2}$ means "less is more". The priority of $\boldsymbol{K_3}$ means "the main thing is to survive, preserving the most valuable until better times". The priority of $\boldsymbol{K_4}$ means that public funds should be used as much as possible to offset the costs of social (non-scientific) programs.

Obviously, the absolute priority of one of $\boldsymbol{K_i}$ cannot be chosen as a solution, however, none of $\boldsymbol{K_i}$ can be completely ruled out. Therefore, the solution will be a balanced set of $\boldsymbol{K_i}$, which determines the priority vector.

© The Authors, published by EDP Sciences. This is an open access article distributed under the terms of the Creative Commons Attribution License 4.0 (http://creativecommons.org/licenses/by/4.0/).

3.3 The priority vector

The method of pairwise comparisons of the analytic hierarchy process is used to determine such a priority vector [11-14]. It can be used to process the judgments of experts on the priorities of the considered alternatives K_i.

Comparing the pairwise presented alternatives K_i, the expert, at his own discretion, determines the preference pattern and assigns a certain point to each of them on a scale of relative importance. As a result, this preference pattern will represent an inverse of symmetric square matrix, its dimension is 4 × 4 (Table 1).

The processing this matrix using the formulas (1)-(5) makes it possible to calculate the priority vector of goals. The components of this vector determine the degree of their importance from the point of view of an expert. In this case, the preferences established by the expert will be as follows:

K_1 = 0,088 – the number of the RMC employees;
K_2 = 0,255 – the scientific and technical potential level;
K_3 = 0,540 – the qualification of the RMC key personnel;
K_4 = 0,117 – the conversion of spending on the RMC development.

Table 1. The preference pattern.

	The number of the RMC employees, K_1	**The scientific and technical potential level, K_2**	**The qualification of the RMC key personnel, K_3**	**The conversion of spending on the RMC development, K_4**
The number of the RMC employees, K_1	1	1/3	1/4	1/2
The scientific and technical potential level, K_2	3	1	1/3	3
The qualification of the RMC key personnel, K_3	4	3	1	5
The conversion of spending on the RMC development, K_4	2	1/3	1/5	1

The consistency check of these expert judgments shows that CI equals 0.0533 (CI<10%). Having determined the priority vector of the 1st level, construct the priority vectors of the 2nd level. To achieve this, define the degree of influence of each of the expense items on each component of the target priority vectors of the previous level. Since in this case the target priority vectors contain four components and five items of expenditure are considered (the R&D (k_j^1), advanced vocational training (k_j^2), the RMC staff costs (k_j^3), procurement of innovative equipment (k_j^4), infrastructure development (k_j^5), it is possible to build and process 4 matrices of pairwise comparisons (its dimensions will be 5×5).

1. The vector of the significance of cost item for the number of the RMC employees.

The degree of influence of cost item on the target factor "The number of the RMC employees" based on the results of processing the matrix is determined by the following components of the normalized eigenvector of the matrix:

k_1^1 = 0,100 – the item of "R&D expenditures";
k_1^2 = 0,302 – the item of "advanced vocational training";
k_1^3 = 0,490 – the item of "the RMC staff costs";

© The Authors, published by EDP Sciences. This is an open access article distributed under the terms of the Creative Commons Attribution License 4.0 (http://creativecommons.org/licenses/by/4.0/).

k_1^4 = 0,133 – the item of "procurement of innovative equipment";
k_1^5 = 0,045 – the item of "infrastructure development".
The consistency index CI_1 equals 0,033.

2. The vector of the significance of cost item for the scientific and technical potential level.

The degree of influence of cost item on the target factor "The scientific and technical potential level" based on the results of processing the matrix is determined by the following components of the normalized eigenvector of the matrix:
k_2^1 = 0,234 – the item of "R&D expenditures";
k_2^2 = 0,302 – the item of "advanced vocational training";
k_2^3 = 0,061 – the item of "the RMC staff costs";
k_2^4 = 0,292 – the item of "procurement of innovative equipment";
k_2^5 = 0,109 – the item of "infrastructure development".
The consistency index CI_2 = 0,028.

3. The vector of the significance of cost item for the qualification of the RMC key personnel.

The degree of influence of cost item on the target factor "The qualification of the RMC key personnel" based on the results of processing the matrix is determined by the following components of the normalized eigenvector of the matrix:
k_3^1 = 0,259 – the item of "R&D expenditures";
k_3^2 = 0,250 – the item of "advanced vocational training";
k_3^3 = 0,259 – the item of "the RMC staff costs";
k_3^4 = 0,158 – the item of "procurement of innovative equipment";
k_3^5 = 0,074 – the item of "infrastructure development".
The consistency index CI_3 = 0,021.

4. The vector of the significance of cost item for the conversion of spending on the RMC development.

The degree of influence of cost item on the target factor "The conversion of spending on the RMC development" based on the results of processing the matrix is determined by the following components of the normalized eigenvector of the matrix:
k_4^1 = 0,126 – the item of "R&D expenditures";
k_4^2 = 0,055 – the item of "advanced vocational training";
k_4^3 = 0,155 – the item of "the RMC staff costs";
k_4^4 = 0,235 – the item of "procurement of innovative equipment";
k_4^5 = 0,429. – the item of "infrastructure development".
The consistency index CI_4 = 0,087.

4 Conclusions

Having received the values of K_i and k_j^i, the weight coefficients in linear form can be determine and the objective function could be written (6):

$$F\,(x1, x2, x3, x4, x5) = \sum_{i=1}^{5} b_j \cdot x_j \;\rightarrow\; max \tag{6}$$

$$b_j = \sum_i K_i \cdot k_j^i, \quad 0 < x_j < 1$$

In this case, the objective function becomes as follows (7):

$$F\,(x_1, x_2, x_3, x_4, x_5) = 0.21{\cdot}x_1 + 0.24{\cdot}x_2 + 0.19{\cdot}x_3 + 0.24{\cdot}x_4 + 0.12{\cdot}x_5 \tag{7}$$

x_j – the percentage of satisfaction of needs under the relevant item of expenditure.

© The Authors, published by EDP Sciences. This is an open access article distributed under the terms of the Creative Commons Attribution License 4.0 (http://creativecommons.org/licenses/by/4.0/).

It is possible to determine the best values of x_1, x_2, x_3, x_4, x_5 to achieve the maximum of the objective function using mathematical programming methods or, for example, using the MS Excel program (via the Solver Tool add-in).

To conclude, the considered case of reducing financial exposure and determining the level of funding for the development of high-tech industries while limiting the state budget of the RMC illustrates the practical application of this approach. The proposed model is universal and can be used by any country to plan budgetary expenditures for the development of science-intensive industries.

References

1. N. Astashova, Int. J. Management Theory and Practice, **8**, 19 (2011)
2. N. M. Baranova, N. A. Shevtsova, E. G. Dmitrieva, GCPMED2018, 1, 1877, (2019)
3. O. S. Sukharev, Financial J., **1**, 17 (2015)
4. I. P. Sydor, L. Koval, Scientific-Theoretical J. «Science and Economics», **3**, 18 (2014)
5. O. I. Tishutina, A. V. Mikhailov, J. Problems of Modern Economics, **4**, 194 (2012)
6. Z. B. Tedeeva, A. Z. Kulova, J. Science: Society Economy Law, **1**, 62 (2020)
7. K. A. Zimarin, Yu. A. Zakirova, J. Science. Education. Economy. Series: Economy, **3**, 6 (2013)
8. I. A. Canay, A. M. Shaikh, Advances in economics and econometrics: Eleventh world congress, **2**, 271 (2017)
9. E. Yu. Khrustalev, I. A. Strelnikova, J. Finance and Credit, **7**, 13 (2011)
10. A. I. Pilipenko, V. I. Dikhtiar, N. M. Baranova, Z. A. Pilipenko, Modeling Economic Growth in Contemporary Russia, 63 (2019)
11. T. L. Saaty, K. Kearns, Analytical Planning. The Organization of Systems (1985)
12. T. L. Saaty, Int. J. Services Sciences, **1** (2008)
13. T. L. Saaty, *Mathematical Principles of Decision Making* (2010)
14. T. L. Saaty, Cloud of Science, **1** (2015)
15. O. Blanchard, G. Gopinath, K. Rogoff, IMF Econ Rev, **69**, 258 (2021)
16. L. Zhang, J. Econ, **115**, 73 (2015)
17. M. A. Gómez, J. Econ., **117**, 37 (2016)
18. D. Crainich, L. Eeckhoudt, O. Le Courtois, J. Math. Econ., **88**, 153 (2020)
19. M. Menegatti, R. Peter, SSRN (2020)
20. C. Gollier, Toulouse School of Economics Working Paper, **19**, 1003 (2019)

© The Authors, published by EDP Sciences. This is an open access article distributed under the terms of the Creative Commons Attribution License 4.0 (http://creativecommons.org/licenses/by/4.0/).

Impact of Covid-19 on Iran startups at biotech, pharmaceutical, engineering and other innovative industries

Ismael Rezaeinejad*, and *Sergey U.* Chernikov

Peoples' Friendship University of Russia (RUDN University), Miklukho-Maklaya Street, 6, 117198 Moscow, Russian Federation

Abstract. This research reviews Corona's spread that caused many governments to take serious actions; limitations in many aspects of society and economic life as an exogenous shock for many economic actors, especially innovative ones. This review examines an entrepreneurial ecosystem to deal with the problem facing them during the crisis and how to use them to deal with that. As a response to Corona's expansion, limitations threaten many innovative startups. Our research response first represents entrepreneurs' challenges as a result of the situation. Second, we show how entrepreneurs support their investment by considering the crisis's effects and what they do. Finally, we will provide measures that policymakers can use to help entrepreneurs who face a challenge. After reviewing the previous research on the crisis and entrepreneurship, this research investigates the main challenges of high-tech startups in a heavily sanctioned developing economy like Iran. The research shows that while the SMEs successfully use their available resources as the first response to the crisis, their growth and innovation potential are at risk. Therefore, policy measures can provide immediate help to innovation by reducing the pressures resulting from limited cash circulation and long-term actions taken by the more expanded entrepreneurship ecosystem and support it to ensure rapid recovery and growth.

1 Introduction

This research is an overview of the study conducted in industrial and modern countries, with the discovery of acute respiratory distress syndrome (sars) in the late 1990s and recently, with all subsequent directions. However, not all of the infections caused by the new virus are not surprising; control measures such as distance working, which are conducted to slow the spread of the Covid-19 virus, enormous pressures on large parts of a country. Most of the leading players to shape the economy admit that the current orientation is a metaphor of the black killer, a surprise, unpredictable, and severe consequences that can significantly change the political and economic situation [25].

While such events can be interpreted as an opportunity or an unprecedented limitation in many parts of the society due to the Covid-19 crisis shows the current situation as an acute

* Corresponding author: mailto:1032205254@pfur.ru

© The Authors, published by EDP Sciences. This is an open access article distributed under the terms of the Creative Commons Attribution License 4.0 (http://creativecommons.org/licenses/by/4.0/).

crisis [19]. The Covid-19 created many problems on the shoulders of health systems worldwide. It has fuelled an economic crisis to limit a comprehensive volume of economic activity. Furthermore, as many of the past problems have been developed at a particular point since time and region or in a given time range have been developed, the Covid-19 disease has dramatically expanded globally. The cross-sectional measures that have suddenly been applied have unexpectedly harmed the economy [8].

Therefore, the quick response has a particular attention to the section that has been firmly ignored in the current general negotiations: Startups of innovation [13]. It seems that the majority of policy initiatives to support the economies during the Covid-19 crisis, established companies, have targeted existing industry sectors and economics and through these measures are aimed at helping the employment and continued economic activity [19]. At present, the focus is on supporting the nowadays situation, while future economic activities attract less attention [25].

However, the Startups of innovative that form future economic activity is among the most vulnerable players of any economy, even encounter creative and new responsibilities and being small, which threaten the continuation of their lives [11]. This situation is likely to worsen in critical cases. The expansion of Covid-19 leads to a decrease in the enormous potential accumulated in recent years to produce economic and possibly social value innovation.

We investigated individuals involved in an entrepreneurial ecosystem in Iran to determine their reaction to the crisis and their view of the measures taken to reduce the economic impact. When writing this paper, the Iranian government's standards to protect companies and innovations affected by the Covid-19 crisis included tax support, government support for working at short hours, improved measures in guaranteed banks, and the loans and programs offered by banks.

These loans also offer additional special support programs for Startups, such as growth loans and investment schemes, which are not explicitly created to address the Covid-19 crisis but may still be beneficial. Besides, a significant undertaking of the national aid package for large enterprises, business people, and prominent entrepreneurs is presented. However, many of these measures, such as the new loan, are not available because they can not quickly meet the traditional criteria for obtaining a loan, especially in the initial stage (they cannot promptly meet the conventional standards for getting a loan). Equally, many investors are expected to did not get profit in a reasonable timeframe.

2 Methodology

After reviewing the previous research on the crisis and entrepreneurship, this research investigates the main challenges of high-tech startups in a heavily sanctioned developing economy like Iran. An ever-growing startup ecosystem was shaped during the past decade. The main research question is then: "What are the main challenges faced by the Iranian startups during the coronavirus pandemic?". To answer this, we first reviewed the existing literature on the topic. Then, we investigate co-founders of startups, which experienced a monthly growth rate of more than ten percent in the past year. Afterward, the author analyzed the gathered data, and the findings are presented.

3 Results and Discussion

© The Authors, published by EDP Sciences. This is an open access article distributed under the terms of the Creative Commons Attribution License 4.0 (http://creativecommons.org/licenses/by/4.0/).

3.1 Managing the entrepreneurial crisis of innovative startups during covid-19

Aside from the covid-19 pandemic, the virus has also increased local economies and decreased global economies. Fear of the effects of unpredictability, the disease currently affects the world's dominant economies [5]. The limited number of studies on the flexibility in entrepreneurship and crises mainly focus on pre-crisis and skills or sources that entrepreneurs and organizations create to resist or adapt to crisis events [3].

In general, specific features of innovative startups should enable them to be better prepared than other firms to deal with the Covid-19 crisis. Innovation is a prerequisite for resilience, as innovative businesses continually face a broad range of troubles [22]. However, companies do not always recognize the real threat that a potential crisis event poses, and most startups have not been prepared for the events of the past year. We have learned from research on Hurricane Katrina's effects and its consequences that lack of preparation can have significant consequences, especially for small businesses disconnected from cash flows, lack of funds for recovery, and severe infrastructure [20].

Given that data related to the immediate effects of the crisis, Covid-19 and subsequent closure are not available in many economies. The previous researches shows that the startups face the immediate and tangible consequences of the Covid-19 prevalence, especially in sales reduction, while fixed costs remain. The combination threatens the liquidity and extended-term survival of budding companies [2].

Besides, the founders of startups reported that the economic situation is now undesirable for innovation [16]. The main partners, customers, and investors themselves are involved in responding to the crisis and uncertainty about how the problem is developed and prevents any response.

When markets face a problem, startups have to compatible with obstacles to achieving a balanced budget that will negatively affect their subsequent growth. The startups were aware that the crisis caused by the situation could also create opportunities and threats because it causes external pressure to be adapted [3].

Startups also noted that they were suddenly forced to suddenly adjust their organizational priorities because value generation processes are changed. The supply chain is affected by this crisis. Simultaneously, some of the co-founders reported that their positions had not been affected by the problem because their jobs, despite the situation, continued to relate to this issue or because they are flexible before the crisis [12]. However, such flexibility is likely to be limited. This not only describes the organization's ability to continue functioning during a disruptive event, but the conceptual aspect of the is also intended to be used before a crisis and then deployed across it [24].

Crisis management is used to strengthen the resistance in this situation, and it is of paramount importance. Managing crisis To minimize the effects of a crisis and in case of doing well, conducted studies on crisis management in entrepreneurship research mainly measures that entrepreneurs or organizations do to mitigate the potential negative consequences of a crisis, including those involved in changes in sales, marketing, and employment practices[1].

In particular, small businesses tend to be willing and flexible, and we should expect them to show such a thing in response to the Covid-19 crisis.

Therefore, crisis management in the field of entrepreneurship has a close relationship with the concept of sustainability. Rather than offering tough and rigid processes to solve borrowers' problems, it seems appropriate for innovative strategies to accept iterative and flexible approaches such as practical logic[10]. The research findings on the earthquake of Emilia in Italy indicated that the resistant entrepreneurs were those with the resources at that time, change and opportunity, so they follow a fundamental principle and respond to how innovative solutions are used to counter the Covid-19 crisis.

© The Authors, published by EDP Sciences. This is an open access article distributed under the terms of the Creative Commons Attribution License 4.0 (http://creativecommons.org/licenses/by/4.0/).

The budding companies reported that they rely heavily on communication capabilities on this basis.

Moreover, co-founders reported that by gathering funds through domestic measures and asking the government to support them, they had strengthened their financial ability. However, given the support of the government, the founders reported the lack of conformity between the support services provided by the government policies and the characteristics of their organizations; therefore, they are excluded from the policy measures because, for example, they are not payable or are supported programs, and the bureaucratic faults are more than that[8].

Therefore, the first responders who responded to the crisis have not been created to obtain immediate government support. Finally, the founders reported that their customer needs change because of the Covid-19 situation. According to the above conditions, they have been using the bricolage crisis response to solve new problems. They identify and pursue new entrepreneurial opportunities and set new directions for their companies. Nevertheless, the situation is more complicated in emerging economies; Since the pandemic is "dramatically worsening the economic outlook for emerging economies," like Iran, India, Brazil, Russia, Singapore, Turkey, and Mexico are experiencing successive waves the pandemic. As these countries prioritize improving their limited public health capabilities and have tighter mobility restrictions, most of their resources were allocated to their basic needs. Besides, the IMF is expecting the emerging world to enter a severe recession in 2021. Issues such as a significant decrease in capital inflows, fewer systematic markets, higher investment risks, and their reliance on export and tourism made these countries more susceptible than developed ones. All of the mentioned issues led to paying less attention to SMEs in general and specifically to startups in countries like Iran.

Moreover, Iranian policymakers are mostly trying to help medium-sized and large firms survive to save the national employment rate. Therefore, startups suffered more from these contextual factors of the heavily sanctioned economy. According to the data published by Iran's Statistical Research and Training Center, thirty-eight percent of the Iranian firms were inactive for at least two months in the last six months. Also, the data showed a significant drop in liquidity, human resources, and production rates of the startups in general. A considerable number of startups were shocked in the first months of the pandemic. Some of them took precautionary cuts, some froze their activities, and some halted their services. Besides, Iranian startups do not have access to international customers, and therefore their markets are generally limited to the national customers, which is a significant issue. Consequently, it seems necessary for Iranian startups to understand better and address their challenges as a preliminary stage in managing crisis as Jack Ma [the founder of Alibaba] mentions: "If I want to write a book that will be "Alibaba 1001 mistakes"." because startups could learn from each other's challenges and mistakes more than the ways they survive, as their survival depends on quite a lot of different elements.

3.2 Economic policy responses to the constraints of covid-19

While the crisis management at the micro-level is mainly the job of entrepreneurs itself, policymakers are asked to support entrepreneurs in their efforts to cope with crises which means that policymakers are doing crisis management at the macro level because they aim to strengthen the flexibility of jobs, including options, and to support their crisis management practices. Research has shown that areas with a pre - entrepreneurial level align exogenous strategies[9]. Its flexibility characterizes an entrepreneurial region, and entrepreneurial activity can contribute to restructuring and adaptation after the crisis [11]. Followed by hurricanes and tornadoes, such individuals are motivated to join their positions, target business, and social goals. It is very likely that after the crisis, we will see areas that deal with

© The Authors, published by EDP Sciences. This is an open access article distributed under the terms of the Creative Commons Attribution License 4.0 (http://creativecommons.org/licenses/by/4.0/).

the consequences of harsh actions on a different limit and several entrepreneurial potentials. The feasibility of this will depend on how entrepreneurs are innovations [22].

From the policymakers' perspective, we must rely solely on the entrepreneurial initiative to overcome the economic damage caused by limitations. Instead, it should be a key priority for policymakers in the short - term during the Covid-19 crisis to ensure that budding companies can benefit from adequate resources. External assistance for the affected areas before, during, and after limits of Covid-19 is essential.

3.3 The Iranian startups situation

The Iranian startup ecosystem has grown in quality and quantity in the past decade. The enterprising communities and the entire startup ecosystem have taken several steps and became more mature during this period. This evolutionary trend was well-presented in some previous studies. They have described how traditional firms have turned to startups and new technology-based firms within four decades. According to this study, the emergence of entrepreneurship in its modern sense went back to the late 1990s, followed by an ICT-based startup boom in the early 2000s. During the 2000s, entrepreneurship was expanded as an academic field, and the government introduced several initiatives.

In the 2010s, this trend was accelerated by all the players of the Triple Helix, i.e., universities, industries, and the government. Several leading startups, such as Snapp, Digikala, and the like, have grown dramatically and changed the whole startup ecosystem. Currently, several vital players exist at different levels, including international (e.g., international venture capitalists/venture capital funds; international startup studios, and non-resident Iranian experts and entrepreneurs), national (e.g., vice-presidency for science and technology; ministries; universities; funds; national elites foundation; Iran lab expo; national technomart of Iran; Iranian knowledge-based leaders federation; export development & technology exchange corridor; Labs' net; the center for progress and development of Iran), industrial (e.g., innovation centers; innovation factories; sector-specific accelerators, venture capital funds, and incubators), firm (e.g., new technology-based firms (NTBFs)- knowledge-based firms; startups; creative firms), the team (e.g., independent and supported teams), and individual (e.g., freelancers; intended/potential entrepreneurs; young talented individuals) players. However, as mentioned previously, the Iranian entrepreneurial ecosystem has changed dramatically after the coronavirus pandemic. The number of failed startups has increased. There have been limited options for startups to fund their entrepreneurial journeys, such as the state-backed venture capital funds, the Vice-presidency for Science and Technology, and the National Innovation Fund. Moreover, some of the accelerators offered customized services, which led to a limited number of success stories, most of which dealt with health-related solutions.

3.4 Iranian startups challenges

Startups generally face several types of challenges in both local and global economies: On the one hand, several indigenous and exogenous factors could turn into challenges for startups. On the other hand, the coronavirus pandemic has already affected the national and global economies [7].

Both developing and developed economies are still suffering from such a crisis and are forecasting new types of recession. Issues such as threatening their performance and highly turbulent economic situations made these entities more vulnerable during such a global crisis [13]. Therefore, problem management has become a critical issue to be taken into account by startups. For managing such a crisis, startups should get prepared. To do so, they must know where they stand before choosing which actions to take or how to go [21]. Thus,

© The Authors, published by EDP Sciences. This is an open access article distributed under the terms of the Creative Commons Attribution License 4.0 (http://creativecommons.org/licenses/by/4.0/).

understanding their challenges has become mandatory for these innovative entities. Nevertheless, just a few recent studies have already published in this domain, most of which were investigating the challenges in the world's top economies, such as Germany (e.g., see Maritz et al. 2020; Kuckertz et al. 2020), and a few studies have previously investigated the challenges faced by startups in emerging economies like Iran. Based on the mentioned above issues, we categorized these challenges into two groups: (1) general challenges and (2) crisis-related challenges.

3.4.1 General challenges of Iranian startups

These challenges are a function of the business environment's general conditions [15]. For instance, according to previous studies, Iranian startups faced difficulties related to financial issues, human resources, and support mechanisms. They considered a lack of access to various types of financing, such as IPOs, bank credits, venture capital funds, and seed finance, as the most critical financial challenges of startups.

Hopefully, during the last five years, the situation has improved dramatically, and only the possibility of IPO has remained a financial challenge. It is noteworthy that the government has recently announced that five leading startups will be entered into the Tehran Stock Exchange until 2021. As a human resources management issue, the lack of team management skills remained a challenge for Iranian startups. This might be due to the country's cultural aspects, called the "Land of Individual Achievers." In the last category, the ecosystem still lacks enough support for securing intellectual property rights. Another study highlighted the government's challenges and mentioned that lack of knowledge about startup activities has led to the enactment of irrelevant rules and regulations. Generally, this was a significant issue and still is in some domains.

Nevertheless, we witnessed that the Vice-presidency for Science and Technology has paved the way for startups to grow more quickly by removing irrelevant rules and regulations and dedicating special offices for handling their insurance and tax-related concerns. Moreover, it conducted a gap analysis to measure the challenges of ICT-based startups in Iran. According to the study, startups have legal and infrastructural problems. In sum, a few studies have already paid attention to the general challenges of Iranian startups.

3.4.2 Crisis-related challenges of Iranian startups

The few studies on the challenges of startups in emerging economies such as Iran have mainly focused on the pre-crisis period. Given that information on the immediate impact of the crises on Iranian startups is not available, the author reviewed some relevant publications. They listed the main challenges through a qualitative research design in the following sections of this manuscript.

As a highly sanctioned emerging economy, Iranian startups have faced several challenges, even before the coronavirus pandemic. For instance, as the sanctions increased, several challenges were imposed on Iran's entrepreneurial or Startup ecosystem. Iranian startups still face a series of challenges due to a lack of access to international markets, providing them with several profitable opportunities. As these sanctions were an integral part of the economy in the past four decades, the previous Global crisis did not affect Iranian SMEs and startups as much as this pandemic did. However, Issues such as the depreciation of the local currency, limited access to international funds and markets, and challenges faced by female entrepreneurs were among the most significant challenges which affected the startup ecosystem in the past decades [17].

Nevertheless, several opportunities existed exploited by startups during these crises. For example, improvements in the Blockchain and Fintech industries were outputs of the pivots

© The Authors, published by EDP Sciences. This is an open access article distributed under the terms of the Creative Commons Attribution License 4.0 (http://creativecommons.org/licenses/by/4.0/).

made by Iranian startups during these crises. In sum, given the lack of data on how Iranian startups were acting during the pandemic, limited evidence exists regarding their challenges. Therefore, this study resorts to qualitative research design to shed light on those significant challenges.

4 Conclusions

Human life is undoubtedly more valuable than economic activity, and this study should not be termed as a critique of the measures taken to control the spread of Covid-19. We have shown how the economic crisis affects infection control actions - and, in particular, the limitation of economic activities on innovative initiatives. For example, Covid-19 threatens the innovation potential that can be sustained at regular times, and it can be claimed this was a tremor during which the Irrecoverable business models were lost. The crisis of Covid- 19 seems different. It is not just about government intervention and the support of the initiative. It identified seven factors related to inconsistency and coping strategies. However, the first step is to understand startups' response to the crisis in general, especially the dire crisis.

Future research should determine how these factors interact, and it will be interesting to decide how to use undesired outcomes to produce positive results. Entrepreneurs, who respond to the crisis, use their approach as part of their initial response to transforming the conflict resulting from the situation. Resilience accompanies the quality of the organization's response to the crisis, and it depends on the capacity to increase improvisation, harmony, flexibility, and endurance. This is also possible to ensure that creative jobs remain alive. Many entrepreneurs have their roles in encouraging the creation of changes and creating conflicts with existing resources. Studies show that the crisis can enhance new products' growth (i.e., the development of new products). Because the crises can also stimulate new opportunities, they can promote innovation and development of productions and alternative services, particularly for lifelong jobs. The Covid-19 pandemic consequences are not yet predictable, but there seem to be more opportunities. This research indicates that coping with uncertainty and failure is a standard part of a business, even if a crisis is like a pandemic. Accordingly, entrepreneurs are expected to show flexibility in response to a situation and adapt their business models. This suggests that Startups are better prepared than other economic actors for the problem. Undoubtedly, some commentators have said that the flexibility of startups and the relatively small number they work with means eliminating them from government aid programs will not substantially impact the economy. However, the complete elimination of Startup could potentially jeopardize the government's future innovation. Therefore, long-term political action, which is unlikely to be the first response to such a crisis, seems necessary. This finding is consistent with prior research suggesting that entrepreneurial responsiveness against concerns is characterized by entrepreneurial culture and knowledge diversity, which cannot be dealt with by short-term measures, but the result is consistent.

Just as the various governments' health systems present their primary ability to respond to a pandemic, the countries that created resistance entrepreneurial ecosystems will be able to resume their level of activity faster than those who cannot afford it.

The analysis shows the Startups strongly rely on their entrepreneurial ecosystem for crisis management. Therefore, policy measures can only be successful if they are successful with a broad sense of entrepreneurship. Encourage investors to fund the startups, be an excellent way to combine both short-term cash delivery to pay bills and develop future improvements.

Startups in emerging economies generally are facing a series of challenges that could affect their success or failure. Crises could also change these challenges and priorities. The coronavirus pandemic has affected almost every single building block of their evolving business models. Then, startups should be more focused on different aspects of their business

© The Authors, published by EDP Sciences. This is an open access article distributed under the terms of the Creative Commons Attribution License 4.0 (http://creativecommons.org/licenses/by/4.0/).

and their connection to the business environment. In other words, startups in emerging economies face some general challenges: endogenous and exogenous.

Moreover, they might face a series of specific challenges during a crisis like a coronavirus pandemic. Although a few scholars marginally studied the first type of challenges, the second type is rarely investigated, especially in emerging economies such as Iran. Therefore, this study attempted to identify the main challenges of Iranian startups in this coronavirus pandemic. The findings of this research could be considered a cornerstone for startups to understand their real challenges better to become able to face the coronavirus pandemic and manage the impacts of the crisis on their startups. The author concluded that startups were facing six types of challenges: financial challenges, human resources management challenges, support measures and mechanisms, marketing challenges, crisis management challenges, and further challenges (Table 1).

Table 1. Implications for startups and policymakers

Challenge	Implications for startups	Implications for policymakers
Financial	- Considering creative fundraising techniques such as crowdfunding - Revising their business plans according to the contingencies ofthe crisis - Cutting extra costs by adjustingtheir business models - Changing their priorities by focusing on profitability insteadof growth	- Supporting venture capital funds,accelerators, and incubators - Providing enough resources for state-backed venture capital funds - Offering working capital loans with lower interest rates - Delivering delayed payments and direct payments
Human resources and management	- Reconfiguring their teams -Their roles to cope with firing extra team members - Hiring new team members with multiple skills and abilities	- Offering wage subsidies - Supporting temporal downsizing - Suggesting talented memberslisted in databases such as the National Elites Organization

© The Authors, published by EDP Sciences. This is an open access article distributed under the terms of the Creative Commons Attribution License 4.0 (http://creativecommons.org/licenses/by/4.0/).

Support measures and mechanisms	- Preparing different scenarios to convince the support mechanisms - Gathering information about the governmental initiatives and applying for the most relevant ones - Receiving legal consultancy services about discontinued or canceled contracts	- Providing rescue packages for support mechanisms - Considering motivating policies such as tax exemptions for investors - Providing information about governmental initiatives and offering support services - Providing legal services to help startups
Market and marketing changes -	- Scanning the market -Revising their marketing plans - Considering creative marketing tools and techniques	- Providing relevant information about markets, e.g., through preparing databases - Devising relevant policies regarding the market failures - Offering creative thinking and marketing programs
Crisis management	- Improving their general crisis management skills - Improving their conflict management skills - Improving their entrepreneurial skills	- Improving the crisis recognition systems - Implementing critical decision-making systems - Considering contingency planningand training for future crises
Further challenges models	- Revising the business - Learning how to manage multiple tasks - Negotiating with key players in their supply chain - Monitoring the relevant governmental responses	- Providing consultancy services for startups - Analyzing results and removingirrelevant responses to the pandemic

4.1 Proposals

There are important issues due to the studies and reviews related to entrepreneurship and crisis, especially for the startups; the international press has shown many opportunities to protect beginners' innovation potential. Although it is not clear that specific measures have the most significant impact, the programs that specifically target innovation must be forced. It is an error that policymakers have to avoid, assuming that actions are also aimed at putting SMEs out of innovative projects. Unlike many other crises, and the Covid-19 crisis has not harmed all countries at the same time. Many European countries learned from China and Italy's experiences and reacted favorably, at least in health care. While the long period of the economic crisis that has engaged nations may disappear in the long run, policymakers can

© The Authors, published by EDP Sciences. This is an open access article distributed under the terms of the Creative Commons Attribution License 4.0 (http://creativecommons.org/licenses/by/4.0/).

see how the measures are taken to protect innovation are emerging, taking them as an appropriate way to improve the critical situation's knowledge. However, it can also be carried out to understand the effect of short response time to economic crises. It will be essential to follow the impact of actions taken during the Covid-19 crisis to prepare for future comparisons.

References

1. D. Alesch, J. Holly, E. Mittler, R. Nagy, Organizations at risk: What happens when small businesses and not-for-profits encounter natural disasters? (2001)
2. T. Baker, R. E. Nelson, Administrative Science Quarterly, **50**, 329 (2005)
3. P. Bishop, Entrepreneurship & Regional Development, **31**, 496 (2019)
4. P. Bishop, D. Shilcof, Entrepreneurship & Regional Development, **29**, 215 (2017)
5. A. Boin, Review of Policy Research, **26**, 367 (2009)
6. R., Brown, C. Mason, Small Business Economics, **49**, 11 (2017)
7. D. A. Buchanan, D. Denyer, Int. J. of Management Reviews, **15**, 205 (2013)
8. C.G. Bush, D. J. Ceru, R. Blackburn, Business Horizons, **52**, 481 (2009)
9. M. Cowling, W. Liu, A. Ledger, Int. Small Business J., **30**, 778 (2012)
10. R. Doern, N. Williams, T. Vorley, Entrepreneurship & Regional Development, **31**, 400 (2019)
11. GDA (Global Data Analysis), Coronavirus (COVID-19) Executive Briefing, Global Data (2020)
12. R. J. Glass, L. M Glass, W. E. Beyeler, H. J. Min, Emerging Infectious Diseases, **12**, 1671 (2006)
13. L. E. Grube, V. H Storr, Entrepreneurship and Regional Development, **30**, 800 (2018)
14. B. Herbane, Int. Small Business J., **28**, 43 (2010)
15. JHCRC (Johns Hopkins Coronavirus Resource Center), Coronavirus COVID-19 Global Cases by the Center for Systems Science and Engineering (CSSE) at Johns Hopkins University (2020)
16. A. Köhn, Management Review Quarterly, **68**, 3 (2008)
17. S. Korber, R. B. McNaughton, Int. J. of Entrepreneurial Behaviour & Research, **24**, 1129 (2018)
18. M. K. Linnenluecke, Int. J. of Management Reviews, **19**, 4 (2017)
19. PWC (PricewaterhouseCoopers GmbH), Financial support in the Covid-19 crisis - factsheet for startups and SMEs (Financial support in the Covid-19 crisis - factsheet for start-ups and SMEs) (2020)
20. R. C.Runyan, J. of Contingencies and Crisis Management, **14**, 12 (2006)
21. D. Smallbone, D. Deakins, M. Battisti, J. Kitching, Int. Small Business J., **30**, 754 (2012)
22. J. Spillane, M. Hough, European Management J., **21**, 398 (2003)
23. T. A. Williams, D. A. Gruber, K. M. Sutcliffe, D. A. Shepherd, E. Y. Zhao, Academy of Management Annals, **11**, 733 (2017)
24. T. A. Williams, D. A. Shepherd, Academy of Management J., **59**, 2069 (2016)
25. A. Winston, Is the COVID-19 Outbreak a Black Swan or the New Normal? MIT Sloan Management Review (2020)

© The Authors, published by EDP Sciences. This is an open access article distributed under the terms of the Creative Commons Attribution License 4.0 (http://creativecommons.org/licenses/by/4.0/).

26. A. Wyns, How our responses to climate change and the coronavirus are linked (2020)

© The Authors, published by EDP Sciences. This is an open access article distributed under the terms of the Creative Commons Attribution License 4.0 (http://creativecommons.org/licenses/by/4.0/).

Income inequality measures and the middle class

Oleg I. Pavlov[1,*], and *Olga Yu.* Pavlova[2]

[1]Peoples' Friendship University of Russia (RUDN University), Miklukho-Maklaya Street, 6, 117198 Moscow, Russian Federation
[2]All-Russian correspondence multidisciplinary school, B-234 Vorob'evy Gory, MGU, 119234 Moscow, Russian Federation

Abstract. We study how the presence of the middle class in the sense of Gevorgyan-Malykhin affects the value of income inequality measures including the Gini coefficient J and the Hoover index H. It is proved that in the presence of the middle class (1) $J \le \frac{1}{2} - \frac{L'(0)}{2}$ (where L is the Lorenz function), (2) $H \le \frac{1}{2}$, (3) the longest vertical distance between the diagonal and the Lorenz curve (which is equal to H) is attained at $z_0 < \frac{3}{4}$. A tight upper bound for P90/P10 ratio is found assuming $L'(0) > 0$. Tight upper and lower bounds for the differential deviation in terms of the Gini coefficient are found as well.

1 Introduction

Roughly speaking the middle class is some middle part of a society in social or socio-economic hierarchy. There is no precise well-established definition of the middle class but most often it is defined in terms of income. According to Ravallion [1], "typically, being "middle class" is defined as having an income within some interval that includes the median and the interval has typically been symmetric in the income space around the median". The middle class is believed to play an important stabilizing role in modern society [2]. In order to do so effectively, obviously, it must constitute a sizable portion of a society and possess a sizable portion of its income.

Recall that the Lorenz function $L(z)$ is the fraction of the total income or wealth, owned by the poorest fraction z of the population.

*Corresponding author: pavlov_oi@rudn.ru

© The Authors, published by EDP Sciences. This is an open access article distributed under the terms of the Creative Commons Attribution License 4.0 (http://creativecommons.org/licenses/by/4.0/).

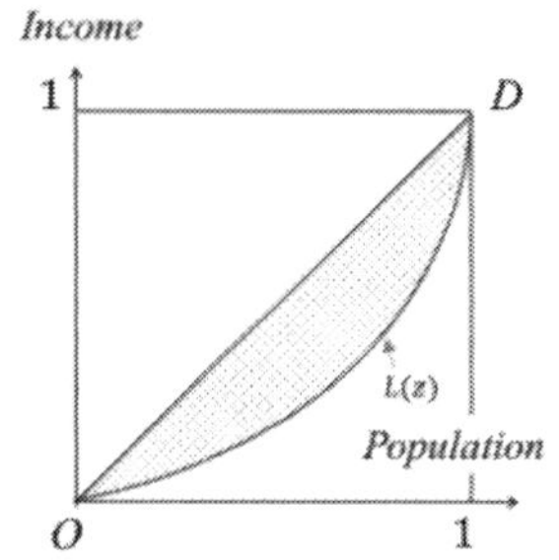

Fig. 1. The Lorenz curve $L(z)$. The doubled area of the shaded region is the Gini coefficient.

How much wealth may belong to the middle part (i.e. the second and third quartiles) of the population?

In case of the perfect equality $L(z) = z$ so that $L\left(\frac{3}{4}\right) - L\left(\frac{1}{4}\right) = \frac{3}{4} - \frac{1}{4} = \frac{1}{2}$. In the example below the richest 25% own more than 50% of wealth so $L\left(\frac{3}{4}\right) < \frac{1}{2}$, hence $L\left(\frac{3}{4}\right) - L\left(\frac{1}{4}\right) < \frac{1}{2}$.

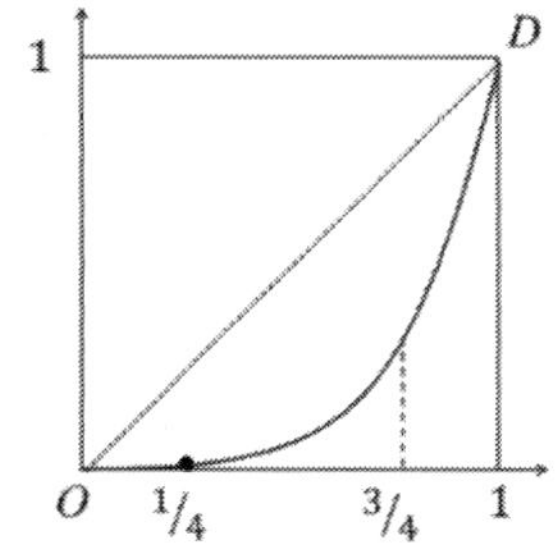

Fig. 2. $L\left(\frac{3}{4}\right) < \frac{1}{2}$ implies the absence of the middle class.

Gevorgyan and Malykhin [3] gave the following mathematical definition: there is the middle class if $L\left(\frac{3}{4}\right) - L\left(\frac{1}{4}\right) \geq \frac{1}{2}$. Equivalently, there is the middle class if the middle part of the population (if ordered by wealth) possesses at least one-half the total wealth. Having the middle class prohibits the richest quarter of the society to have over a half of the total wealth. For example, if $L(z) = z^3$, then $L\left(\frac{3}{4}\right) = \left(\frac{3}{4}\right)^3 = \frac{27}{64} < \frac{1}{2}$, hence there is no middle class in a such a society. Similarly, there is no middle class in a society with the Lorenz curve from Figure 2, but there is a middle class in case of the perfect equality. The authors proved in [4] that in the class of two-segment polygonal Lorenz curves (with the vertices (0;0), (a;b), and (1;1)) there is the middle class if and only if $a \leq \frac{1}{2}$.

It is clear that the presence of the middle class does influence the distribution of wealth in a society. In order to measure such an influence numerically we shall consider several inequality indices (which can be interchangeably applied both to income and savings). The most popular one, *the Gini coefficient*, was defined in [5], also see [6]. Geometrically it can be described as the fraction of the area between the Lorenz curve and the diagonal AD (which is shaded in Figure 1) relative to the area of the triangle under that diagonal. The closer is the Lorenz curve to the diagonal, the smaller is the Gini index. Given a Lorenz curve L, we denote the correspondent Gini coefficient by $J(L)$ or simply by J.

The Hoover index [7] is the largest deviation of $L(z)$ from the identity, that is, the maximal value of $z - L(z)$. The Hoover index is also known as the Pietra index and the Schutz index; it is also called the Robin Hood index because it equals to the amount of wealth needed to be redistributed from the rich to the poor in order to attain the perfect equality. We also consider the P90/P10 ratio which is the ratio of the upper bound value of income of the

© The Authors, published by EDP Sciences. This is an open access article distributed under the terms of the Creative Commons Attribution License 4.0 (http://creativecommons.org/licenses/by/4.0/).

90% of people with highest income (i.e. of the ninth decile) to that of the 10%. Yet another inequality measure – differential deviation – was introduced in [8] as $\int_0^1 (L'(z)-1)^2 dz$.

The main research question considered in this paper is: How can the presence of a middle class affect the value of the Gini coefficient and other income inequality measures? Throughout the paper we assume $L(z)$ to be continuous.

2 Results and discussion

2.1 The Gini coefficient

Gevorgyan and Malykhin proved that if there is the middle class, then $J \leq \frac{1}{2}$. But the maximal value $J = \frac{1}{2}$ is attained at $L(z)$ which is equal to zero on $[0; 0.5]$ (in such a society the poorest one half of the population one does not have any income whatsoever). It is unlikely that such a distribution of wealth is possible in a modern society! So we will assume that every initial segment of the population does have some income. As $L'(z)$ is proportional to the income of person/household corresponding to z, it is feasible to consider the case $L'(0) > 0$.

What are the possible values of the Gini coefficient given $L'(0) > 0$ in the presence of the middle class? We answer this question in the following Theorem 1:

Theorem 1. If there is the middle class, then $J \leq \frac{1}{2} - \frac{L'(0)}{2}$.

Proof of theorem 1. Denote $L'(0)$ by k. Since $L(z)$ is concave up (convex), the Lorenz curve lies on or above the line $y = kx$. Since $L\left(\frac{3}{4}\right) - L\left(\frac{1}{4}\right) \geq \frac{1}{2}$ and $L\left(\frac{1}{4}\right) \geq \frac{k}{4}$, we have that $L\left(\frac{3}{4}\right) \geq \frac{1}{2} + \frac{k}{4}$. Due to convexity, the Lorenz curve is on or above some straight line l_1 which goes through the point $\left(\frac{3}{4}, L\left(\frac{3}{4}\right)\right)$. ($l_1$ coincides with the tangent line to the Lorenz curve at $\left(\frac{3}{4}, L\left(\frac{3}{4}\right)\right)$ if the latter exists.) By shifting l_1 by $L\left(\frac{3}{4}\right) - \left(\frac{1}{2} + \frac{k}{4}\right)$ units down, we obtain a straight line l_2 which goes through the point $\left(\frac{3}{4}, \frac{1}{2} + \frac{k}{4}\right)$ and such that the Lorenz curve is on or above l_2. Also, the vertex D is on or above l_2 since the Lorenz curve is convex. Then the upper envelope of the lines $y = kx$ and l_2 is a two-segment polygonal chain OAB (see Figure 3) which contains the points $(0; 0)$ and $\left(\frac{3}{4}; \frac{1}{2}\right)$ such that no point of the Lorenz curve lies below it. Hence the area between the diagonal AD and the Lorenz curve is not greater than the area of the quadrilateral OABD. This means that $J(L)$ is not greater than twice the area of OABD which is further denoted by S_{OABD}.

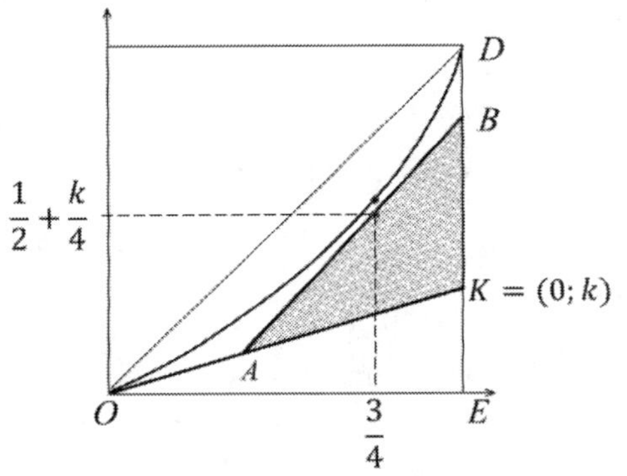

Fig. 3. The line l_2 (not shown) goes through the points A, $\left(\frac{3}{4}, \frac{1}{2} + \frac{k}{4}\right)$, and B.

© The Authors, published by EDP Sciences. This is an open access article distributed under the terms of the Creative Commons Attribution License 4.0 (http://creativecommons.org/licenses/by/4.0/).

Denote the point $(1; k)$ which is the intersection of the line $y = kx$ and the segment ED by K. Since $S_{OABD} = \frac{1}{2} - S_{OEK} - S_{ABK}$ and S_{OEK} is equal to $\frac{k}{2}$, it is enough to estimate S_{ABK}. Denote the point $\left(\frac{3}{4}, \frac{1}{2} + \frac{k}{4}\right)$ by K'. Let segments $K'B'$ and $K'A'$ be parallel to AK and BK respectively, see Figure 4. The distance between the lines $K'A'$ and $B'K$ is equal to $\frac{1}{4}$ since both are vertical and the second coordinates of K' and B are $\frac{3}{4}$ and 1 respectively. The second coordinate of A' is equal to $\frac{3}{4}k$ since A' is on the line $y = kx$ and the first coordinate of A' is $\frac{3}{4}$ (the same as that of K'). Therefore $|K'A'| = \frac{1}{2} + \frac{k}{4} - 3\frac{k}{4} = \frac{1}{2} - \frac{k}{2}$. This means that $S_{A'K'B'K} = \frac{1}{4}\left(\frac{1}{2} - \frac{k}{2}\right) = \frac{1}{8} - \frac{k}{8}$.

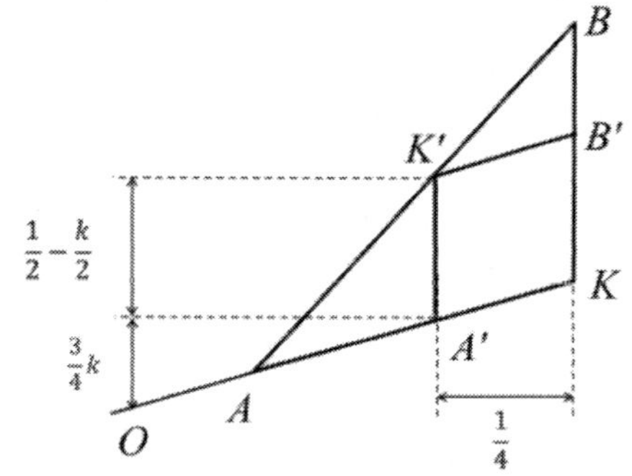

Fig. 4. Calculating the area of the parallelogram A'K'B'K.

It is proved in [9] that if a triangle KAB is cut off an angle K by a line, which goes through an interior point K' of the angle, then this triangle has the minimum area if $AK' = K'B$. In order to consider this case, we need to check whether B belongs to the segment ED. We have that $BB' = B'K$ since $K'B = AK'$ and $K'B' \parallel AK$. $|B'K| = |K'A'| = \frac{1}{2} - \frac{k}{2}$, therefore $|BE| = BB' + B'K + KE = \left(\frac{1}{2} - \frac{k}{2}\right) + \left(\frac{1}{2} - \frac{k}{2}\right) + k = 1$. This means that B coincides with D, hence $S_{OABD} = \frac{1}{2} - S_{OEK} - S_{ABK}$ in the considered case indeed. Both triangles $K'BB'$ and $AK'A'$ are similar to the triangle ABK and are in proportion $\frac{1}{2}$ to it. Therefore, $S_{K'BB'} = S_{AKA'} = \frac{1}{4}S_{ABK}$. This implies that the area of the parallelogram $A'K'B'K$ is one half of S_{ABK}, so $S_{ABK} \geq 2S_{A'K'B'K} = 2\left(\frac{1}{8} - \frac{k}{8}\right) = \frac{1}{4} - \frac{k}{4}$. So $S_{OABD} = \frac{1}{2} - S_{OEK} - S_{ABK} \leq \frac{1}{2} - \frac{1}{2}k - \left(\frac{1}{4} - \frac{k}{4}\right) = \frac{1}{4} - \frac{k}{4}$. Thus $J(L) \leq 2S_{OABD} \leq 2\left(\frac{1}{4} - \frac{k}{4}\right) = \frac{1}{2} - \frac{k}{2} = \frac{1}{2} - \frac{L'(0)}{2}$. The proof is complete.

The estimate for the Gini coefficient given by Theorem 1 is sharp. Indeed, for the two-segment polygonal Lorenz curve L_k with vertices $(0; 0)$, $\left(\frac{1}{2}; \frac{k}{2}\right)$, and $(1; 1)$, $L'(0) = k$ because the first segment is a part of the line $y = kx$. The estimate for Gini coefficient, given by the theorem, is $\frac{1}{2} - \frac{k}{2}$. The actual value of $J(L_k)$ is equal to the difference of the coordinates of the midpoint $\left(\frac{1}{2}; \frac{k}{2}\right)$ (see [10]), which equals $\frac{1}{2} - \frac{k}{2}$ too. In the worst case $L'(0) = 0$ Theorem 1 reduces to the theorem of Gevorgyan and Malykhin.

Note that without a requirement for the middle class the Gini coefficient may reach 1-$L'(0)$ (this value is attained at the two-segment polygonal Lorenz curve with vertices $(0; 0)$, $(1; k)$, and $(1; 1)$. Therefore, the presence of the middle class cuts the Gini coefficient in half.

© The Authors, published by EDP Sciences. This is an open access article distributed under the terms of the Creative Commons Attribution License 4.0 (http://creativecommons.org/licenses/by/4.0/).

2.2 The Hoover index

Let z_0 be such that $z_0 - L(z_0)$ is maximal among $z - L(z)$ for all $0 \leq z \leq 1$. (In other words, z_0 corresponds to the person/household having the average income. If z_0 is unique, then every person from $[0; z_0)$ has a below average income, while very person from $(z_0;1]$ has an above average income.) In general, z_0 can be any single number from the interval (0;1) (this observation can easily be seen from the fact that $(z_0; L(z_0))$ is the point of the Lorenz curve which is the most remote from the diagonal AD), or it can be an interval of numbers. The following Figure 5 shows three Lorenz curves corresponding to the same value of the Hoover index (and the same value of the Gini coefficient as well because for a two-segment polygonal Lorenz curves the Hoover index is equal to the Gini coefficient), but attaining their average income at different z_0.

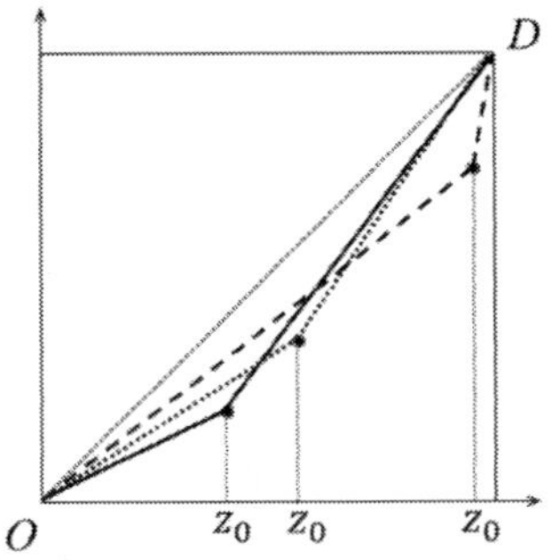

Fig. 5. Three Lorenz curves with the same Hoover index and Gini coefficient but with average income values at various z_0.

For simplicity, we assume that in the following theorem the Lorenz function is differentiable on (0;1).

Theorem 2. If there is the middle class, then

a. There is $z_0 < \frac{3}{4}$.

b. $H \leq \frac{1}{2}$.

Proof of theorem 2.

a. According to the mean value theorem there exists $z' \in \left[\frac{1}{4}; \frac{3}{4}\right]$ such that $L'(z') = \left(L\left(\frac{3}{4}\right) - L\left(\frac{1}{4}\right)\right) : \left(\frac{3}{4} - \frac{1}{4}\right)$. Then $L'(z') \geq 1$ since the denominator in the latter formula is equal to $\frac{1}{2}$ and because $L\left(\frac{3}{4}\right) - L\left(\frac{1}{4}\right) \geq \frac{1}{2}$ in the presence of the middle class. $L'(z)$ is a non-decreasing function because $L(z)$ is concave up. Hence the set of all $z \in [0; 1]$ such that $L'(z') < 1$ is an initial interval of [0;1] which does not contain $z' \leq \frac{3}{4}$ by the above argument. Clearly the right endpoint of this segment is the required z_0 such that $z_0 \leq z' \leq \frac{3}{4}$.

b. The presence of the middle class implies that $L\left(\frac{3}{4}\right) \geq \frac{1}{2}$. Then since $z_0 \leq \frac{3}{4}$ by item **a.** and due to convexity of $L(z)$, the point $\left(\frac{3}{4}, \frac{1}{2}\right)$ should be on or below the segment with the endpoints $P = (z_0; L(z_0))$ and $(1; 1)$. This means that P is on or above the line which goes through the points $\left(\frac{3}{4}, \frac{1}{2}\right)$ and $(1; 1)$ (see Figure 6).

© The Authors, published by EDP Sciences. This is an open access article distributed under the terms of the Creative Commons Attribution License 4.0 (http://creativecommons.org/licenses/by/4.0/).

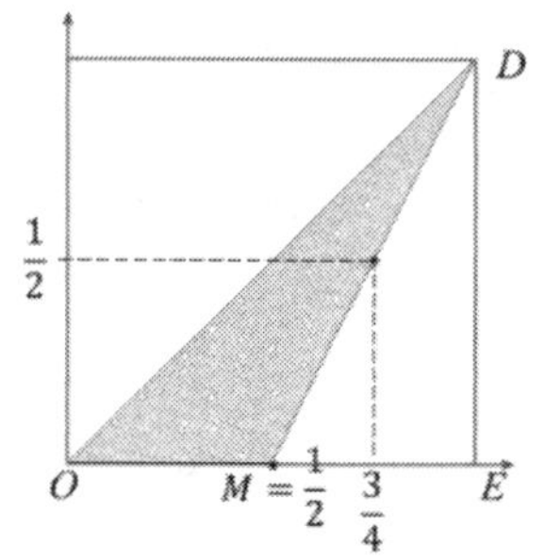

Fig. 6. Possible position of P.

Therefore, P belongs to the triangle ODM (to its part which is to the left of the vertical line $z = \frac{3}{4}$). However, the longest vertical segment lying in this triangle contains M and is of length $\frac{1}{2}$. So $H(L) \leq \frac{1}{2}$. Theorem 2 is proved.

Note that the set of all z_0 is not necessarily bounded by $z = \frac{3}{4}$. For example, in case of the perfect equality ($L(z) \equiv z$) this set is exactly $[0; 1]$. Theorem 2 states that this set, if not singleton, necessarily contains $z_0 < \frac{3}{4}$.

The maximal possible value of the Hoover index in the presence of the middle class, which is $\frac{1}{2}$, is attained at the two-segment polygonal Lorenz curve with vertices $(0; 0)$, $\left(\frac{1}{2}; 0\right)$, and $(1; 1)$.

2.3 The P90/P10 ratio

Since $L'(z)$ is proportional to income/prosperity of a person, the P90/P10 ratio which is the ratio of the ninth decile to tenth decile is also equal to $L'(0.9)/L'(0.1)$ (in this section we assume $L(z)$ to be differentiable on (0;1)). Clearly the P90/P10 ratio ≥ 1; it is well-known that the P90/P10 ratio can be *any* number ≥ 1. [11] noted that the presence of the middle class does not affect the P90/P10 ratio, i.e. it can still be any number ≥ 1.

Theorem 3. Let $L'(0)>0$.

a. $P90/P10 \leq \frac{10-9L'(0)}{L'(0)}$.

b. If there is the middle class, then $P90/P10 \leq \frac{3.5-2.5L'(0)}{L'(0)}$.

Proof of theorem 3. Denote $L'(0)$ by k. Throughout the proof we shall use the mean value theorem according to which $L(b) - L(a) = L'(t)(b-a)$ for some $t \in (a,b)$. Then $L'(t) \geq 0$ implies that $L(b) - L(a) \geq L'(t)(b-a)$. $L'(z)$ is a non-decreasing function because $L(z)$ is concave up, therefore $L'(t) \geq L'(a)$ and

$$L(b) - L(a) \geq L'(a)(b-a) \quad (1)$$

whenever b$\geq$ a. Now the proof splits in two parts.

a. It then follows from (1) that $L(0.9) \geq L'(0) \cdot 0.9 = 0.9k$. Due to convexity of $L(z)$, $L'(0.9)$ does not exceed the slope of the secant line which goes through points $(0.9; L(0.9))$ and $(1; 1)$. This slope is equal to

$\frac{1-L(0.9)}{0.1} \leq \frac{1-0.9k}{0.1} = 10 - 9k$.

Hence $L'(0.9) \leq 10 - 9k$. Now the inequality $L'(0.1) \geq k$ implies that $P90/P10 \leq \frac{10-9k}{k} = \frac{10-9L'(0)}{L'(0)}$ as required.

© The Authors, published by EDP Sciences. This is an open access article distributed under the terms of the Creative Commons Attribution License 4.0 (http://creativecommons.org/licenses/by/4.0/).

b. Similarly to the proof of the item **a**, we have that $L(0.25) \geq L'(0) \cdot 0.25 = 0.25k$.

$L(0.75) - L(0.25) \geq \frac{1}{2}$ due to the presence of the middle class.

$L'(0.75)$ is not less than the slope of the secant line which goes through points $(0.25; L(0.25))$ and $(0.75; L(0.75))$, thus $L'(0.75) \geq 1$.Consequently, $L(0.9) - L(0.75) \geq L'(0.75) \cdot 0.15 \geq 0.15$. Therefore,

$L(0.9) = L(0.25) + (L(0.75) - L(0.25)) + (L(0.75) - L(0.25)) \geq 0.25k + 0.5 + 0.15 = 0.25k + 0.65$.

As in the previous item, $L'(0.9)$ does not exceed the slope of the secant line which goes through points $(0.9; L(0.9))$ and $(1; 1)$, which is equal to

$\frac{1-L(0.9)}{0.1} \leq \frac{1-0.25k-0.65}{0.1} = \frac{0.35-0.25k}{0.1} = 3.5 - 2.5k$.

Finally, $P90/P10 \leq \frac{3.5-2.5k}{k} = \frac{3.5-2.5L'(0)}{L'(0)}$ as required.

In both items of Theorem 3 the upper estimates for the P90/P10 ratio are of the form $a + \frac{b}{L'(0)}$. We note that neither a, nor b can be decreased (independently of the other). Indeed, consider the following Lorenz curves: the two-segment polygonal Lorenz curve L_1 with vertices $(0; 0)$, $(1; k)$, and $(1; 1)$ and at the three-segment polygonal Lorenz curve L_2 with vertices $(0; 0)$, $(0.25; 0.25k)$, $(0.9; 0.25k + 0.65)$, and $(1; 1)$. The former one can be modified into a smooth curve if we replace its part to the right of the line $z = 0.95$ with a piece of parabola which connects the points $(0.95; 0.95k)$ and $(1; 1)$ and has the necessary derivative (namely, k) at $z = 0.95$. The P90/P10 ratio for this curve is exactly $\frac{10-9k}{k} = \frac{10-9L'(0)}{L'(0)}$. L_2 can also be modified into a smooth curve. For this, shift the left endpoint of the third segment up by a small $\varepsilon > 0$ to obtain the point $(0.9; 0.25k + 0.65 + \varepsilon)$ and connect the second segment with the old first segment and the new third segment by smooth curves (such as an arc of a parabola or a circle). The P90/P10 ratio for the new curve is slightly less than $\frac{3.5-2.5k}{k} = \frac{3.5-2.5L'(0)}{L'(0)}$ and it gets as close to $\frac{3.5-2.5k}{k}$ as desired as ε is getting closer to zero.

2.4 The differential deviation

This index was proposed in [8] as $\int_0^1 (L'(z) - 1)^2 dz$. Because defined in terms of the Lorenz function, this index does satisfy desirable properties of income inequality measures (see [12]). Also, it was shown [8] to be coherent with the natural partial order on the set of Lorenz curves. Throughout the rest of the paper the differential deviation of a society with the Gini coefficient J will be denoted by $D(J)$ or simply by D. We will use the following fact from [4]: there is the middle class for a two-segment polygonal Lorenz curve if and only if its middle vertex is on or to the left of the line $z = \frac{1}{2}$.

Theorem 4. For a two-segment polygonal Lorenz curve:

a. If $J \leq \frac{1}{2}$, then $4J^2 + 1 \leq D(J) \leq \frac{1}{1-J}$.

b. If $J > \frac{1}{2}$, then $D(J) \geq \frac{1}{1-J}$ and $D(J)$ can be indefinitely large.

Proof of theorem 4. **a.** It was proved in [13] that $D(J) \geq 4J^2 + 1$ for a two-segment polygonal Lorenz curve whenever $J \leq \frac{1}{2}$.

As derived in [13], $D(J) = J^2 \left(\frac{1}{z_0} + \frac{1}{1-z_0}\right) + 1$, where $(z_0, z_0 - J)$ is the middle vertex of the two-segment Lorenz curve. The presence of the middle class implies that $z_0 \leq \frac{1}{2}$; the fact

© The Authors, published by EDP Sciences. This is an open access article distributed under the terms of the Creative Commons Attribution License 4.0 (http://creativecommons.org/licenses/by/4.0/).

that the vertex $(z_0, z_0 - J)$ is in the unit square means that $z_0 - J \geq 0$. Hence $J \leq z_0 \leq \frac{1}{2}$. But it is known that given segment $I \subset \left[0; \frac{1}{2}\right]$, the largest value of $\frac{1}{z} + \frac{1}{1-z}$ on I is attained at that endpoint of I which is closer to $z = 0$, and for the segment $\left[J; \frac{1}{2}\right]$ this endpoint is $z = J$. Consequently,

$D(J) \leq J^2\left(\frac{1}{J} + \frac{1}{1-J}\right) + 1 = J^2\left(\frac{1}{J(1-J)}\right) + 1 = \frac{J}{1-J} + 1 = \frac{1}{1-J}$.

b. $D(J) \geq \frac{1}{1-J}$ whenever $J > \frac{1}{2}$ according to [13]. Fix J. If z_0 is sufficiently close to 1, then $\frac{1}{1-z_0}$ can be indefinitely large and so will $D(J) = J^2\left(\frac{1}{z_0} + \frac{1}{1-z_0}\right) + 1$.

Theorem 4 is proved.

Corollary 1. If there is the middle class, then $4J^2 + 1 \leq D(J) \leq \frac{1}{1-J}$.

Proof of Corollary 1. If there is the middle class, then $J \leq \frac{1}{2}$ according to Gevorgyan-Malykhin theorem. The minimum value $4J^2 + 1$ is attained at the curve with vertices $(0; 0)$, $\left(\frac{1}{2}; \frac{1}{2} - J\right)$, and $(1; 1)$ (see [13]) for which there is the middle class according to the fact from [4]. The two-segment polygonal Lorenz curve with vertices $(0; 0)$, $(J; 0)$, and $(1; 1)$ at which the maximal value $\frac{1}{1-J}$ is attained has the middle class due to the same fact.

2.5 Question

What are the possible values of $D(J)$ in the presence of the middle class (if the Lorenz curve is not necessarily a two-segment polygonal chain)?

2.6 Further discussion

The study of the middle class in the sense of Gevorgyan-Malykhin can be continued in several directions. One direction is to relax the concept and allow for the difference $L\left(\frac{3}{4}\right) - L\left(\frac{1}{4}\right)$ to be less than $\frac{1}{2}$. In [4] the authors gave a definition of an *ε-middle class* by replacing the inequality $L\left(\frac{3}{4}\right) - L\left(\frac{1}{4}\right) \geq \frac{1}{2}$ with $L\left(\frac{3}{4}\right) - L\left(\frac{1}{4}\right) > \frac{1}{2} - \varepsilon$. This definition is more stable than the original one in the sense that if a given Lorenz curve L satisfies the inequality $L\left(\frac{3}{4}\right) - L\left(\frac{1}{4}\right) > \frac{1}{2} - \varepsilon$, then a sufficiently close $\tilde{L}$ will also satisfy it (for example, in the topology of uniform convergence).

It is possible to quantify to which extent the middle class is absent. We called in [14] the difference $\frac{1}{2} - \left(L\left(\frac{3}{4}\right) - L\left(\frac{1}{4}\right)\right)$ the *middle-class deficit* and proved that $L\left(\frac{3}{4}\right) - L\left(\frac{1}{4}\right) \geq \frac{1}{2} - \frac{2}{3}J$ for every Lorenz curve L. In other words, the middle-class deficit never exceeds $\frac{2}{3}J$.

Also, it is possible to stipulate one or another degree of homogeneity of the middle class itself. For example, Gevorgyan and Malykhin [3] suggested to consider the P75/P25 ratio and they gave an upper estimate for the Gini coefficient provided the middle class is present and there is a perfect equality within the second and third quartiles of the population (the latter condition implies that the Lorenz curve is linear between $z = 0.25$ and $z = 0.75$).

© The Authors, published by EDP Sciences. This is an open access article distributed under the terms of the Creative Commons Attribution License 4.0 (http://creativecommons.org/licenses/by/4.0/).

3 Conclusions

We established that the presence of the middle class in the sense of Gevorgyan-Malykhin strongly influences various income inequality measures. The values of the Gini coefficient and the Hoover index are cut in half. The upper bound for the P90/P10 ratio (provided $L'(0) > 0$) is reduced by almost three times. On the contrary, both the upper and lower bounds on the differential deviation in terms of the Gini coefficient are attained on the middle-class Lorenz curves whenever $J \leq \frac{1}{2}$.

References

1. M. Ravallion, World Development, 38 (2009)
2. N. Vi. Kuznetsova, Ar. Y. Tupitsyn, Asian Social Science, 10 (2014)
3. P. S. Gevorgyan, V. I. Malykhin, Labor and Social Relations, **12,** 90 (2010)
4. O. I. Pavlov, O. Yu. Pavlova, Innovations in life, **4,** 153 (2016)
5. C. Gini, Bologna (1912)
6. L. Ceriani, P. Verme, J. Econ, Inequal, 10 (2012)
7. E. Hoover, The Review of Economics and Statistics, 18 (1936)
8. O. I. Pavlov, O. Yu. Pavlova, Management of economic systems, 12 (2017) http://uecs.ru
9. Line through a Point in an Angle, https://www.cut-the-knot.org
10. A. Frank, Farris, The American Mathematical Monthly, 117 (2010)
11. A. N. Astashenko, V. I. Malykhin, *Indicators of social differentiation* (LAP LAMBERT Academic Publishing, 2012)
12. Hugh Dalton, The Economic Journal, 30 (1920)
13. O. I. Pavlov, O. Yu. Pavlova, Russian Economics online-journal, 4 (2018) https://readera.org
14. O. I. Pavlov, O. Yu. Pavlova, Management of economic systems, 12 (2016) http://uecs.ru

© The Authors, published by EDP Sciences. This is an open access article distributed under the terms of the Creative Commons Attribution License 4.0 (http://creativecommons.org/licenses/by/4.0/).

Mathematical Model with Feedback Loop in Company Control

Arkadiy Plotnikov, *Tatyana* Goryacheva*, *Flyura* Kazakova, and *Ekaterina* Zakharchenko

Yuri Gagarin State Technical University of Saratov, 77 Politekhnicheskaya St., 77, 410054 Saratov, Russia

Abstract. This article is devoted to the further development of feedback loop models. They are used in the management of a single-industry firm. The development of the studying consists in the mathematical modeling of transients and assessing their impact on economic indicators. To solve this problem, the following was done in the article. A structural model of the functioning of a single-industry firm (SIF) is presented. It is built in accordance with the theory of automatic control (TAC) and includes elements: a management unit, a production unit, a sales unit, information nodes, as well as a feedback loop (FL). Equations and relationships describing the logic of the functioning of the company as a production system are presented. They allowed us to derive dynamic relationships and differential equations that reflect feedback loops on revenue and production costs. The system of expressions in operator form is presented, which describes the contour of the FL SIF. It has the form of a system of differential equations. It forms the basis of the mathematical model of SIF in the control system. This model made it possible to obtain a graphical interpretation of transients with closed and open FL based on the use of the Mathcad editor. Transients are fluctuations in sales volumes and production costs in the presence of disturbing influences. This is the scientific result and determines the novelty of the article.

1 Introduction

The presence of feedback loop in the management system is a prerequisite for ensuring the effectiveness and sustainability of the functioning and development of the company. It is difficult to monitor the compliance of actually achieved parameters of functioning and development with the given ones without fulfilling this condition. It is impossible to make managerial decisions on the adjustment of strategy elements, planned indicators, timelines, methods and means of management, etc. Therefore, the feedback loop principle has a certain distribution in the mathematical modeling of economic objects and aspects of their activities. However, the analysis of scientific works on this issue showed: a mathematical model was not built that allows us to interpret the nature of transients. Fluctuations in sales volumes and production costs under disturbing influences are considered as transients with a closed and open feedback loop. Such an interpretation is of both scientific and practical interest. It

* Corresponding author: tvgsgtu@rambler.ru

© The Authors, published by EDP Sciences. This is an open access article distributed under the terms of the Creative Commons Attribution License 4.0 (http://creativecommons.org/licenses/by/4.0/).

contributes to the further development of the methodology of mathematical modeling in the management of economic systems. It is aimed at increasing management efficiency by building a stable feedback loop. The feedback loop provides the necessary operational information for making and monitoring the implementation of management decisions. Thus, the purpose of the article is to build a mathematical model that allows you to perform a graphical interpretation of the nature of transients with a closed and open feedback loop.

2 Theoretical analysis

Mathematical modeling of the activity of economic systems and management systems is reflected in a number of studies. This formed an independent field both in mathematics and in management science. Of particular interest, according to the authors of the article, are models that take into account the influence of feedback, which is justified in the introduction. We will briefly review and analyze developments in this area. D. Forrester at work [1], examined the dynamic modeling of the company. He described the theory of feedback loop information system management. Introduced cybernetic models of dynamic systems with feedback. His research laid the foundation for systematic research in this area. One of the largest specialists in the field of TAC, academician V.A. Trapeznikov, in his work [2], considered the possibility of using theory in economics. The study was deployed in a book [3]. This approach in a generalized form was presented in the work [4]. It briefly analyzes new research in this direction. Proceedings of Sirazetdinova T.K. [5], Sio K.K. [6], Bonilla M. [7], Dooren V. [8], Klir G. [9], Checkland P. [10], Tolk Andreas [11], Lychkina N.N.[12], Slivitsky A.B. [13] devoted to various aspects of the construction and application of cybernetic and mathematical models based on the use of the feedback. These studies cover a wide range of issues and problems of the theory, methodology, methodology and practice of modeling. In particular, the principles, conditions and limitations of modeling, functioning, development and management of various economic systems are presented. Methodological foundations and methodological tools for constructing and using models with feedback loop are developed. A systematization is proposed, the possibilities and limitations of using models with feedback loop in the management of economic objects and their areas of activity are analyzed.

In addition to Forrester J, Jennings N R [14, 15], Wooldridge M.J [16] and others contributed to the formation of data of multi-agent resource conversion processes and dynamic liner modeling systems. Developments on distributed automated systems for optimizing enterprise processes are presented in the works of K.A. Aksenov with co-authors [17,18,19], as well as works [20, 21] on the development of data preparation modules in the framework of mathematical modeling of enterprise management systems, however, these works did not conduct a mathematical study of the effect of feedback loop on the parameters of the state and dynamics of economic systems. Structural schemes and models were built in articles [22-24]. Equations of the functioning of the enterprise as a production and economic system are also derived using the scientific apparatus of TAC. The possibilities of applying the proposed methodology to assess the sustainability of innovative development of enterprises in the innovation management system are substantiated. A number of points on the application of the TAC methodology in economic systems will be presented below. It should be noted that these schemes and models have received practical testing. This has resulted in the improvement of economic analysis and operational planning by the activity of a single-industry instrument-making enterprise (single-industry is an enterprise that produces homogeneous products). The proposed approach in the form of a virtual system for automated control of the activity of an instrument-making enterprise received practical implementation in the form of a patent of the Russian Federation. This proves the adequacy of the model.

© The Authors, published by EDP Sciences. This is an open access article distributed under the terms of the Creative Commons Attribution License 4.0 (http://creativecommons.org/licenses/by/4.0/).

In these studies, an extensive scientific apparatus and the practice of using models with feedback loop in the management of activities and the development of economic systems have been developed. But until now, a mathematical model has not been built that allows us to obtain a graphical interpretation of the nature of transients with a closed and open feedback loop. This is necessary to identify the impact of the feedback loop on the economic performance of the company. This determined the scientific task and focus of the research article.

3 Study results

None of the economic and mathematical models can fully reflect the complexity and diversity of the processes of functioning and development of the company. This is due to the difficult formalizability of the description and presentation of economic objects and processes in an optimal form for modeling. We offer some simplifications and assumptions. What is manifested in the following initial ideas.

1. To solve this scientific problem, the article considers just a single-industry firm (a company that produces homogeneous products). Diversified production requires much more complex multi-level bulky models, which is extremely difficult and impractical. The structure and logic of the SIF functioning are presented in the form of a structural diagram (Fig. 1). Two feedback loops are marked on the diagram - by revenue and cost price (which distinguishes it from [25]), taking into account inertia (time delays). The structural model demonstrates the logic of the process of production `TAC.
2. The planned value of the volume of output is introduced into the SIF activity model. The model includes blocks of production and sales of products with feedback loops. This approach allows you to evaluate deviations from the planned cost, production volume and sales, affecting the stability of the company. In the model SIF continuously produces and sells products. SIF has a steady volume of production and sales.
3. It should be understood that in the exact sciences, methods are used to derive differential equations of motion of mechanical systems based on Lagrange equations of the 2nd kind, based on the kinetostatic method, Maxwell's equations, etc. The derivation of economic and mathematical models involves the use of structural analysis methods. Blocks have transfer functions, which are obtained by the experimental-statistical method. Based on the constructed block diagram, the transfer functions of the blocks are determined. The transfer functions and ratios for individual circuits consisting of blocks are derived. Next, two differential equations are derived in the time domain for cost (4th order equation) and revenue (1st order equation).

Under the transition process is understood the process of restoring the values (rates of change) of economic indicators after exposure to certain disturbances.

© The Authors, published by EDP Sciences. This is an open access article distributed under the terms of the Creative Commons Attribution License 4.0 (http://creativecommons.org/licenses/by/4.0/).

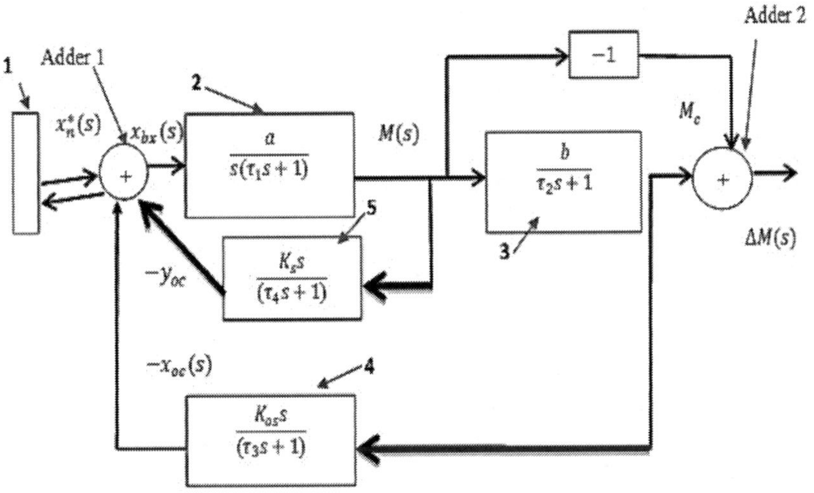

Fig. 1. Block diagram of SIF with feedback loops.

Designations: 1 - management unit; 2 - production unit; 3 - block sales of products; 4 - feedback loop on revenue from product sales; 5– cost feedback loop; summing blocks (in circuitry TAC-adders) 1 and 2 - information nodes in which the comparison of planned and actual indicators is carried out.

In the unit 1, a comparison is made of the rate of change in the planned volume of production $x_n^*(s)$ (in accordance with the notation system adopted in TAC - the Laplace image of the input impact on the system) with the actual. And accordingly, the planned and actual rate of sales, presented in the form of a derivative of revenue $M_c(s)$. The unit 1 generates input management actions on the system;

$M_n(s)$ - image of the planned cost of production;

K_{os} - feedback coefficient on revenue from sales of products, the value of which is derived from the ratio:

$sK_{os}M_c(s) = x_{os}(s)$,

where $x_{os}(s)$ – the rate of change in the volume of sales;

K_s – feedback coefficient on the cost of production, entered in the transfer function (4a), where the product $K_s s$ appears;

$M(s)$ - Laplace image of the cost of production;

b – the level of profitability of the product;

$M_c(s)$ – image of the actual revenue of the enterprise (volume of sales);

$\Delta M(s)$ – profit from sales of products (Rub.);

τ_1, τ_2, τ_3, τ_4 - time constants characterizing the inertia (delay processes) of the corresponding blocks.

y_{oc} (s) – the feedback signal of the circuit at cost.

Let us derive the differential equations of the firm's functioning according to the structural diagram of Fig. 1 taking into account the time inertia (the presence of time lags M), which take place in real activity. Differential equations from [22], [23] are introduced. Additional members are introduced into them, taking into account the following representations.

For simplicity of further calculations, we introduce the notation:

$$\frac{K_s s}{\tau_4 s+1} = K_\alpha. \quad (1)$$

© The Authors, published by EDP Sciences. This is an open access article distributed under the terms of the Creative Commons Attribution License 4.0 (http://creativecommons.org/licenses/by/4.0/).

- transfer function of the production unit: $\frac{a}{s(\tau_1 s+1)}$; (1.a)

- transfer function of the sales unit: $\frac{b}{\tau_2 s+1}$; (1.b)

- transfer function of the revenue feedback loop: $\frac{K_{os}s}{\tau_3 s+1}$; (1.c)

- the transfer function of the feedback loop at cost $\frac{K_s s}{\tau_4 s+1}$ (1.d)

After that we derive relations for interconnection of blocks.

For adder 1, we have the sum of the input variables

$$x_{\text{вх}}(s) = x_n^*(s) - x_{os}(s) - y_{os}(s) = x_n^*(s) - K_s M(s) - \frac{K_{OS} s M_c(s)}{\tau_3 s+1}. \quad (2)$$

For product sales unit 3, respectively:

$$M_c(s) = x_{\text{вх}}(s) * W_2(s) * W_3(s) \quad . \quad (3)$$

Where $W_2(s)$, $W_3(s)$ - transmission functions of production unit 2 and sales unit (realization) 3.

Substitute (5) in (6), get when taking into account expressions

$$M(s) = \frac{a x_{\text{вх}}(s)}{s(\tau_1 s+1)} \quad , \quad M_c(s) = \frac{bM(s)}{\tau_2 s+1}, \quad (4)$$

$$M_c(s) = \frac{a}{s(\tau_1 s+1)} * \frac{b}{\tau_2 s+1} * (x_n^*(s) - K_s M(s) - \frac{K_{OS} s M_c(s)}{\tau_3 s+1}). \quad (5)$$

The resulting expressions describe the process of functioning of the company. They make it possible to obtain dynamic finite relations in the form of differential equations of the system for feedback options for revenue, cost and their totality.

The derivation of these differential equations implies obtaining a system of expressions in operator form:

$$W_{3s} = \frac{W(S)*W_n(S)}{1+W(S)*W_n(S)W_{os}(S)}, \qquad W_{os} = \frac{K_{os}S}{\tau_3 S+1} \quad (6),$$

Then:

$$W_{3s} = \frac{ab(\tau_4 s+1)*[\tau_1\tau_4 S^2 + (\tau_1+\tau_4)S + 1 + K_s a]*(\tau_2 s+1)*(\tau_3 s+1)}{[\tau_1\tau_4 S^2 + (\tau_1+\tau_4)S + 1 + K_s a]S*(\tau_2 s+1)\{[...]\}},$$

$$W_{3s}(S) = \frac{ab(\tau_4 s+1)(\tau_3 s+1)}{S\{[...]\}},$$

$$W_{3s}(S) = \frac{M_C(S)}{X_n{}^*(S)} = \frac{ab(\tau_4 s+1)(\tau_3 s+1)}{S[D_4 S^4 + D_3 S^3 + D_2 S^2 + D_1 S + 1]}, \quad (7)$$

$$M_c(S) = \frac{bM(S)}{\tau_2 s+1}; \qquad \{[\ldots]\} = D_4 S^4 + D_3 S^3 + D_2 S^2 + D_1 S + 1 \quad (8)$$

Based on (7) and (8), we get:

$$M(S) = \frac{(\tau_2 s+1)}{b} * M_c = \frac{(\tau_2 s+1)}{b} * W_{3s}(S) * X_n * (S), \quad (9)$$

$$[D_4 S^4 + D_3 S^3 + D_2 S^2 + D_1 S + 1]S * M(S) = a(\tau_3 S+1)(\tau_2 S+1)(\tau_4 S+1) * X_n.$$

The terms of sustainability of the enterprise are expressed by the next system of equations:

$$\left.\begin{array}{l} D_i > 0; D_3 D_2 > 2.25 D_1 D_4; \\ D_1 D_2 > 2.25 D_3 \end{array}\right\} \quad (10)$$

Next, the system is transformed into differential equations in the time domain.

© The Authors, published by EDP Sciences. This is an open access article distributed under the terms of the Creative Commons Attribution License 4.0 (http://creativecommons.org/licenses/by/4.0/).

The equation for the description of the dynamics of cost:

$$\ddddot{M}_c + d_3\dddot{M}_c + d_2\ddot{M}_c + d_1\dot{M}_c + d_0 M_c = AX_n^* \quad (11)$$

$$A = \frac{ab}{\tau_1\tau_2\tau_3\tau_4}, \quad \tau = 0.02 \quad (12)$$

The equation for the description of revenue dynamics:

$$\dot{M} = -\frac{b}{\tau}M + \frac{\tau_2}{\tau}M_c + \frac{1}{\tau}\dot{M}_c \quad (13)$$

Enter the designations:

$$\left.\begin{aligned} M_c &= x_0 \\ \dot{M}_c &= x_1 = \dot{x}_0 \\ \ddot{M}_c &= x_2 = \dot{x}_1 \\ \dddot{M}_c &= x_3 = \dot{x}_2 \\ \ddddot{M}_c &= \dot{x}_3 \\ M &= x_4 \\ \dot{M} &= \dot{x}_4 \end{aligned}\right\} \quad (14)$$

The system of differential equations in a normal form takes the form of:

$$\left.\begin{aligned} \dot{x}_0 &= x_1 \\ \dot{x}_1 &= x_2 \\ \dot{x}_2 &= x_3 \\ \dot{x}_3 &= -d_3x_3 - d_2x_2 - d_1x_1 - d_0x_0 + -AX_n^* \\ \dot{x}_4 &= -\frac{b}{\tau}x_4 + \frac{\tau_2}{\tau}\dot{x}_1 + \frac{1}{\tau}x_0 \end{aligned}\right\} \quad (15)$$

In this case, di are displayed on the basis of D_i (i=0...3).
For graphic interpretation, print: x_1, x_4, $x_0 - x_4$
$x_0 = M_c$, $x_4 = M$, $\Delta M = M_c - M$.

For simulation, parameters are set:

$$\begin{aligned} x_n^* &= 10^6 \\ a &= 1 \\ b &= 1.05 \\ \tau_1 &= \tau_2 = \tau_3 = \tau_4 = 1 \\ K_s &= 0.3 \\ K_{os} &= 0.4 \\ \tau &= 0.02 \end{aligned} \quad (16)$$

Using this model and the Mathcad editor allowed us to obtain a graphical interpretation of transients. They are shown for the cost of production, revenue from sales of products and profits of the company, as well as the dynamics of indicators with open on K_s (parameters 16) and closed loops (parameters 16a) feedback in the control system SIF. Presented in figures 2, 3.

© The Authors, published by EDP Sciences. This is an open access article distributed under the terms of the Creative Commons Attribution License 4.0 (http://creativecommons.org/licenses/by/4.0/).

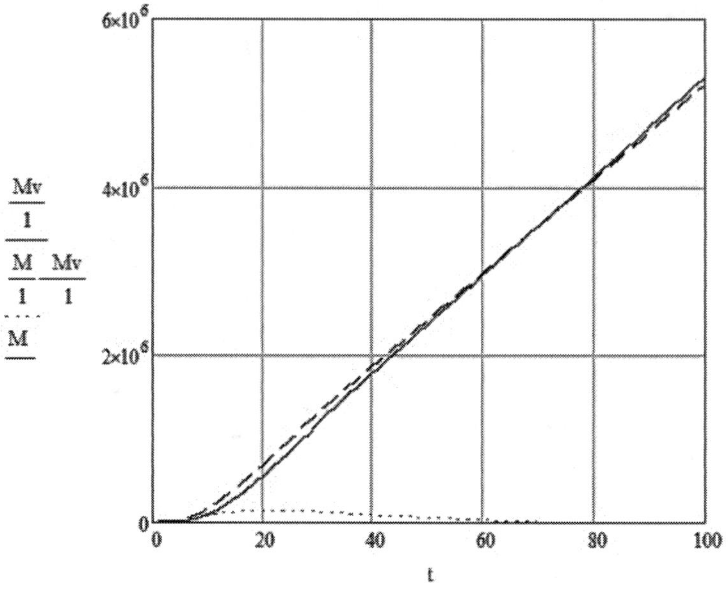

Fig. 2. Transients on the cost of production, revenue from sales of products and profits of the company.

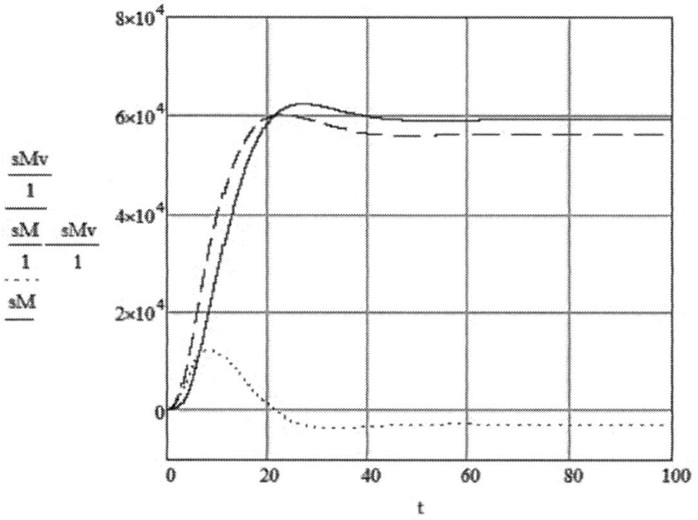

Fig. 3. Transients dynamics of economic indicators of the company.

$$K_c = 0.6 \quad K_{oc} = 0.5 \qquad (16\text{ a})$$

© The Authors, published by EDP Sciences. This is an open access article distributed under the terms of the Creative Commons Attribution License 4.0 (http://creativecommons.org/licenses/by/4.0/).

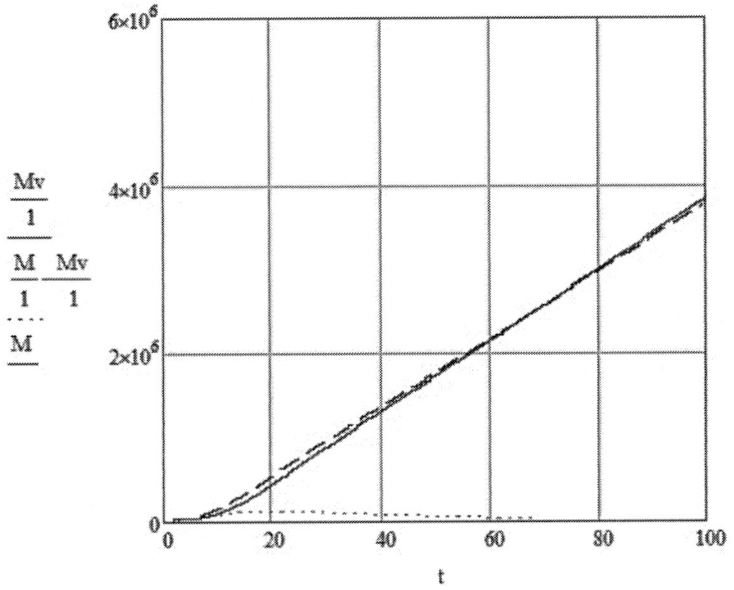

a) Transients in the rate of change in the volume of revenue from sales and production costs.

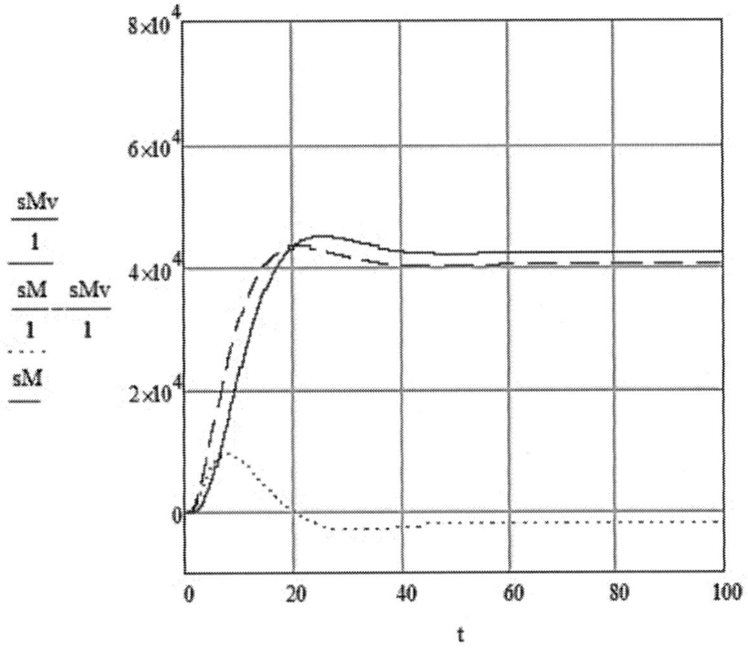

b) Transients in the rate of change of profit of the company.

The graphs show the simulation results in the presence of an operating system (the control loop of the company is closed) and its absence (the control loop of the company, respectively, is open). The process is considered under conditions of perturbations determined by changes in the economic parameters of modeling.

© The Authors, published by EDP Sciences. This is an open access article distributed under the terms of the Creative Commons Attribution License 4.0 (http://creativecommons.org/licenses/by/4.0/).

In fig. 3. a) it is shown that with the open FL at cost, the rate of increase in cost is higher than with closed feedback, the deviation from the planned values increases, and the duration of the transition process, which leads to a decrease in management efficiency and an increase in the instability of the company's functioning. Fig. 3. b) shows a similar dependence of the rate of change of the firm's profit and the duration of the transition process on the state of the feedback loop in the SIF control system. Thus, mathematical modeling made it possible to clearly demonstrate the influence of the feedback state on the economic performance of the enterprise and transients.

4 Conclusions

The article provides a brief review and analysis of scientific papers on the issues and problems of mathematical modeling of activity and management of economic systems using the feedback principle. He made it possible to determine one of the directions in the development of modeling related to the interpretation of the nature of transients with a closed and open feedback loop in the company management system. Transients are determined by fluctuations in sales volumes and production costs in the presence of disturbing influences. The formulated initial ideas of the article made it possible to build a structural model of the functioning of a single-discipline firm, based on the scientific apparatus and ideology of TAC. The structural model clearly demonstrates the logic and content of the process of controlling the functioning of the SIF. Equations mathematically describing this process were derived. The study performed the transformation of the obtained expressions into a system of differential equations. This made it possible to build a mathematical model of transients. It is presented by revenue from sales and production costs, company profits, their totality with a closed and open loop operating system in the company management system. Using the Mathcad editor led to the desired graphical interpretation of the transients. This was a scientific task and determines the novelty of the article. Modeling clearly demonstrated the dependence of the dynamics of the main economic indicators of the company and the duration of the transition processes on the state of the feedback loop in the SIF control system. It reflected the mismatch of actual and planned indicators with an open FL. There is the positive role of feedback loops on the dynamics of the firm's functioning. At the same time, from a mathematical point of view, the direction of increasing the rates and profits of small businesses is the introduction of positive feedback loops in the range of values (0 ... - 0.5), as well as the combination of positive and negative feedback loops in terms of cost and revenue. The proposed mathematical model contributes to the further development of the methodology of mathematical modeling in the management of economic systems. It is aimed at increasing management efficiency by building a stable feedback loop. The circuit provides the necessary operational information for making and monitoring the implementation of various management decisions.

References

1. J. W. Forrester, Management Review, **9(2)** (1968)
2. V. A. Trapeznikov, Automation and telemechanics, **1**, 5 (1966)
3. V. A. Trapeznikov, Management and technological advances (1983)
4. V.L. Epstein, Management problems, **1**, 2 (2006)
5. T. K. Sirazetdinov, *Dynamic modeling of economic objects* (1996)
6. K. K. Sio, Management economy (2000)
7. M. Bonilla, Linear Algebra and its Applications, **425 (2–3)**, 345 (2007)

© The Authors, published by EDP Sciences. This is an open access article distributed under the terms of the Creative Commons Attribution License 4.0 (http://creativecommons.org/licenses/by/4.0/).

8. V. Dooren, Linear Algebra and its Applications, **351–352**, 219
9. G. Klir, Architecture of Systems Problem Solving, 322 (1999)
10. P. Checkland, Systems Thinking, Systems Practice, 126 (1999)
11. A. Tolk, Ontology, Epistemology, and Teleology for Modeling and Simulation, 372 (2013)
12. N. N. Lychkina, Management sciences in the modern world, **2(1)**, 233 (2015)
13. A. B. Slivitsky, Scientific Cooperation, **14**, 544 (2019)
14. A. Greenwald, Intelligent Systems, **18**, 12 (2003)
15. R. Dash, Intelligent Systems, **18**, 40 (2003)
16. R. Bordini, M. Fisher, W. Visser, M. Wooldridge, Intelligent Systems, **18**, 40 (2003)
17. K. Aksyonov, A. Antonova, CEUR Workshop Proceedings, **2109**, 1 (2018)
18. S. N. Medvedev, K. A. Aksyonov, 24th Int. Crimean Conf. "Microwave & Telecommunication Technology" (CriMiCo'2014), 1, 433
19. K. Aksyonov, E. Bykov, O. Aksyonova, N. Goncharova, A. Nevolina, Proceedings of the 5th IASTED International Conference on Modelling, Simulation and Identification (MSI) (2014)
20. A. Borodin, Y. Kiselev, S. Mirvoda, S. Porshnev, Proceedings of the 11th Int. Conf. BDAS: Beyond Databases, Architectures and Structures: Communications in Computer and Information Science, 505 (2015)
21. A. Borodin, S. Mirvoda, I. Kulikov, S. Porshnev, Proceedings of the Int. Conf.: Beyond Databases, Architectures and Structures, 224 (2017)
22. A. P. Plotnikov, Vestnik SSTU, **3(34)**, 32 (2008)
23. P. A. Plotnikov, Economic and humanitarian research in the regions, **6**, 263 (2011)
24. A. P. Plotnikov, S. F. Nahov, Izvestiya of Saratov University. New Series. Series: Economics. Management. Law, **3**, 285 (2015)
25. S. F. Nahov, P. K. Plotnikov, A. P. Plotnikov, *Virtual system for managing the process of producing homogeneous products of the company*, 36

© The Authors, published by EDP Sciences. This is an open access article distributed under the terms of the Creative Commons Attribution License 4.0 (http://creativecommons.org/licenses/by/4.0/).

Model of socio-economic factors in the spread of SARS-CoV-2 across Russian regions

E.V. Sinitsyn[1,*], *A.V.* Tolmachev[1, 2], and *V.M.* Laptev[1]

[1]Ural Federal University, Mira St., 19, 620002 Ekaterinburg, Russia
[2]Datatel-Ural LLC, Sibirsky trakt, 12/8, 620100 Ekaterinburg, Russia

Abstract. The worldwide spread of a new infection SARS-CoV-2 makes relevant the analysis of the different factors that lead to the vulnerability of modern civilization to previously unknown diseases. In this regard, the development of mathematical models describing the evolution of epidemics like COVID-19 and the identification of socio-economic factors affecting the epidemiological situation in regions is an important research task. The paper proposes a probabilistic mathematical model for the spread of the COVID-19 epidemic, which allows to analyze the evolution of the main characteristics of the disease and to assess main factors influencing them. The study is based on the official statistical data on the spread of the COVID 19 presented on coronavirus sites in the Russian Federation and other countries, the Yandex Data Lens dataset service, as well as the data from the Federal State Statistics Service. In the research some data mining methods were used for evaluation the model's parameters. The model equations allow to predict the evolution of the disease and estimate the confidence interval of such prognosis. We estimated the ratio of detected and hidden cases of the disease, the distribution of the disease's duration probability and its average value for different regions. It has been mathematically proven that the vaccination is the necessary and sufficient condition of achievement a stationary stable state - the cessation of a pandemic. The regions of Russian Federation were clustered by the course of the disease COVID-19 on the base of k-means method. The analysis of the most important socio-economic factors affecting the epidemiological situation was provided separately for each cluster.

1 Introduction

The worldwide spread of the new SARS-CoV-2 infection raise a question about socio-economic problems that make modern civilization vulnerable to new, previously unknown diseases. To prevent situations like the current in the future, it is worth to model pandemic, analyze it possible development, distinctive features. It is important to assess the influence of various socio-economic factors on the rate of the disease's spread and its course (duration of the disease, mortality). The results of this analysis can be used to make the decisions to constrain epidemic. This study is the second work of the authors devoted to this problem. In a previous work [1] we presented a review of works (see, for example [2-9]) devoted to

* Corresponding author: sinitsyn_ev@mail.ru

© The Authors, published by EDP Sciences. This is an open access article distributed under the terms of the Creative Commons Attribution License 4.0 (http://creativecommons.org/licenses/by/4.0/).

modeling of the epidemics. In particular, the last of the mentioned works laid the foundations of the so-called SIR epidemics models (from the English Susceptible - Infected - Recovered). The approach considered in our first work was also close to SIR model. On its basis, we analyzed the initial phase of the COVID-19 disease (until the end of the so-called "first" wave in July 2020), mainly in the regions of the Russian Federation, with regional comparisons. Contemporary development of the epidemic situation in the Russian Federation and in the world, the emergence of new data, the start of the vaccination company, requires adapting both the model and the analysis results to the current data.

In connection with the above, the main goals of our present research are:

— The adaptation of a probabilistic mathematical model [1] describing the course of the epidemic to its current phase.

— Regional comparisons of the current epidemic situation in the regions of the Russian Federation based on the base of adapted model.

— The official statistical data on the spread of the epidemic presented on coronavirus sites in the Russian Federation and other countries, the Yandex Data Lens dataset service, as well as data from the Federal State Statistics Service, were used.

2 Theoretical foundations of the analysis of the COVID-19 spread

As in the previous work of the authors [1] for modeling the COVID-19 epidemic, we divide the population into the following clusters (groups):
I – infected individuals in whom the disease has not been identified: those who are ill without clinical symptoms, who did not apply to medical institutions or were mistakenly diagnosed with the absence of the disease.
H – healthy and susceptible to disease individuals.
A – detected active cases of the disease.
D – deaths.
U – disease-resistant (disease-resistant individuals: recovered, including implicitly, as well as vaccinated).

Hereinafter, the size of each group will be denoted by a letter N with the corresponding index $N_I \ldots N_U$. The relative share of each group in the population of the analyzed region N will be denoted as *n* with an index, denoting the group: $n_I \ldots n_U$.

$$N_I + N_{\mathrm{H}} + \cdots + N_U = N$$

or (1)

$$n_I + n_{\mathrm{H}} + \cdots + n_U = 1$$

The graph corresponding to the model used herein is shown in Fig. 1.

© The Authors, published by EDP Sciences. This is an open access article distributed under the terms of the Creative Commons Attribution License 4.0 (http://creativecommons.org/licenses/by/4.0/).

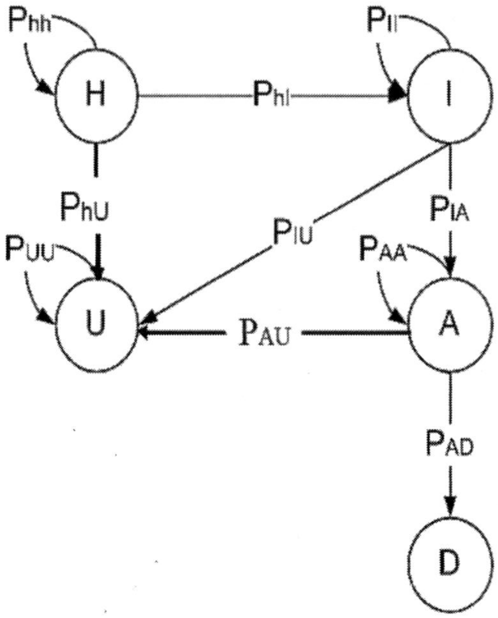

Fig. 1. The graph of possible transitions between groups of participants in the epidemic. The indicators on the arrows are transition probabilities.

New statistical data on the course of the disease caused by the SARS-CoV-2 virus, that appeared since the publication of our previous work, led to some transformations of the graph in Fig. 1, in comparison with that one in [1]:

— The beginning of the vaccination campaign is considered (transition H → U).

— Statistics, at least for RF, shows that a severe course of the disease in group I, finishing by a death, is usually accompanied by an appeal to medical attention and the isolation of the patient (falling into group A); exceptions are quite rare, in this regard, the transition (I → D) is not considered because of it negligible effect on the spread of the epidemic.

— Cases of re-infection have been identified so few that they can be neglected in modeling (only about 50 cases of re-infection are officially announced around the world, and most of them are associated with the specific of the patient's health).

The listed features of the transitions between the vertices of the graph transform it into the tree with one initial (H) and two final states (U, D) The transitions shown in Fig. 1 can be described using a 5 × 5 matrix $\hat{T}$ which has the form:

$\hat{T}$ =

	H	**I**	**A**	**D**	**U**	
	P_{HH}	*0*	*0*	*0*	*0*	**H**
	P_{HI}	P_{II}	*0*	*0*	*0*	**I**
	0	P_{IA}	P_{AA}	*0*	*0*	**A**
	0	*0*	P_{AD}	P_{DD}	*0*	**D**
	P_{HU}	P_{IU}	P_{AU}	*0*	P_{UU}	**U**

(2)

The columns of the matrix (2) contain the probabilities of transitions from the group indicated in the top row to the group presented in the rightmost column, which implies that the sum of all elements of each column is equal to 1. Such matrices are usually called stochastic [10].

© The Authors, published by EDP Sciences. This is an open access article distributed under the terms of the Creative Commons Attribution License 4.0 (http://creativecommons.org/licenses/by/4.0/).

Using the matrix (2), it is easy to formulate the equation for the probability $P(n_I, n_H, \dots, n_U;t) \equiv P(\vec{X};t) \equiv P$ – the fraction of participants in groups $I-U$ at the time t will be $n_I, n_H, \dots, n_U$. Here $\vec{X} = \{n_I, \dots, n_U\}$ – is a vector, with the sum of the components equal to 1. After simple transformations performed by the analogy with the approach described in our previous work and, for example, in the work of Feller (1964) one can get:

$$\frac{\partial P}{\partial t} = P \cdot \{T_{kk}N_k - N\} + T_{k\neq j}(N_k + 1)P(\dots, N_k + 1, \dots, N_j - 1, \dots; t) \quad (3)$$

It is implied that repeated indices are summed.

The equation (3) allows to study various aspects of the changes in the probability $P(\vec{X};t)$, in particular, the scenario analysis of the epidemic evolution. Let's start from the deriving the equations for mathematical expectations $n_I, n_H, \cdots, n_U$ on the base of (in the equations (4a, d), for brevity and simplicity of notation, we assumed that a positive effect starts immediately after vaccination and its duration is longer than the duration of analyzing time periods. Now there is not enough statistical data about the time of the immunity formation and its duration for more general consideration) (3):

$$\frac{dn_H}{dt} = -(P_{HI}+P_{HU})n_H(t) \qquad (4a)$$

$$\frac{dn_I}{dt} = -\int_0^t f_{IA}(t-\tau)\cdot n_I(\tau)\cdot d\tau - \int_0^t f_{IU}(t-\tau)\cdot n_I(\tau)\cdot d\tau + P_{HI}n_H(t) \qquad (4b)$$

$$\frac{dn_A}{dt} = -\int_0^t f_{AD}(t-\tau)\cdot n_A(\tau)\cdot d\tau - \int_0^t f_{AU}(t-\tau)\cdot n_A(\tau)\cdot d\tau +$$

$$+\int_0^t f_{IA}(t-\tau)\cdot n_I(\tau)\cdot d\tau \qquad (4c)$$

The equations for the derivatives $\frac{dn_D}{dt}, \frac{dn_U}{dt}$ are:

$$\frac{dn_D}{dt} = \int_0^t f_{AD}(t-\tau)\cdot n_A(\tau)\cdot d\tau \qquad (4d)$$

$$\frac{dn_U}{dt} = \int_0^t f_{AU}(t-\tau)\cdot n_A(\tau)\cdot d\tau + \int_0^t f_{IU}(t-\tau)\cdot n_I(\tau)\cdot d\tau + P_{HU}n_H(t). \quad (4e)$$

Here time $t=0$ corresponds to the beginning of the pandemic, and t to the considered current moment. The main feature of integra-differential equations (4) is their temporal nonlocality (delay) associated with the following factors:

— $f_{IA}(t-\tau)$ – is connected with incubation (latent) period, due to which the symptoms of the disease and its detection occurs with a delay on some time distributed near it most probable value.

— $f_{AD}(t-\tau), f_{AU}(t-\tau), f_{IU}(t-\tau)$ – is caused by the finite time of the disease's course (in general, different for its various outcomes). These times are also distributed in the neighborhoods of the most probable values.

The features of the functions $f\ (t-\tau)$, characterizing the time delay, will be described in more details in the next section. Here we only note that the distribution of the delay times mentioned above is rather "narrow", so the integrands in (4) are different from zero in narrow intervals near current time t. These intervals are small compared to the characteristic scales of significant changes in the unknowns n_I, n_H, n_U. So, let's expand the unknowns $n_A(\tau)$, $n_I(\tau)$ in integrands (4) into the Taylor series in powers $(t-\tau)$ and left only the linear terms. This gives:

$$\frac{dn_H}{dt} = -(P_{HI}+P_{HU})n_H(t) \qquad (5a)$$

© The Authors, published by EDP Sciences. This is an open access article distributed under the terms of the Creative Commons Attribution License 4.0 (http://creativecommons.org/licenses/by/4.0/).

$$\frac{dn_I}{dt}(1 + P_{IA}\overline{T_A}+P_{IU}\overline{T_C}) = -(P_{IA} + P_{IU})n_I(t) + P_{HI}n_H(t) \tag{5b}$$

$$\frac{dn_A}{dt}(1 + P_{AD}\overline{T_C}+P_{AU}\overline{T_C} - P_{IA}\overline{T_A}) = -(P_{AD} + P_{AU})n_A(t) + P_{IA}n_I(t) \tag{5c}$$

$$\frac{dn_D}{dt}(1 + P_{AD}\overline{T_C}) = P_{AD}n_A(t) \tag{5i}$$

$$\frac{dn_U}{dt}(1 + P_{AU}\overline{T_C} +P_{IU}\overline{T_C}) = (P_{AU}+P_{IU})n_A(t)+P_{HU}n_H(t) \tag{5f}$$

Note that from equations (4) and (5), it follows:

$$\frac{dn_H}{dt}+\frac{dn_I}{dt}+\frac{dn_A}{dt}+\frac{dn_D}{dt}+\frac{dn_U}{dt} = 0 \tag{6}$$

In equation (the extension of the limits of integration in (7) and (8) to ∞ is possible due to the aforementioned "narrowness" of the distributions of the corresponding times) (5)

$$P_{IA} = \int_0^\infty f_{IA}(\tau)\cdot d\tau \tag{7a}$$

$$P_{IU} = \int_0^\infty f_{IU}(\tau)\cdot d\tau \tag{7b}$$

$$P_{AU,D} = \int_0^\infty f_{AU,D}(\tau)\cdot d\tau \tag{7c}$$

As will be shown in the next section, the functions $f_{AU}(\tau)/P_{AU}$ and $f_{AD}(\tau)/P_{AD}$ are practically identical. It is reasonable to assume that the function $f_{IU}(\tau)$, which characterizes the distribution of recovery times for group I, also slightly differs from $f_{IA}(\tau)$.

The values <T> in (5) are the corresponding average times:

$$\overline{T_A} = \frac{1}{P_{IA}}\int_0^\infty f_{IA}(\tau)\cdot\tau\cdot d\tau \tag{8a}$$

$$\overline{T_C} = \overline{T_U} \approx \overline{T_D} = \frac{1}{P_{AU,D}}\int_0^\infty f_{AU,D}(\tau)\cdot\tau\cdot d\tau \tag{8b}$$

As it was mentioned in the paper [11], the probability P_{IA} can be represented as:

$$P_{IA} = P_\alpha(d + \frac{T}{N}) \tag{9}$$

Here *d* is the likelihood of an infected person self-referring to a doctor per time unit, *T/N* is the number of tests per capita carried out per time unit to detect COVID-19, and finally, P_α is the reliability of detecting the disease – the probability of a type I error in diagnosis (the disease is present, but not detected).

These values were defined in the paper [1]: $P_\alpha d = 0{,}12$; $P_\alpha = 65.3\%$

To evaluate the probability P_{HI} in (4, 5) we also use the model of the work [11], in which:

$$P_{HI} = \frac{f\cdot n_I(t)\cdot n_H(t)}{n_H(t)+n_I(t)}, \tag{10}$$

where *f* is the average frequency of contacts in the population multiplied by the probability of infection by contact. Thus, the presented formula assumes that a person can be infected with a probability proportional to the fraction of infected persons among the all individuals with whom she or he has contacted. The identified (and therefore isolated) infected persons, as well as persons from group U, do not influence on the probability of infection. In the work [1], the parameter *f* = 0.14 was determined on the base of statistical data on the COVID-19 for the Russian Federation.

© The Authors, published by EDP Sciences. This is an open access article distributed under the terms of the Creative Commons Attribution License 4.0 (http://creativecommons.org/licenses/by/4.0/).

Let us provide a general analysis of equations (5). Firstly, we'll define rigorously the popular term “plateau” (the reports that it is either already achieved, or not yet in the mass media are very diverse and contradictory). We will assume that the "plateau" corresponds to the situation:

$$\frac{dn_A}{dt} = 0, \tag{11a}$$

this, in accordance with (5c), gives:

$$\frac{n_A(t)}{n_I(t)} = \frac{P_{IA}}{P_{AD}+P_{AU}} \tag{11b}$$

Note that evaluation the ratio of the detected cases of the disease to their total number (including latent cases o) is an urgent and important problem in the planning of anti-epidemic measures. The equation (11b) can be used to assess latent morbidity, since all parameters in it can be determined based on data, presented in the official medical statistics. The illustrations will be given in the next section.

It would be tempting to expand definition (11a) by extending the concept of "plateau" to stationary points of the equations (4, 5) defined by the relations:

$$\frac{dn_A}{dt} = \frac{dn_I}{dt} = \frac{dn_H}{dt} = \frac{dn_D}{dt} = \frac{dn_U}{dt} = 0. \tag{12}$$

It is easy to verify that the stationary point defined by (12) is stable [12]. However, from (5d, e) it follows that in this case: $n_A(t) = 0$, and hence: $n_H(t) = 0$. Then from (5b) one can get– $n_I(t) = 0$. Thus, (12) describes not a plateau, but the end of the epidemic. The analysis of (5) "from top to bottom", starting with (5a), shows that the achievement of the condition $n_H(t) = 0$ is necessary and sufficient to reach the stationary point (12) – the end of the epidemic. Moreover, state $n_H = n_A = n_I = 0, n_D + n_U = 1$ is the only stationary point of equations (4,5). This means, that the pandemic can be finished only if all the population become either dead (group D) or immune (group U)

We would like to emphasize That this conclusion from the equation (4,5) is especially important in connection with the vaccination company. The stable state (12) (the transition *H→U,D* for every person), is achievable by either the vaccination or the COVID-19 disease. The choice of the way to achieve it – at the discretion of citizens.

3 Methods of analysis

To analyze the entire set of collected epidemic and the socio-economic data we used:

1. methods of statistical analysis [13,14]:

— correlation analysis of parameters characterizing morbidity and factors describing the socio-economic state of the regions.

— multivariate regression.

2. methods of data mining - [10, 15], for example, clustering by k-means method.
3. methods of statistical modeling.

For the analysis, we used data from the website on coronavirus in the Russian Federation[1]and in the World[2], the service with data sets Yandex DataLens[3], as well as data from Rosstat for the constituent entities of the Russian Federation [16].

In accordance with the Data mining standard [17], the initial data underwent preprocessing. We have excluded from the analysis those countries where recoveries were not registered, or the reported values differed significantly from their typical values for most countries. Subjects with a small number (less than 2000) of closed disease cases (recoveries and deaths)- were also excluded. Finally, the errors in data were corrected, (for example:

© The Authors, published by EDP Sciences. This is an open access article distributed under the terms of the Creative Commons Attribution License 4.0 (http://creativecommons.org/licenses/by/4.0/).

negative numbers of new or closed cases). Data analysis was performed by means of scripts, developed by the authors in the Python in Anaconda data analysis package.

4 The analysis of factors affecting the epidemic of COVID-19

In this section, the mathematical model, presented above (section 2), will be used for a comparative analysis of the regions of the Russian Federation by characteristics of the COVID-19 disease.

4.1 The course duration of the disease COVID-19 in the Russian Federation

The initial data for analyzing the probability distribution of recovery (death) times from COVID-19 disease were obtained using the Yandex DataLens service. We analyzed the period from the start of the pandemic to the current moment 13.05.2021.

To determine the time of the disease's, duration we used the following supposition - the person, who entered the group of active cases first was the first who left it (after recovery or death). This is analog of well-known accounting FIFO method (First In - First Out). This method is practically the only possible way who estimate the duration of disease on the base of purely official statistics without detailed analysis of individual disease histories.

To share closed disease cases between groups "Recovered" and "Dead" we applied Monte Carlo procedure: closed cases were distributed between mentioned groups with a probability of recovery or death at the moment when the case of disease was closed.

The obtained probability distributions of the COVID-19 duration up to recovery or death for the Russian Federation are presented in Fig. 2. For comparison, we presented the same probability distribution evaluated near one year ago in our work [1] (Fig. 2b)

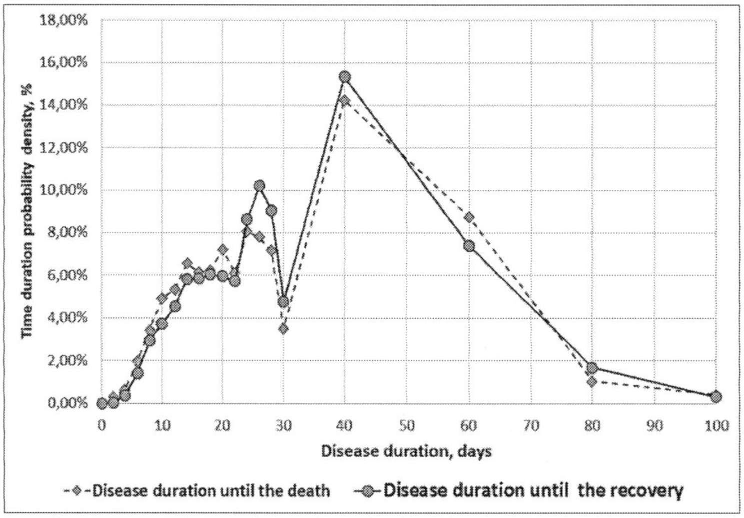

a)

© The Authors, published by EDP Sciences. This is an open access article distributed under the terms of the Creative Commons Attribution License 4.0 (http://creativecommons.org/licenses/by/4.0/).

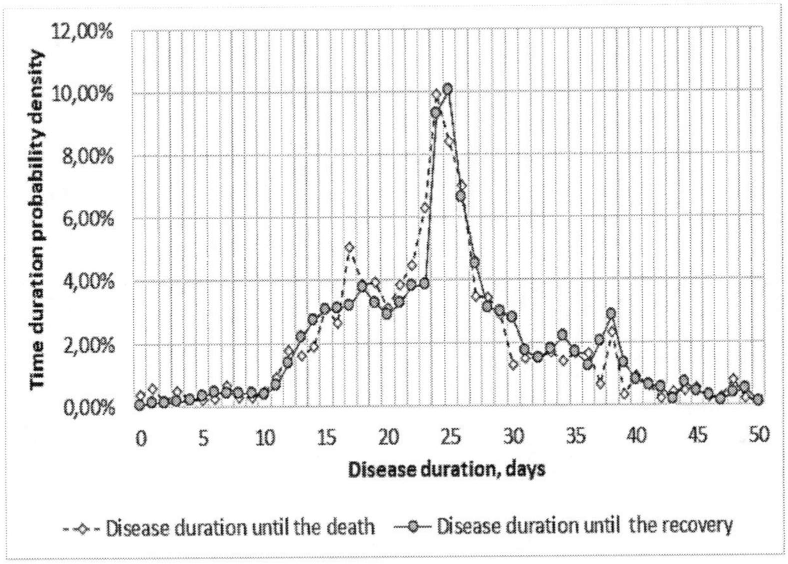

b)

Source: the authors' calculations based on the dataset from the Yandex DataLens service.

Fig. 2. Time duration (recovered and dead cases of COVID-19) probability density, Russian Federation: a) – on the date 13.05.2021, b) – 18.07.2020.

Note that Fig. 2a shows an additional peak of the probability density at the point near 40 days. Highly likely this reflects the experience gained during the year passed in the treatment of severe cases of the disease, which have a relatively long duration.

We calculated also shows the same probability distributions as in Fig. 2, but for the whole World (Fig. 3).

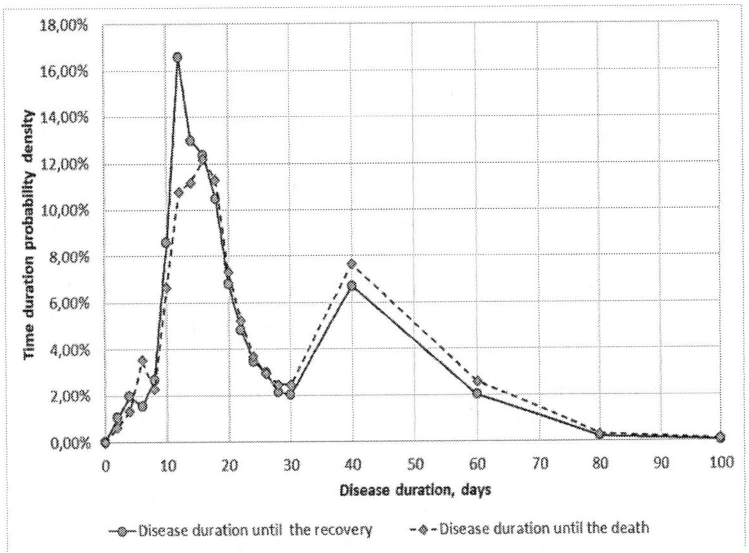

Source: the authors' calculations based on the dataset from the Yandex DataLens service.

Fig. 3. Time duration (recovered and dead cases of COVID-19) probability density, the whole World.

We can also see (Fig. 3) the presence of additional peak in the region of 40 days (like that in Fig. 2), that was absent one year ago [1].

© The Authors, published by EDP Sciences. This is an open access article distributed under the terms of the Creative Commons Attribution License 4.0 (http://creativecommons.org/licenses/by/4.0/).

It can be seen from the presented figures (Fig. 2, Fig. 3) that the probability distributions of the disease durations up to recovery or death do not differ statistically significantly. So, one can use only one time – T_C instead of two times of recovery and death. The data on the time of illness for different countries and different regions of Russia are significantly different. Note that, as was mentioned in Section 3, the probability distribution of these times really can be treated as "narrow" in comparison with characteristic times of pandemic course.

The duration of the COVID-19 disease may depend on a number of factors, in particular: the virus strain, that is predominantly common in the region, the level of medical care, the organization of disease detection, measures to limit it spread, etc.

It seems useful to cluster the regions of the Russian Federation and countries of the world (Fig. 4, Fig. 5), according to such parameters as the average time of disease – T_C and the coefficient of its variation:

$$C_v(T_C) = \frac{\sigma(T_C)}{<T_C>} , \quad (13)$$

here $\sigma(T_C)$ is the standard deviation, $< T_C >$ the mathematical expectation (average) of the time of disease course.

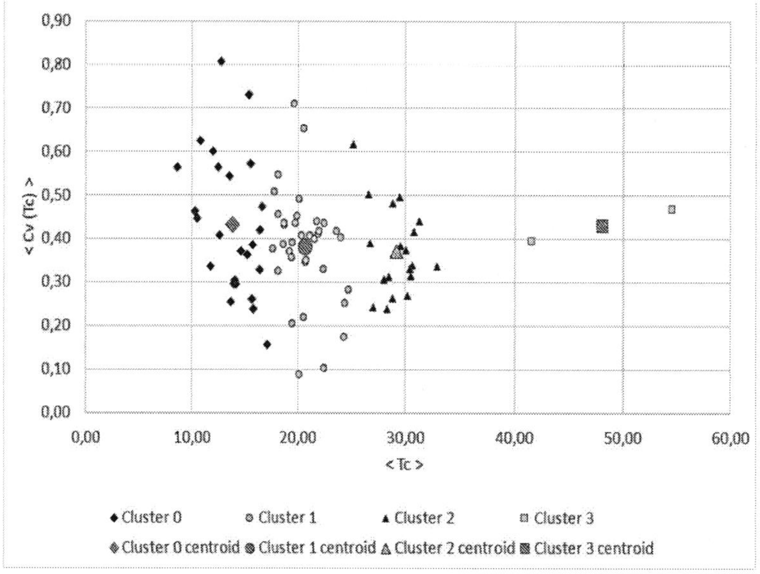

Source: the authors' calculations based on the dataset from the Yandex DataLens service.

Fig. 4. Clustering the regions of the Russian Federation by the parameters $< T_C >$ and $C_v(T_C)$ (13).

There are 4 clusters in the RF. Typical representatives of cluster 0 are: the Komi Republic, Volgograd region, Tula region, Altai Territory, Belgorod region. Cluster 1: Kursk Region, the Republic of Chuvashia, Karelia, Oryol, Kursk and Sverdlovsk Regions, Perm Territory. Cluster 2 is represented by such subjects as: Primorsky Territory, the Republic of Tatarstan, Karachay-Cherkessia, the Chelyabinsk region, the cities of St. Petersburg and Moscow. Finally, cluster three is represented by only two subjects: the Moscow and Pskov regions. Some indicators describing the socio-economic state in the above-mentioned clusters 1-4 are presented below (Table 1). These indicators have statistically significant correlation coefficients with the average duration of COVID-19 disease in one or more clusters.

© The Authors, published by EDP Sciences. This is an open access article distributed under the terms of the Creative Commons Attribution License 4.0 (http://creativecommons.org/licenses/by/4.0/).

Table 1. Some socio-economic indicators for clusters 0-3.

Indicators for clusters	Average values				Variation coefficients			
	0	1	2	3	0	1	2	3
Population density, people / sq. Km	40,1	34,6	26,4	11,5	2,4	0,8	1,0	7,9
The fraction of the employed people (average annual number) in the population	0,5	0,5	0,5	0,4	0,1	0,1	0,2	0,01
Gross regional product, mln. RUB per capita	0,4	0,4	0,7	0,4	0,9	0,7	1,0	0,5
Population per hospital bed, people	118,1	125,0	123,8	131,6	0,1	0,2	0,2	278,0
Population per doctor, people	219,9	216,8	209,7	0,1	0,2	0,2	0,2	0,1
Number of diseases PER 1000 people	830,6	771,9	772,9	713,1	0,2	0,2	0,2	0,02

For the whole World "k-means" method gives four clusters also (Fig. 5). The most numerous cluster is 0: representatives – Kazakhstan, Slovakia, Armenia, Russia. Next cluster 1, represented by Bulgaria, Ukraine, Egypt, Hungary, Spain. The remaining clusters have few members: cluster 2 (Cyprus, Honduras, France) and 3 (Netherlands).

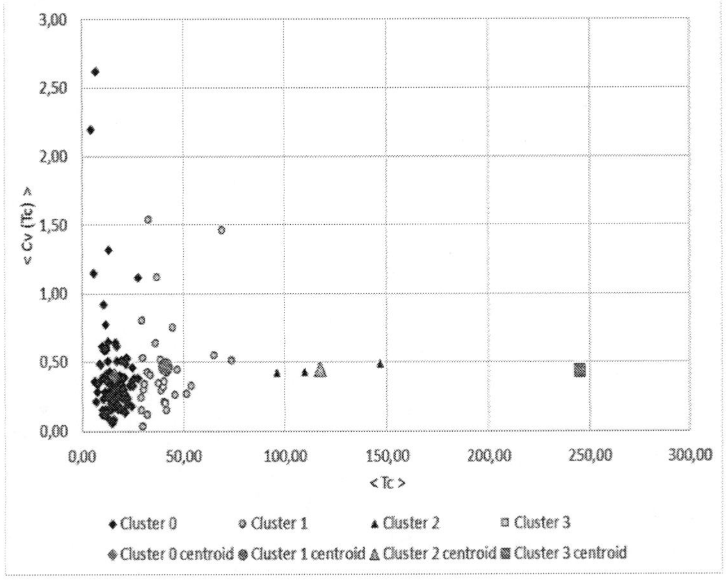

Source: the authors' calculations based on the dataset from the Yandex DataLens service.

Fig. 5. Clustering the countries of the World by the parameters $< T_C >$ и $C_v(T_C)$ (13).

© The Authors, published by EDP Sciences. This is an open access article distributed under the terms of the Creative Commons Attribution License 4.0 (http://creativecommons.org/licenses/by/4.0/).

The performed clustering was used for the correlation analysis of the influence of socio-economic factors on the course of the disease in the Russian regions. Such analysis was carried separately for each cluster, over the regions included into it. The subjects of Cluster 3 were excluded from the correlation analysis.

4.2 Assessment the ratio of detected and latent cases of COVID-19 disease in the Russian Federation

Let's use the theoretical model of Section 3, namely, formula (11b) to assess the ratio between detected (active) and latent cases of COVID-19 that did not registered by the official statistics, for the Russian Federation. For this purpose, the we'll use the data on active cases and tests on SARS-CoV-2 performed (Fig. 6).

The horizontal lines (Fig. 6) mark the dates when $\frac{dn_A}{dt} = 0$. With the help of (9) and the data [1], on the parameters included in it, the value of P_{IA}. Probabilities P_{AU}, P_{AD} for the disease COVID-19 in the Russian Federation, were calculated on the base of data presented in the Yandex DataLens service[3]. The results for the dates marked in Fig. 5 are presented in Table 2.

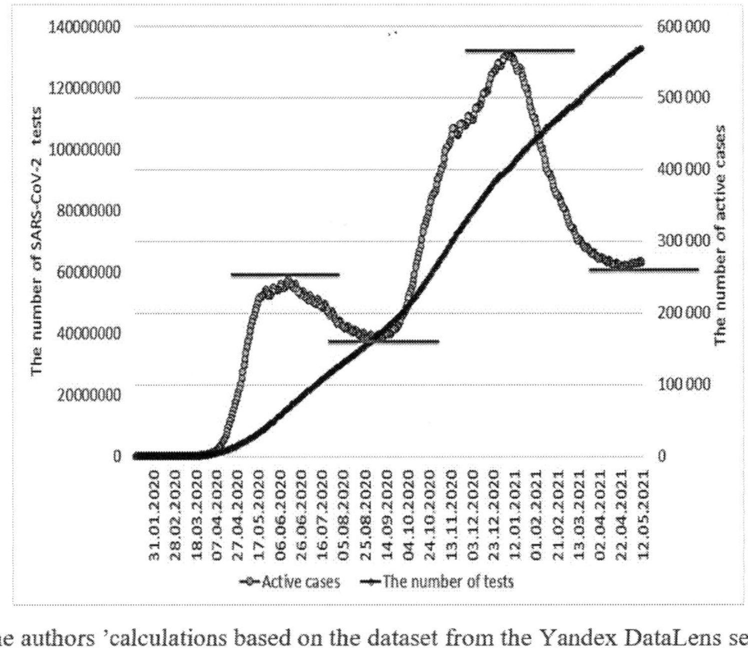

Source: the authors 'calculations based on the dataset from the Yandex DataLens service.

Fig. 6. The Data for calculating the ratio between detected and latent cases of the disease COVID-19 according to the formula (11b).

Table 1. The ratio between detected and latent cases of COVID-19, calculated according to (11b).

Dates	22.04.2021	01.01.2021	03.09.2020	12.06.2020
Value	19,7	15,1	8,3	5,3
Confidence interval (Significance10%)	±9,5	±7,3	±4,0	±2,6

Unfortunately, the data on the number of tests in the Russian Federation and other countries contain many omissions and inaccuracies, which reduces the reliability of the estimations. Nevertheless, the estimations (Table 2), even for the lower limit of the

© The Authors, published by EDP Sciences. This is an open access article distributed under the terms of the Creative Commons Attribution License 4.0 (http://creativecommons.org/licenses/by/4.0/).

confidence interval, are impressive. In RF we have near five million detected cases of the COVID -19 disease, most of them has finished by recovery. In addition, according to table 2, there are about 50 million of latent cases of the disease, and about 21 million citizens have get the first vaccination. Thus, the total number of people immune to SARS-CoV-2 infection is approaching to 75 million. This is already close to the notorious 70% of the population.

4.3 Discussion. Correlation and regression analysis of socio-economic factors affecting the characteristics of the epidemic

The modeling of the spread of COVID-19 disease proposed in this work shows that there is the only one way to finish this pandemic. It is the complete immunization of the population, either through vaccination or through disease. Significant differences in socio-economic conditions, quality of medical care, cultural traditions in different regions, as well as mutations of the virus and the emergence of new strains, will lead to the fact that the end of the pandemic will not be instantaneous. The most likely scenario is the migration of disease flashes from region to region, and a gradual decrease in the number of cases with the increasing of the immunized population.

In this regard, the analysis of socio-economic factors affecting the spread of the epidemic in its initial phase provided in [1], remains relevant. We have updated, its results by treating the new data on the Covid-19 pandemic and the socio-economic state of the regions of the Russian Federation. This allows to confirm the conclusions made in [1]. The main of them are listed below.

The most significant factors affecting the spread of the COVID-19 epidemic (these indicators have the greatest positive correlation coefficients with the number of cases, the rate of increase in incidence, and the duration of the course of the disease) are:

— Average annual number of employed, thousand people.
— Gross regional product.
— Retail turnover.
— Number of unemployed.
— The number of foreign citizens who have a patent for employment.
— The number of enterprises operating in the field of culture, sports, leisure and entertainment, organizations.
— The number of students in educational programs of primary, basic, and secondary general education.
— Passenger transportation by public buses.
— Passenger traffic by public buses.

These factors indicate the adequacy of the measures taken to prevent the explosive spread of the pandemic, although they may not have been consistent and timely everywhere. On the other hand, the listed risk factors should be used as the basis for providing the standard anti-epidemic measures in the cases of the flashes of SARS-CoV-2 - type diseases.

5 Conclusions

The mathematical model proposed in this work gives logically consistent description of the COVID-19 pandemic spread. It should be emphasized that the model relies on data that can be obtained by means of national system of macro-medical statistics and does not require the analysis of individual case histories. The last data is much more difficult to obtain.

The analysis of the various factors influencing on the spread of the disease indicates that it is worth to develop the methods of mathematical probability modeling of epidemics, especially since there are good traditions of such researches in the Russian Federation. This

© The Authors, published by EDP Sciences. This is an open access article distributed under the terms of the Creative Commons Attribution License 4.0 (http://creativecommons.org/licenses/by/4.0/).

will make possible to develop a reliable scientific base for various anti-epidemic measures and decisions. It could be recommended to analyze from this point of view the strategies for the socio-economic development of Russian regions for the period up to 2035 and make the necessary adjustments. Such activities and adjustments should be based on the models similar to the model outlined in this paper.

References

1. E.V. Sinitsyn, A.V. Tolmachev, A. S. Ovchinnikov, R-Economy, **6(3)**, 129 (2020)
2. L. A. Rvachev, Experiment on machine forecasting of influenza epidemic, **198(1)**, 68 (1971)
3. P. Whittle, Biometrika, **42(1-2)**, 116 (1955)
4. N. T. J. Bailey, The Mathematical Theory of Infectious Diseases and Its Application (1975)
5. N. S. Barlow, S. J. Weinstein, Physica D: Nonlinear Phenomena, **408**, 132540 (2020)
6. F. Ndairou, I. Area, J. J. Nieto, D. F. M. Torres, Chaos, Solitons & Fractals, **135**, 109846 (2020)
7. M. S. Abdo, K. Shah, H. A. Wahash, S. K. Panchal, Chaos, Solitons & Fractals, **135**, 109867 (2020)
8. X. Zhang, R. Ma, L. Wang, Chaos, Solitons & Fractals, **135**, 109829 (2020)
9. T. Chakraborty, I. Ghosh, Chaos, Solitons & Fractals, **135**, 109850 (2020)
10. J. Leskovec, A. Rajaraman, J. D. Ullman, Mining of Massive Datasets (2020)
11. Yu. M. Svirezhev, Nonlinear waves, dissipative structures and catastrophes in ecology (1987)
12. L. E. Elsgolz, Differential Equations (2020)
13. G. I. Ivchenko, Yu. I. Medvedev, Mathematical Statistics (2014)
14. N. N. Amosova, B. A. Kuklin, S. B. Makarova, Yu. D. Maksimov, Probabilistic sections of mathematics (2001)
15. A. A. Barsegyan, M. S. Kupriyanov, V. V. Stepanenko, I. I. Kholod, Data analysis technologies, Data mining, Visual mining, Text mining, OLAP (2007)
16. *Regions of Russia, Socio-economic indicators*, 32 (2018)
17. P. Chapman, J. Clinton, R. Kerber, T. Khabaza, T. Reinartz, C. Shearer, R. Wirth, *CRISP-DM 1.0 Step-by-step data mining guide* (2000)

© The Authors, published by EDP Sciences. This is an open access article distributed under the terms of the Creative Commons Attribution License 4.0 (http://creativecommons.org/licenses/by/4.0/).

Modeling the activities of financial institutions in changing macroeconomic conditions as a stability factor of the economy

Ayur Ayurzanain*, and *Oksana* Darmaeva

East-Siberian state University of technology and management, Klyuchevskaya St. 40B, p. 1, 670033 Ulan-Ude, Russia

Abstract. This article is devoted to the study of the activity of financial market institutions in modern conditions, the main feature of which is the constant change in the conditions of the institutional environment and macroeconomic stability. Scientists have proved the increasing importance of financial institutions in ensuring the economic growth of the economies of States in the XXI century. The authors consider and systematize the approaches of different authors and representatives of different directions of institutionalism and economic science to the definition of the role and essence of financial institutions. Also, the main conclusions about the impact of the global financial and economic crisis of 2008-2009 and sanctions of Western countries on the activities of the institutions of the Russian financial market in 2018 are formulated, based on the collection of statistical data, calculations are made and a model is formed, as a result of which a number of conclusions are obtained. As a result of calculations of the mathematical model made in the MS Excel program on the basis of statistical data of the Central Bank of Russia characterizing a condition of the financial market in 2008, 2009 and 2018 it is revealed that in economy of Russia there were considerable changes. The proposed model of assessing the institutional impact of financial institutions on economic growth as a practical tool can use a larger number of statistical indicators for calculations and obtaining more accurate data to solve more complex scientific and practical problems.

1 Introduction

The modern financial market is a very complex and multi-faceted system, in which the main element is formed institutions. the processes of digitalization and the intensive development of innovative technologies accelerate the processes of money exchange many times, offer new investment mechanisms, including those using electronic and cryptocurrency.

A number of significant institutional changes in the financial market have taken place in Russia over the past decade, and how effective they are for the current financial market system in the country remains an open question.

Macroeconomic instability, as a global phenomenon for the entire world economic system, is now seen as a challenge for any market, and the Russian financial market is no

*Corresponding author: ayurza9@yandex.ru

© The Authors, published by EDP Sciences. This is an open access article distributed under the terms of the Creative Commons Attribution License 4.0 (http://creativecommons.org/licenses/by/4.0/).

exception. A number of economic sanctions, including in the financial sphere, directly affected Russia's financial institutions, which had to adapt to the new economic realities by implementing significant economic transformations, both at the state level and significant institutional changes on the ground.

This article attempts to give a qualitative assessment of the impact of financial institutions on economic growth, as a result of their activities within the system. However, at the same time, there is a need to systematize quantitative and qualitative indicators in order to be able to monitor, including control over the quality of implemented institutional changes in order to develop a balanced economic policy and maximize efficiency.

The beginning of the new century marked the growing role of financial institutions in the economy. Innovative technologies and high rates of technological development currently require additional development of investment institutions, which today, in order to ensure maximum efficiency, are intensively implementing significant structural and institutional changes.

The investment process in the economy largely contributes to the support of high rates of economic growth. Financial institutions play a key role in the implementation of investment processes in the economy, because they represent a multi-level system of contractual relations, a huge number of investment transactions that are associated with the distribution of a large number of assets both in national economies and around the world.

The very development of financial institutions in the most developed countries plays a key role today and largely ensures the high competitiveness of national economies.

Thus, it can be recognized that financial institutions are the most important factor of economic growth in modern conditions [1].

In Russia today there is a situation in which we can observe the underdevelopment of the financial market, low level of financial literacy of the population and institutional trust in financial institutions. In this case, it becomes a limiting factor of commodity-money exchange. The population simply does not know how to use the available money to make investments based on the use of financial instruments. At the same time, a number of crises and bankruptcies of banks have formed a consistently low level of institutional confidence in the financial sector as such and financial institutions in particular [2].

The institutional direction of economic theory considers the functioning of financial institutions from a different angle: from the point of view of the impact on socio-economic relations, the development of commodity-money and investment processes. Classical institutionalists (T. Veblen, J. Galbraith, J. Commons, and others) believed that institutions are the basis of any economic system, and they are constantly evolving through the implementation of institutional changes.

In turn, the process of institutional change was considered in the works of a new direction of economic thought - a new institutional economic theory, the main ideologists of which are O. Williamson, R. Coase, D. North, M. Olson, V. A. Schmidt, G. M. Hodgson, E. G. Furubotn and others.

Among Russian scientists in the field of institutional research include the following: L. I. Abalkin, R. I. Kapelyushnikov, Yaroslav Kuzminov, D. S. Lvov, V. I. Mayevsky, R. M. Nureev, V. A. naishul, A. N. Oleinik, V. V. Radaev, V. L. Tambovtsev, A.E. Shastitko, etc.

The aim of the study is to identify the features and trends of financial institutions in the context of economic development in changing macroeconomic conditions using the proposed author's model, which can be used to assess the state of the economy on quantitative statistical data on the activities of financial institutions (for example, such as Deposit, credit operations and investments in securities) [3].

© The Authors, published by EDP Sciences. This is an open access article distributed under the terms of the Creative Commons Attribution License 4.0 (http://creativecommons.org/licenses/by/4.0/).

2 Theory

The modern financial market is a very complex and multifaceted system, in which the main element is formed institutions. the processes of digitalization and the intensive development of innovative technologies accelerate the processes of money exchange many times, offer new investment mechanisms, including those using electronic and cryptocurrency.

A number of significant institutional changes in the financial market have taken place in Russia over the past decade, and how effective they are for the current financial market system in the country remains an open question.

Macroeconomic instability, as a global phenomenon for the entire world economic system, is now seen as a challenge for any market, and the Russian financial market is no exception. A number of economic sanctions, including in the financial sphere, directly affected Russia's financial institutions, which had to adapt to the new economic realities by implementing significant economic transformations, both at the state level and significant institutional changes on the ground.

This article attempts to give a qualitative assessment of the impact of financial institutions on economic growth, as a result of their activities within the system. However, at the same time, there is a need to systematize quantitative and qualitative indicators for the possibility of monitoring, including monitoring the quality of implemented institutional changes in order to develop a balanced economic policy and maximize efficiency.

In the study of approaches to the definition of a financial institution, it is very important to note that within the framework of the works of scientists from the institutional direction, the term «financial institution» is vague. There is no doubt that financial institutions should be classified as economic institutions allocated within the institutional framework. It should be noted that in the new institutional economic theory, with the actual identification of the concept of Institute, firm and organization; the activities of banking, investment, insurance and non-profit institutions are often considered on the example of specific relations mainly for the purpose of disclosing the category of transaction costs and analyzing the impact on the economy. The main views of scientists on the concept of a financial institution in the modern economy are presented in table 1.

Table 1. Concepts «financial institution» and «economic Institute» in the works of scientists-institutionalists.

Author	The definition of a financial institution
D. North	Formal economic institutions in the scientific literature are considered most often in the same context with property rights, since they «establish property rights, that is, a bundle of rights to use and receive income from property, and alienation of other persons from the use of property or resources».
John. Commons.	Features of «financial capitalism» are manifested in the strengthening of the role of banks and other financial institutions and in the formation of developed collective social groups – trade unions, corporations and political parties.
W. J. Samuels.	Financial institutions include companies that provide financial services, markets, intermediaries, that is, those institutions, by which the counterparty of the market realize they have taken financial decisions.
V. V. Zotov, V. F. Presnyakov, V. O. Rosenthal	Economic institutions are socially recognized functional and organizational forms of collective economic activity, through which the system functions of the economy are realized.
A. E. Shastitko	The institutional response to the costs of settlements has been the development of a system of specialized units within organizations, as well as independent organizations that perform an intermediary function in the calculations. Such organizations have become banks. However, due to the increase in the volume of operations and their frequency, further specialization was required. Among the firms that provide net transaction services, or transaction services, the following should be distinguished: Finance and real estate transactions, banking and insurance.

© The Authors, published by EDP Sciences. This is an open access article distributed under the terms of the Creative Commons Attribution License 4.0 (http://creativecommons.org/licenses/by/4.0/).

Table 1. Continued

O. Williamson	The main purpose and result of the functioning of economic institutions of capitalism is to minimize transaction costs... Although the theory of transaction costs is applicable to the study of economic organizations of all types, the focus is on the economic institutions of capitalism with a focus on firms, markets and «relational» Contracting. Firms are a set of contractual relations.
R. Coase	Such institutions as the firm and the market make up the institutional structure of the economic system... the Main purpose of economic institutions is to save on transaction costs and to reduce risk, and the content is to unite individuals, each of which pursues its own economic benefit.
J. Hodgson	The nature of modern financial institutions, state economic policy, case law clearly favors the traditional firm. The relationship between the financial and industrial sectors of the economy should also be taken into account. The financial environment, including access to Finance and interest rates, has an impact on any business. It can be proved that a stable financial environment focused on long-term growth of the entire economy contributes to the creation of favorable conditions for business and technological improvement.

It should be noted that the concept of a financial institution and its activities are directly related to the economic category of financial assets. Traditionally, the term «asset» refers to property owned by the constituent entities of the economy [4,5].

The category of «financial asset» in the economy is defined as property presented in the form of financial documents, which record the rights and obligations of economic entities regarding cash payments. In addition to money, financial assets include securities and insurance instruments. The most detailed concept of the category of financial assets in the economy can be represented in the diagram (figure 1).

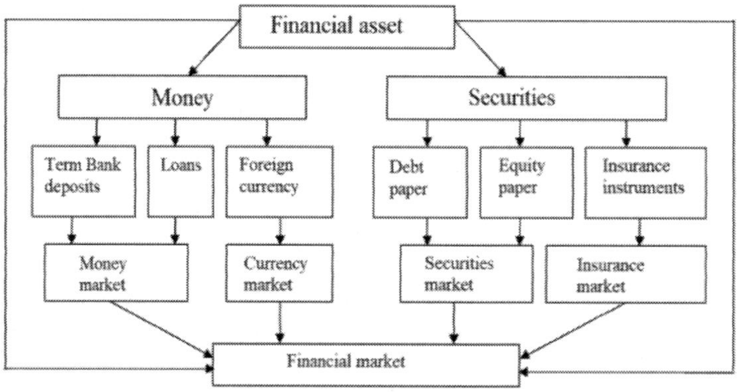

Fig. 1. Types of financial assets.

At the same time, the financial market is considered as a set of different mechanisms, institutions and economic relations through which financial assets are traded [7].

Taking into account the analysis of approaches to the definition of «Institute», «economic Institute», «financial institution» in the framework of the institutional direction, we have formed the following definition:

Financial institutions are a set of contractual relationships in the formation, distribution and use of financial assets in the economy. Banking, insurance, investment, exchange and non-profit financial institutions should be singled out. The fundamental element of the successful functioning of financial institutions in the economy is the institutional environment [6].

The objectives of the study require the gradual disclosure of the basic categories of institutional economy, among which the most important are the concept of transactions, transaction costs and the transaction sector in the economy. The formation of the transaction

© The Authors, published by EDP Sciences. This is an open access article distributed under the terms of the Creative Commons Attribution License 4.0 (http://creativecommons.org/licenses/by/4.0/).

sector in the modern economy is largely carried out at the expense of transaction services provided by financial institutions [8].

3 Data and methods

The basis for the formation of the theoretical prerequisites of this study is a new institutional economic theory, which today offers the necessary tools to solve the questions and problems faced by modern researchers in the field of the functioning of market institutions, one of which is financial institutions.

Today, we can refer financial institutions to the transactional sector of the economy, on which the processes of commodity-money exchange as a basic element of the functioning of the entire system largely depend.

For the tasks of the study, it is necessary to form a certain universal model, which can be formed on the basis of quantitative indicators in order to determine the qualitative state of the functioning of market institutions.

The tools developed in the framework of this study offer a comparison of the resulting market conditions in order to analyze the impact of the institutions of this market on economic growth, which is the main resulting indicator of the market system.

The formation of a mathematical model based on the possibility of simultaneous accounting for an unlimited number of indicators provides an opportunity not only to quantify each indicator of the system, but also to give a short-term forecast of its development, which makes it possible to take into account an additional characteristic-trends.

The next important aspect of the proposed model of the influence of financial institutions on economic growth (in this case) is the ability to track the qualitative state of the market, financial market or other, in which the functioning of the analyzed institutions takes place.

In general, we can say that the proposed model is indicative, a gradual deterioration of macroeconomic indicators, along with a negative trend and an increase in macroeconomic instability, can act as a forecast for the upcoming crisis of the economy or its individual sectors.

The selection of quantitative indicators and the level of their significance is a key component and issue in the formation of this model. To simplify mathematical calculations, we propose to use a model based on four basic indicators of the system of banking institutions, which are regularly monitored by the Central Bank of the Russian Federation.

The tools developed in the framework of this study offer a comparison of the resulting market conditions in order to analyze the impact of the institutions of this market on economic growth, which is the main resulting indicator of the market system.

The basis for the formation of the theoretical prerequisites of this study is the new institutional economic theory, which today offers the necessary tools to solve the questions and problems posed by modern researchers in the field of the functioning of market institutions, one of which is financial institutions.

Financial institutions today can be attributed to the transactional sector of the economy, on which the processes of commodity-money exchange as a basic element of the functioning of the entire system largely depend.

For the tasks of the study, it is necessary to form a certain universal model, which can be formed on the basis of quantitative indicators in order to determine the qualitative state of the functioning of market institutions.

To perform calculations of the model for assessing the institutional impact of financial institutions on economic growth, there are two calculated parameters necessary for the construction of a matrix model in a two-dimensional coordinate system:

1) k – the share of the average statistical indicator of the financial institution in the total volume of statistical indicators during the analyzed period.

© The Authors, published by EDP Sciences. This is an open access article distributed under the terms of the Creative Commons Attribution License 4.0 (http://creativecommons.org/licenses/by/4.0/).

For each statistic parameter k is calculated by the formula:

$$ki = yi/y0 * 100\%, \quad (1)$$

where y0 is the average volume of the indicator in monetary terms for the analyzed period; yi is the total volume of the indicator for the year.

2) t – the contribution of the trend of each indicator to the change in the total trend is calculated for each indicator according to the formula:

$$ti = ai/a0 * 100\%, \quad (2)$$

where ai is the trend coefficient of the i-th indicator during the analyzed period; a0 is the trend coefficient of the total indicator (the sum of private statistical indicators) for the analyzed period.

According to our model, based on the results, we can distinguish four basic states of the financial sector of the economy, on the basis of which the contribution of financial institutions to economic growth is estimated.

1) the state of economic growth - this state is characterized by positive trends in most of the most capital-intensive indicators of financial institutions in the analyzed period.

2) inertial development – this state can be characterized as a temporary uncertainty and reaction time to significant institutional changes or the negative impact of macroeconomic conditions. at the same time, this condition is characterized by multidirectional trends in the basic performance indicators of financial institutions during the analyzed period.

3) Economic instability – this state assumes a multidirectional growth trends of economic indicators of financial institutions, with the largest in quantitative terms indicators have a negative trend in the analyzed period;

4) Crisis economy or economic recession – this state is characterized mainly by negative trends in the growth of economic indicators of financial institutions in the analyzed period.

Thus, these estimated indices ki and ti are data to determine the coordinates of points when building a graphical model, which allows us to trace the resulting impact of institutional change and outcome that is possible in the framework of the proposed model (economic growth, slow economic development, unstable economy or the crisis economy) [9].

The model of assessing the institutional impact of financial institutions on economic growth allows us to trace and identify through which transactions in the transaction sector we have a particular state of the financial system of the state. The coordinate system is divided into 4 conditional quadrants, which characterize economic indicators based on quantitative characteristics and growth trends.

© The Authors, published by EDP Sciences. This is an open access article distributed under the terms of the Creative Commons Attribution License 4.0 (http://creativecommons.org/licenses/by/4.0/).

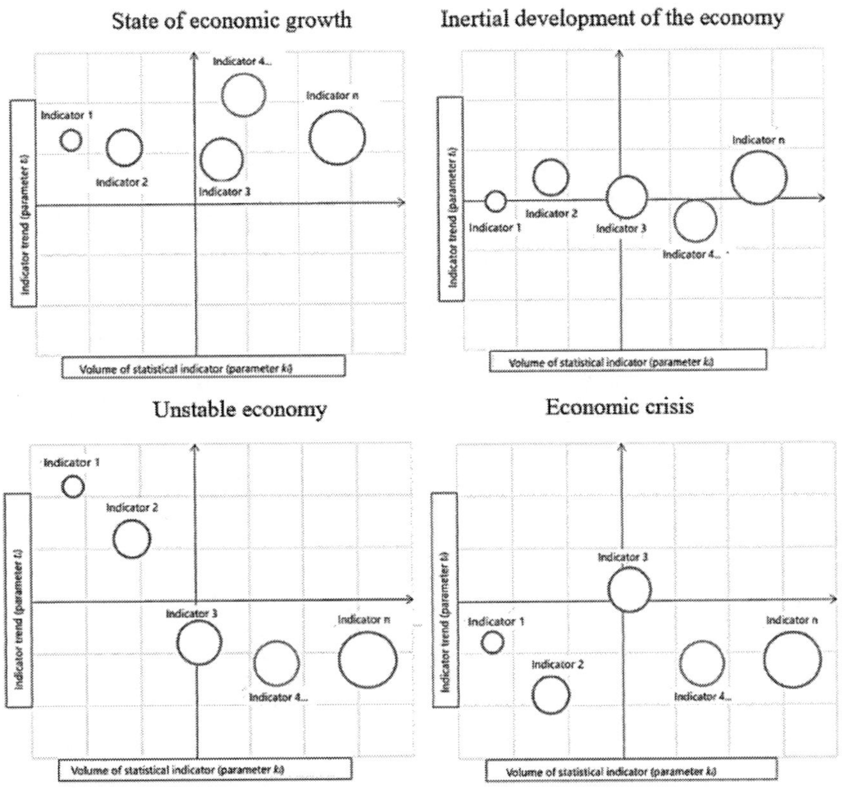

Fig. 2. Graphical representation of the results of the model for assessing the institutional impact of financial institutions on economic growth.

Thus, the indicators of this quadrant will have a positive impact on the final result, and with a large number of indicators in this quadrant, we will have a state of economic growth, according to the conditions of our proposed model.

In the second quadrant of the coordinate system, we have indicators of the transaction sector of the financial market with low quantitative characteristics, that is, in this quadrant, indicators with the lowest result are grouped.

These indicators have a minimal contribution to the development of the economy and have a minimal impact on economic growth, while with a large number of indicators in this quadrant, we will have a situation of inertial development.

In the third quadrant, economic indicators in the trade sector have negative growth trends and relatively low quantitative characteristics. These economic indicators make the least contribution to the formation of the final economic result. Therefore, there are a large number of indicators in this quadrant, and we will be in an unstable economic state.

In the fourth quadrant, there are economic indicators of the trade sector with relatively low quantitative characteristics, but high growth trends. These economic indicators make the least contribution to the formation of the final economic result.

4 Model

Based on the results of empirical statistics of the Central Bank of the Russian Federation, it is possible to construct a matrix model of the impact of financial institutions on economic growth for the period that is characterized by the most serious consequences for economic growth in terms of economic instability, we are talking about the period from 2008 to 2010.

© The Authors, published by EDP Sciences. This is an open access article distributed under the terms of the Creative Commons Attribution License 4.0 (http://creativecommons.org/licenses/by/4.0/).

Table 2. Performance indicators of financial institutions in 2009, billion RUB.

Record\Month	**Jan**	**Feb**	**Mar**	**Apr**	**May**	**Jun**	**Jul**	**Aug**	**Sept**	**Oct**	**Nov**	**Dec**
Loans to organizations	12 8 44 854	13 7 51 695	13 6 83 045	13 4 68 452	13 5 11 745	13 3 24 7 45	13 1 77 962	13 1 29 456	13 1 27 124	13 0 47 9 63	12 9 73 962	13 0 14 745
Loans to individuals	4 01 7 596	4 03 7 547	3 97 1 964	3 87 2 595	3 81 1 567	3 73 8 786	3 69 8 452	3 68 2 458	3 66 0 452	3 61 9 956	3 59 3 458	3 58 6 965
Interbank credit	2 50 1 478	2 86 4 962	2 69 0 785	2 66 5 962	2 44 6 892	2 31 4 487	2 37 7 566	2 80 1 963	3 00 9 789	2 90 8 745	2 63 9 965	2 82 3 874
Volume of investments in securities	585 962	586 964	821 964	848 745	819 581	860 015	859 962	894 852	920 853	994 045	1 07 0 758	1 23 4 364
TOTAL	**19 949 890**	**21 241 168**	**21 167 758**	**20 855 754**	**20 589 785**	**20 238 033**	**20 113 942**	**20 508 729**	**20 718 218**	**20 570 709**	**20 278 143**	**20 659 948**

The analysis of the activities of financial institutions was carried out on a number of parameters and quantitative data, among which the entire model was built on four grouped parameters:

1) loans to organizations;
2) loans to individuals;
3) interbank loans;
4) the volume of investments in securities.

Table 3. Performance indicators of financial institutions in 2008, trln. RUB.

Record\Month	**Jan**	**Feb**	**Mar**	**Apr**	**May**	**Jun**	**Jul**	**Aug**	**Sept**	**Oct**	**Nov**	**Dec**
Loans to organizations	8 73 1 215	10 0 29 412	10 2 29 451	10 6 07 745	5411 0 945	11 2 45 748	11 5 48 654	11 8 40 4 51	12 2 13 451	12 4 19 548	12 6 30 748	12 7 33 985
Loans to individuals	3 24 2 124	3 0123 21	3 0925 42	3 19 9 874	3 3484 63	3 46 5 954	3 59 0 985	3 73 9 748	3 89 0 478	4 01 8 748	4 08 3 654	4 05 5 748
Interbank credit	818 541	1 5106 58	1 7924 32	1 92 2 214	1 82 1 451	1 88 8 456	1 80 2 548	1 76 3 548	1 93 7 984	2 17 0 541	2 38 0 987	2 51 7 547
Volume of investments in securities	1 24 7 648	1 13 5 789	1 0015 12	904 562	923 412	979 324	1 02 3 748	998 654	906 654	887 658	810 213	834 748
TOTAL	14 039 528	15 688 180	16 115 937	16 634 395	60 204 271	17 579 482	17 965 935	18 342 401	18 948 567	19 496 495	19 905 602	20 142 028

© The Authors, published by EDP Sciences. This is an open access article distributed under the terms of the Creative Commons Attribution License 4.0 (http://creativecommons.org/licenses/by/4.0/).

At the next stage, we proceed to build a graphical model and calculate the linear trend equations for each of the presented indicators, as shown in Figure 3.

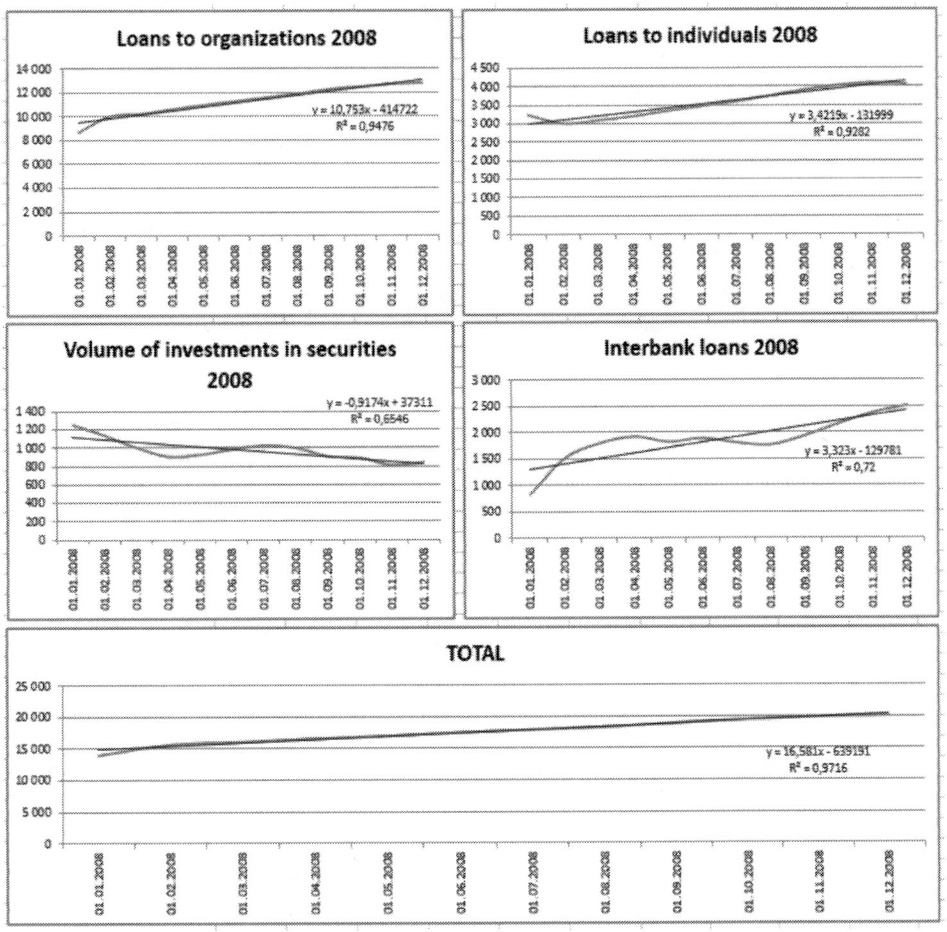

Fig. 3 Dynamics and linear trends in a number of indicators of banking institutions of the Russian Federation, 2008.

© The Authors, published by EDP Sciences. This is an open access article distributed under the terms of the Creative Commons Attribution License 4.0 (http://creativecommons.org/licenses/by/4.0/).

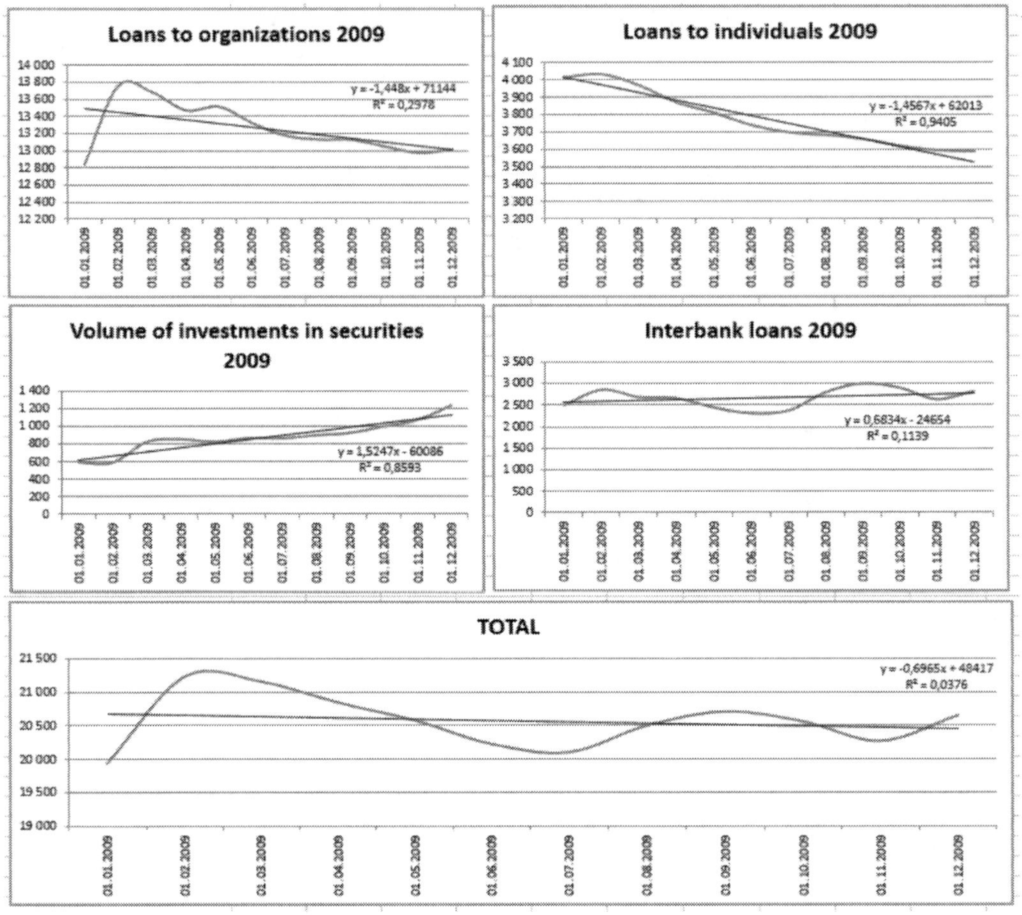

Fig. 4 Dynamics and linear trends in a number of indicators of banking institutions of the Russian Federation, 2009.

The calculation of the linear trend is important and necessary for taking into account the trends of the indicator in the reporting year and constructing points in a two-dimensional coordinate system in order to determine the structure of transaction services of the Russian financial market. Identification of crisis phenomena and trends in the functioning of this sector of the economy.

Table 4. Calculation of indicators Ki and Ti for model construction, 2008.

Record\Month	Total annual average, billion. RUB	Coefficient Ai	Coefficient Ao	Ki = Yi /Y0 * 100% (abscissa)	Ti = Ai /$A_{общ}$ * 100% (axis of ordinates)
Loans to organizations	11 264 145	327929	505632	63,8	64,9
Loans to individuals	3 560 458	104289		20,2	20,6
Interbank credit	1 859 784	101358		10,5	20,0
Volume of investments in securities	970 052	-27944		5,5	-5,5
TOTAL	**17 654 439**	**Agen = 505633**		**100,0**	**100,0**

© The Authors, published by EDP Sciences. This is an open access article distributed under the terms of the Creative Commons Attribution License 4.0 (http://creativecommons.org/licenses/by/4.0/).

In order for the necessary values to be compared most correctly, it is necessary to adjust the value of each indicator by the amount of inflation of the analyzed period.

Table 5. Calculation of indicators Ki and Ti for model construction, 2009.

Record\Month	Total average for the year with inflation 8.8%, billion. RUB	Ai	Coefficient Ao	Ki = Yi /Y0 * 100% (abscissa)	Ti = Ai /$A_{общ}$ * 100% (axis of ordinates)
Loans to organizations	12 088 457	-43 937	505632	64,4	-207,0
Loans to individuals	3 442 879	-44 357		18,3	-208,9
Interbank credit	2 435 412	20 604		13,0	97,1
Volume of investments in securities	797 652	46 461		4,2	218,9
TOTAL	**18 764 400**	**Agen = 21 229**		**100,0**	**-100,0**

Based on the obtained values of the key parameters Ri and Ti, a matrix model of 2 analysis cycles is constructed and changes are tracked. The starting point is determined based on the average value of the parameters Ki and Ti (Figure 5). In this case, the volume of the ball in Figure 5 is equal to the average value of the indicator for the reporting period, also expressed in millions of rubles in the data name of each indicator.

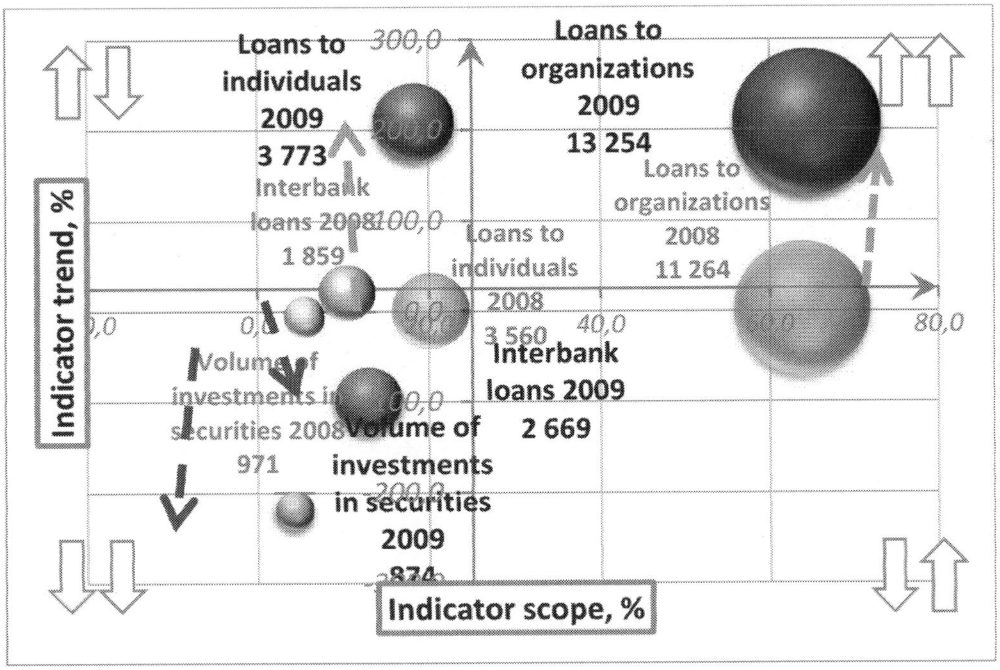

Fig. 5. Model for assessing the institutional impact of financial market institutions on economic growth, 2008-2009.

Consider the situation in the financial sector on the basis of the proposed model in 2018 and assess the situation in the Russian economy at the moment. After all, experts now largely

© The Authors, published by EDP Sciences. This is an open access article distributed under the terms of the Creative Commons Attribution License 4.0 (http://creativecommons.org/licenses/by/4.0/).

disagree on how significant the impact of Western sanctions on the financial sector of the country and the economy of the Russian Federation as a whole.

Table 6. Performance indicators of financial institutions in 2018, trln. RUB.

Record\Month	Jan	Feb	Mar	Apr	May	Jun	Jul	Aug	Sept	Oct	Nov	Dec
Loans to organizations	3019 3745	30 296 875	3018 1752	3062 7754	3143 9965	3128 7856	3139 8748	3159 1741	3268 4758	3289 1748	3298 6741	3307 7789
Loans to individuals	12 174 968	12 261 963	1237 1641	1257 2632	1280 1546	1305 6789	1329 6789	1356 1963	1389 3741	1417 3978	1441 6963	1473 4741
Interbank credit	8070 748	6635 741	6448 741	5275 741	5947 963	5335 748	5649 165	5407 745	5861 963	5942 784	6522 748	7 174 325
Volume of investments in securities	9 947 963	9 977 962	10 031 368	10 433 785	10 507 741	10 665 963	10 691 788	10 845 985	10 879 478	10 836 963	10 691 894	11 009 854
TOTAL	60 387 424	59 172 541	59 033 502	58 909 912	60 697 215	60 346 356	61 036 490	61 407 434	63 319 940	63 845 473	64 618 346	65 996 709

Based on the generated data and the calculated values of the basic indicators, the next step is to build a graphical model and calculate the linear trend equation for each indicator.The resulting trend equations and the dynamics of indicators will be reflected in the graphs below.

© The Authors, published by EDP Sciences. This is an open access article distributed under the terms of the Creative Commons Attribution License 4.0 (http://creativecommons.org/licenses/by/4.0/).

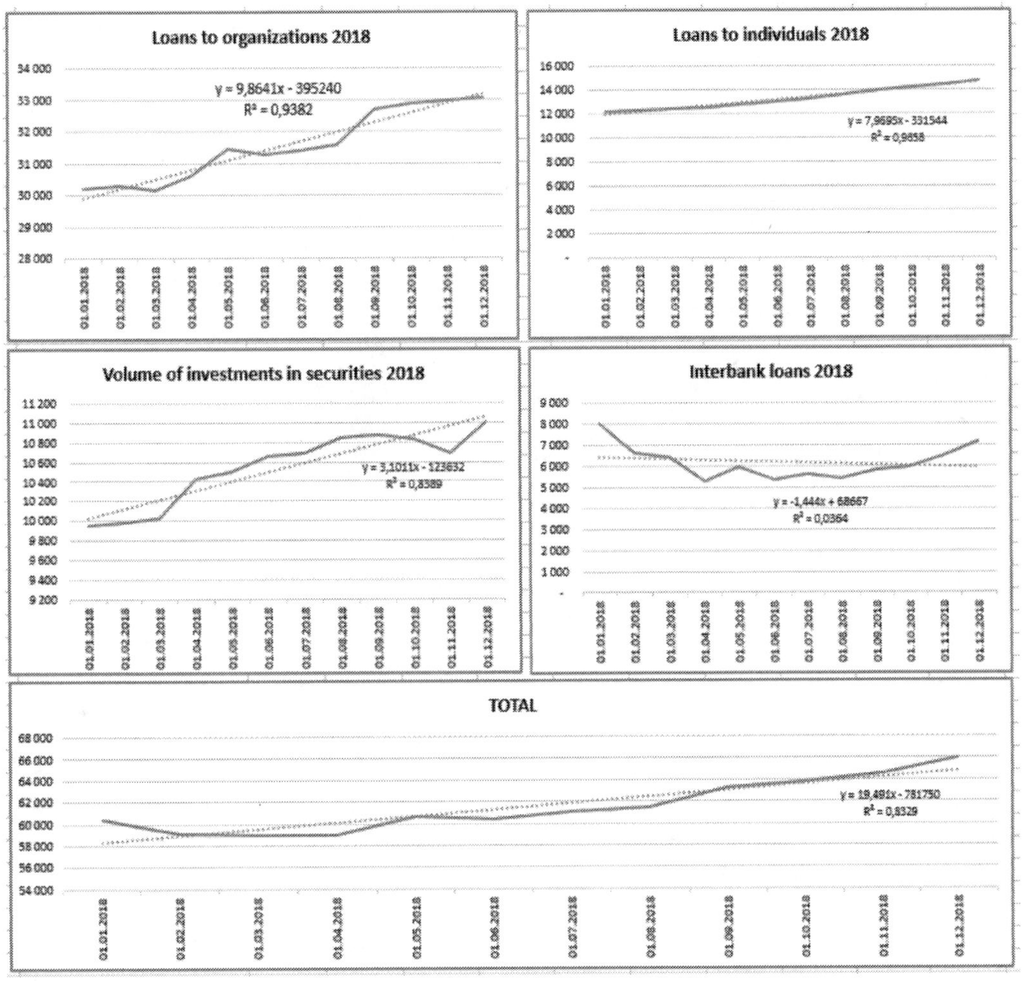

Fig. 6. Dynamics and linear trends in a number of indicators of banking institutions of the Russian Federation, 2018.

Next, we calculate the linear trend equation necessary for us to take into account the trends in the indicator in the reporting year and build points in a two-dimensional coordinate system in order to determine the structure of transaction services of the financial market of the Russian Federation in the modern period of the impact of external economic sanctions.

Table 7. Calculation of indicators Ki and Ti for model construction, 2018

Record\Month	Total average for the year with inflation 8.8%, trillion. RUB	Ai	Coefficient Ao	Ki = Yi /Y0 * 100% (abscissa)	Ti = Ai /$A_{общ}$ * 100% (axis of ordinates)
Loans to organizations	31 554 548	9,86	19,49	51,3	50,6
Loans to individuals	13 276 412	7,97		21,6	40,9
Interbank credit	6 189 521	-1,44		10,1	-7,4
Volume of investments in securities	10 543 652	3,10		17,1	15,9
TOTAL	61 564 133	19,49		100,0	100,0

© The Authors, published by EDP Sciences. This is an open access article distributed under the terms of the Creative Commons Attribution License 4.0 (http://creativecommons.org/licenses/by/4.0/).

Similar to the construction of the model for the previous two periods of 2008 and 2009, we will make a graphical model for 2018.

We construct a model based on the obtained values of the parameters Ki and Ti. The origin point is determined based on the average values of the Ki and Ti parameters. The volume of the ball in the figure will be equal to the average value of the indicator in the reporting period, also indicated in the data signature of each indicator in trillion. as a result, the resulting model will have the following form, shown in the figure

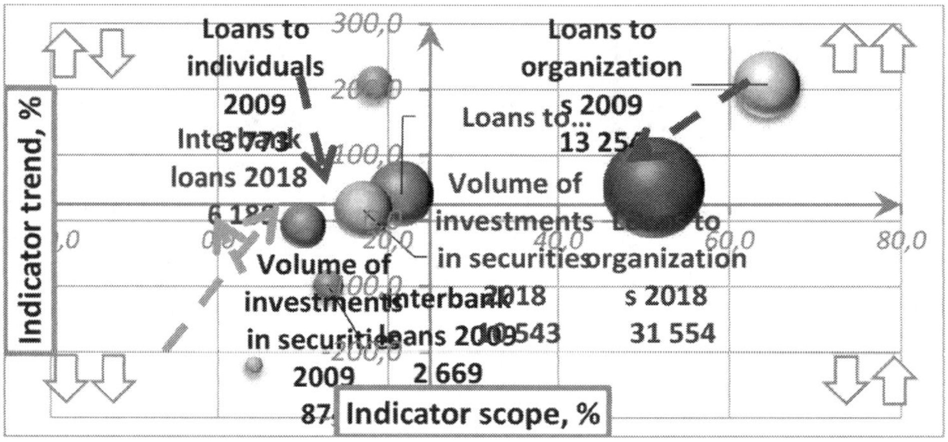

Fig. 7. Model for assessing the institutional impact of financial market institutions on economic growth, 2018.

Based on the results of the model, we see that in 2018 compared to 2009, the institutions of the financial sector of the Russian Federation moved from a state of "unstable economy" to a state of "economic growth".

5 Research result

According to the results of the study, it is possible to form a number of certain conclusions about the development of the financial sector of the Russian economy. The analysis of the activities of financial institutions showed that during the economic crisis of 2008-2009 and until 2018 in the country's economy there were a number of significant structural changes.

According to the results of the analysis of the period of influence of the global financial and economic crisis of 2008-2009 on the Russian economy, the financial sector of the country moved from the state of "inertial growth" to the state of "unstable economy".

Particularly interesting is the analysis of the financial sector in the modern period and the development of the Russian economy under the negative impact of external economic sanctions, which showed that in comparison with the situation in 2008-2009, the situation on the financial market in 2018 can be characterized as a state of "economic growth". According to the presented model, this economic growth in fact has some borderline state with the state of "inertial growth", however, it can be argued that the impact of economic sanctions on the Russian economy from other States is not so significant and does not lead to significant consequences in the form of a total deterioration of financial and economic indicators of the Russian financial market.

© The Authors, published by EDP Sciences. This is an open access article distributed under the terms of the Creative Commons Attribution License 4.0 (http://creativecommons.org/licenses/by/4.0/).

6 Conclusion

The level of development of financial institutions has a significant impact on ensuring high rates and sustainability of economic growth.

The new institutional economic theory with the use of elements of mathematical modeling is currently able to act as a significant tool for monitoring and controlling the effectiveness of implementing institutional changes, or to act as an indicator indicating the need to implement institutional changes in a particular period of time [10].

The new system economic theory in systems research focuses on aspects related to the allocation of resources and the effectiveness of economic development.

In assessing the impact of financial institutions on economic growth, we note that the current analysis of financial institutions is based on a discrete analytical approach that takes into account only the quantitative characteristics of transactional financial services, and not the structural aspects, including the dynamics of this area of economic research, and that it is preferable to use the model to assess the institutional impact of financial institutions on economic growth.

Based on the proposed model, which assesses the institutional impact of financial institutions on economic growth, it is possible to characterize the state of the national financial system and analyze the trade sector of the economy. According to the calculation of the model, the following states of the national financial system should be distinguished: the state of economic growth; inertial development of the economy; economic instability; crisis economy or recession.

As a result of the research, we determine the state of financial institutions, the degree of their development, their interaction, the institutional environment in which they operate, and the structure of their ability to lend, invest, and sponsor certain projects.

The proposed model of assessing the institutional impact of financial institutions on economic growth can also be used to analyze the activities of financial institutions at the meso level, which will allow to regulate and make timely changes in the policy on the operational management of the financial market.

References

1. A. B. Ayurzanain, Economic sciences, **131,** 7 (2015)
2. A. B. Ayurzanain, Economic Sciences, **73(12),** 84 (2010)
3. S. E. Zhelaeva, A. B. Ayurzanain, O. A. Darmaeva, Russian Conference on Digital Economy and Knowledge Management, 523 (Atlantis Press, 2020)
4. D. C. North. *Violence and social orders: A conceptual framework for interpreting recorded human history, 65* (Cambridge University Press, 2009)
5. A. E. Shastitko, I. P. Shabalov, I. N. Filippova, Economic Issues, **2,** 74 (2018)
6. O. E. Williamson, E. Brousseau, J. M. Glachant, *The New Institutional Economics Guidebook New Institutional Economics*, 181 (2008)
7. R. H. Coase, Economics, 386 (1937)
8. G. M. Hodgson, Rethinking Economics, 59 (2017)
9. R. King, R. Levine, Quarterly Journal of Economics, 108 (1993)

© The Authors, published by EDP Sciences. This is an open access article distributed under the terms of the Creative Commons Attribution License 4.0 (http://creativecommons.org/licenses/by/4.0/).

https://doi.org/10.1051/shsconf/202111401023

National systems of technology transfer of BRICS countries: features of formation

Yuliana Solovieva[1,*], and *Mingjun* He[2],

[1] Peoples' Friendship University of Russia (RUDN University), Miklukho-Maklaya Street, 6, 117198 Moscow, Russia
[2] Kunming Polytechnic University, Qingming nanlu str., 727, Chengdong district, Kunming, China

Abstract. The article considers the features of creation and development of organizational systems of technology transfer in the BRICS member countries (Brazil, Russian Federation, India, China, South Africa), interaction of participants in the innovation process. The author identifies the significant role of the regional component in the formation of effective national innovation systems. The purpose of the article is to analyze the national innovation systems of the BRICS member countries, since each of these countries has gone its own way of innovative development, justified by regional characteristics. Based on the analysis of transfer systems operating in the BRICS countries, the author identifies key and most promising areas for the development of the integration of the scientific and educational, production spheres and the state in order to form special mechanisms for organizing innovative processes that ensure effective interaction between all its participants, contributing to the formation of competitive high-tech products.

1 Introduction

The economic development of any state depends on many of socio-economic, political and other factors that affect not only the increase in real production, but also economic growth, in general, its improvement in quality. As economies develop, these factors change, assess and reassess their importance. For most of the technology leaders, the development of the innovation sphere, the knowledge economy, and technological transfer are becoming priority areas. In the context of modern socio-economic transformations, it is the level of development of the national innovation system that creates the world economic superiority of the country.

Technology transfer is one of the main factors in the development of both national economies and the global economy as a whole. The degree of participation of each country in the process of obtaining and implementing the results of intellectual activity largely determines both the prospects for the country's scientific and technological development and its competitiveness. In modern conditions, it is the level of innovation and technological development, based on the production and transfer of high-tech products, that contributes to the world economic superiority of the country.

* Corresponding author: jouliana_sol@mail.ru

© The Authors, published by EDP Sciences. This is an open access article distributed under the terms of the Creative Commons Attribution License 4.0 (http://creativecommons.org/licenses/by/4.0/).

Economic organizations, unions, various integration groups occupy a special place in the world market of high technologies. So, the BRICS member countries have considerable potential based on an effectively developing economy and a large resource base of world importance: «Brazil - the 7th economy of the world at par of GDP, is rich with agricultural production, is close to position of great powers; Russia - the 5th-6th economy of the world in terms of GDP parity, rich in mineral resources, has the largest territory in the world, one of the two largest nuclear powers in the world; India - the 3rd economy of the world at par GDP, cheap intellectual resources, one of two countries with more than milliard population, the nuclear power, is close to position of great powers; China - the 1st economy of the world in terms of GDP parity and the 1st exporter in the world ("world factory"), the owner of the world's largest foreign exchange reserves, has the world's largest population, a nuclear power; The Republic of South Africa - the 25-29th economy of the world at par GDP, various natural resources, the main African regional power» [10].

The purpose of the article is to analyze the national innovation systems of the BRICS member countries, since each of these countries has gone its own way of innovative development, justified by regional characteristics (climatic, cultural, socio-economic, etc.).

2 Methodology

As basic research methods we used system approach, comparative and statistical analysis of indicative values of innovation and technological development of the BRICS member countries. So, the systematic approach is reflected in the application of all elements of the innovation process (state, science, industry, etc.), which allows to identify trends in the technological development of the economies of the BRICS countries. Investigating problems of technological transfer in the countries of the BRICS, the authors analyse both statistical data, and the legislative base which is the cornerstone of formation of national innovative systems of member countries of BRICS. Methods of the comparative and statistical analysis of data are for this purpose applied.

3 Results

The Global Innovation Index (GII) is calculated annually by the Economist Intelligence Unit to assess the innovative development potential of the world. So, in 2020, the study covered 131 countries [11], which account for 99.4% of world GDP (Table 1). Russia, which was in 62nd place in 2013, reached 47th place in 2020.

Table 1. GII for the BRICS member countries, 2020.

Country / Economy	Score (0-100)	Rank	Income	Rank	Region	Rank
China	53.28	14	UM	1	SEAO	4
Russian Federation	35.63	47	UM	6	EUR	32
South Africa	32.67	60	UM	14	CSA	2
India	35.59	48	LM	3	CSA	1
Brazil	31.94	62	UM	16	CSA	4
World Bank Income Group Classification: LI - low income; LM - lower-middle income; UM - upper-middle income; HI - high income. Regions are based on the United Nations Classification: EUR - Europe; NAC - Northern America; LCN - Latin America and Caribbean; CSA - Central and Southern Asia; NAWA - Northern Africa and Western Asia; SEAO - South East Asia, East Asia and Oceania [11]						

© The Authors, published by EDP Sciences. This is an open access article distributed under the terms of the Creative Commons Attribution License 4.0 (http://creativecommons.org/licenses/by/4.0/).

Each of the BRICS countries has gone its own way of innovative technological development, the formation of national innovation systems (NIS) and technology transfer systems, the main features of which we will consider. As note Edler& Fagerberg [3], "There is a narrow perspective, considering invention only, and there is a broader, more holistic perspective, which emphasizes the importance of looking at the entire innovation cycle from the creation of novel ideas to their implementation and diffusion".

Brazil. The construction of the NIS of Brazil began in the 1900s. in the conditions of relatively closed micro- and macroeconomic processes, accompanying the state-proclaimed industrialization policy. In the 1940s, the creation of institutes implementing research, technological and industrial projects began. So, in 1947, the Center for Aerospace Research was created, which played a significant role in the development of Brazilian aircraft engineering. In 1951, the National Council of Scientific and Technological Research was formed, initially promoting atomic programs. The Council is now playing a significant role in funding public research, especially in university research centers. In 1953, the Development Bank of Brazil was established, which remains the main institution of long-term financing in Brazil. In 1964, two funds were established through the Bank to finance technological innovations - FUNTEC (finances training of personnel associated with the activities of university research centers) and FINAME (finances the development of mechanical engineering and the improvement of industrial equipment). In 1965, the National Agency for Financing Education and Research (FINEP) was formed, which until now is an important state enterprise for financing innovative processes.

In the 1970s there is a phased implementation of three Plans for the Development of Science and Technology. These plans were aimed at increasing the financial base of support for the development of technologies in the field of new energy sources, microelectronics and the aerospace industry, as well as at reorienting horizontal links between private and public entities.

In 1972, the Ministry of Industry and Trade of Brazil established the Secretariat of Industrial Technologies, whose activities are aimed at regulating the protection of intellectual property, coordinating research and technology programs, technological development of private and public companies, and technology transfer.

In 2006, the Government adopted the Act on Innovation, aimed at developing a technology transfer system from a research center to mass production. For its implementation, the Development Bank of Brazil creates a special program for financing entrepreneurs and enterprises wishing to increase R & D spending, introducing new productive business processes or new high-tech products.

At present, the Ministry of Science and Technology (established in 1985), which includes the National Council of Scientific and Technological Research and FINEP, is a key part of Brazil's national innovation system.

India. The NIS of India is a research centre with an extensive infrastructure, with both modern facilities for research and development in various fields (biotechnology, medicine, electronics, communication technologies and others) and interacting with enterprises to implement research results.

Having elected the US Massachusetts Institute of Technology in the early 1950s as the original model, India begins the creation of technological institutes that have received the status of technology parks (in New Delhi, Mumbai, Kampur and Madras), with the aim of forming centers of advanced knowledge and technology with the rapid introduction of the latter into production.

The first technological park - Softmare Technology Park of India (STPI) – was founded by the government of India in 1990 for development of the industry of software products and an entry into the world market. Priority problems of its functioning were: assistance and development of potential in export of software products; innovative activity; rendering full

© The Authors, published by EDP Sciences. This is an open access article distributed under the terms of the Creative Commons Attribution License 4.0 (http://creativecommons.org/licenses/by/4.0/).

support to the companies releasing software products and the equipment for information technologies; rendering telecommunication and infrastructure services to the companies occupied with development and export of software products and which are a part of science and technology park [10].

STPI currently has 36 branches nationwide. Its headquarter is in Delhi. Technopark centers are located in Bangalore, Pune, Bhubaneshwar, Hyderabad, Gandinagar, Thiruvananthapuram, Chennai, Mohali, Jaipur, Navi Mumbai, Hubli and other cities. More than 7,500 companies specializing in export-only software development are members of STPI.

Technology parks in India have a number of benefits. So, they are completely exempt from import tax, and for five years - from paying domestic taxes and fees.

China. The formation of the modern NIS of China was preceded by the reforms of 1970-80 and the national development programs adopted on their basis. Approved in 1986, the state program for the development of science and high technologies "Program 863" identified priority industries (microelectronics, informatics, space, fiber-optic technologies, genetic engineering and biotechnology, energy-saving technologies and medicine). The implementation of the Programme has been quite effective. So, literally in the 10 first years of its operation, over a thousand of the most important scientific and technical achievements were registered, of which 560 developments received worldwide recognition, 73 were awarded state prizes, 266 were patented abroad [8].

Two years later, China launched a new research and production program (Torch), which aimed at the commercialization and industrialization of knowledge-intensive technologies. In 1988, by a resolution of the State Council of China, the first technopark was established - the Beijing High-Technology Industry Development Experimental Zone Zhongguancun (in 1999 Zhongguancun has become the "Zhongguancun Science & Technology Zone" or Z-park).

The integration component in China has its own territorial organization. It is based on the division into the High technology and development zones (HTDZ), formed in the mid-1980s, which are scientific and technological parks.

In 2020, China had 168 HTDZ at various levels, including 53 for strategic purposes (China Development Zones Association, 2020). Among Chinese HTDZ, one can distinguish zones located in the central regions (Beijing, Shenyang), as well as in the coastal regions (Shanghai, Hainan). During the existence of the Shenyan technopark zone, about 600 new types of high-tech products were developed and introduced into production.

China's state policy is aimed at fully supporting enterprises of new and high technologies, technology park structures, and the effective development of the country's economy, which focuses on its own scientific and technical potential. According to the national program adopted in 2006, the state authorities are obliged to allocate a certain share of their expenses for the products of only Chinese innovative companies, despite the profitability/disadvantage of such purchases. Under the new regulations, public authorities can only purchase foreign products if there is no alternative in China [8].

Republic of South Africa. In 2013, Bloomberg Rankings included the Republic of South Africa for the first time. According to the World Intellectual Property Organization published following the results of 2019 in the rating of the countries of the world by the number of patents of the Republic of South Africa rose to the 24th place, having overtaken Austria, Sweden, Switzerland and a number of other countries [12].

The Republic of South Africa is experiencing a steady increase in R & D spending (so, the public expenditures grew from 300 million dollars in 1997 to 5,266 million dollars in 2020). Analyzing the structural component of innovation costs, we note that mainly funds are spent on innovations in industrial production and technology, scientific research of the

© The Authors, published by EDP Sciences. This is an open access article distributed under the terms of the Creative Commons Attribution License 4.0 (http://creativecommons.org/licenses/by/4.0/).

Earth and outer space, in activities related to the use of computer technology and information technology.

The Republic of South Africa is currently striving to establish an effective national innovation system, create sustainable linkages between the participants in the innovation process and the technology transfer system being created. To achieve this goal, the South African Association for Scientific Research and Innovation Management (SARIMA) was founded in 2002. SARIMA has taken the lead in this process. This Association is funded by both the South African Government and various academic organizations, as well as philanthropists from the USA and Europe.

Russia. The technology transfer system that has developed at the moment in Russia is characterized by a number of features characteristic of the transition period. Thus, «an effectively developing economy is characterized by the predominance in the structure of imports of new knowledge-intensive technologies, and in the structure of exports - on the contrary, the marketing of mature technologies» [10].

According to the Federal State Statistic Service of Russia [4], in the structure of exports in technology trade, only scientific research was the predominant of all transaction objects. For other objects of transactions, the share of imports showed a significant excess over the share of exports (see tab. 2). In the system of transfer of advanced production technologies with foreign partners (see tab. 3), there is a noticeable significant excess of imports over exports as a whole [5], which corresponds to the preferential import of technologies that do not have a sufficient degree of novelty from the position of the global technology market. Thus, 86.6% of advanced technologies developed are new for Russia, of which only 13.4% are brand new in the global market.

Table 2. Trade in Technology with foreign countries by the subject of agreement, 2020

Subject of agreement	**Exports**			**Imports**		
	Agreements	**Cost of subject of agreement, mln US dollars**	**Receipts for year, mln US dollars**	**Agreements**	**Cost of subject of agreement, mln US dollars**	**Payments for year, mln US dollars**
Total	4196	66565	3520	5518	12323	4837
including by subject of agreement:						
patent on invention	8	2.4	1.4	34	89.0	4.6
patent on selection achievement	1	0.0	0.0	19	1.3	0.8
patent license on invention	163	141	34.0	277	742	220
utility model	6	1.3	1.3	12	8.7	7.2
know-how	75	22.7	4.7	235	955	486
trademark	95	13.4	4.1	416	1017	580
industrial design	5	1.6	0.1	22	4.1	4.1
engineering services	1144	64302	2588	2382	7326	2824
research and development	1142	1230	397	512	157	120
others	1558	851	490	1628	2023	591

© The Authors, published by EDP Sciences. This is an open access article distributed under the terms of the Creative Commons Attribution License 4.0 (http://creativecommons.org/licenses/by/4.0/).

Table 3. Advanced production technologies developed by groups, 2015-2019 (units).

Advanced production technologies	**Total number of technologies**	**Category**		
		New for Russia	**Brand new**	**Number of technologies using patented inventions**
Total				
2015	1398	1223	175	589
2019	1620	1403	217	530
including:				
design and engineering				
2015	359	323	36	145
2019	456	403	53	168
production, processing and assembly				
2015	548	471	77	247
2019	510	447	63	173
automated uploading / downloading actions; materials and parts transporting				
2015	12	10	2	6
2019	29	25	4	10
automated surveillance (monitoring) devices				
2015	117	82	35	53
2019	159	98	61	88
communication and control				
2015	232	218	14	92
2019	316	295	21	55
production informational system				
2015	84	78	6	33
2019	81	73	8	18
integrated control and monitoring systems				
2015	46	41	5	13
2019	69	62	7	18

The basis of the currently formed technology transfer system was laid in the USSR in the 1950s. The integration processes taking place then in the academic and university environment led to the interaction of scientific organizations with production, with sectoral ministries and departments. In the 1970s, inter-university complexes appeared in the country, uniting scientific teams of various universities in order to solve joint research and scientific and technical problems.

In the 1990s Russia is forming a number of new integrated scientific and educational structures (university complexes, scientific and educational and other centers). However, this is accompanied by a sharp reduction in the network of design and design organizations, experimental plants, scientific and technical services of enterprises (that is, structures designed to ensure the transfer of scientific results to the innovative sphere). Thus, until the early 2000s, there was an increase in the number of research institutes through the creation of new or fragmentation of existing organizations. This was often not done by strengthening the research base of universities and enterprises (the basis of innovative systems in developed market economies), but by the formation of new legal entities.

Everything that happened led to institutional irregularities in the structure of science: if in 1992 there were 3437 independent scientific research institutes, the construction, design and exploration organizations, then in 2018 their number was reduced to 1848 organizations, and from them the number of the design and survey organizations was reduced by 24.8 times

© The Authors, published by EDP Sciences. This is an open access article distributed under the terms of the Creative Commons Attribution License 4.0 (http://creativecommons.org/licenses/by/4.0/).

[5]. At the same time, there are noticeable trends towards an increase in the number of educational organizations and experienced enterprises.

The technological transfer system established in Russia was formed in the conditions of socio-economic transformation. In this regard, there is a need to create an effective mechanism for organizing innovative processes aimed at the interaction of all its participants - the state, production enterprises, institutes of science and education. The foreign experience we reviewed shows the need to create an appropriate innovative infrastructure, which includes: an innovative entrepreneurship-oriented legislative framework that guarantees the protection of intellectual property rights; scientific and educational centers, scientific and technological parks, technology transfer and commercialization centers and other specialized organizations that contribute to the transformation of promising scientific ideas into product and other innovations; financing R&D through the business sector, development of a network of various funds, venture capital firms, small-scale enterprises, other elements of innovation infrastructure.

4 Conclusions

Although the theoretical basis of research on innovative processes is very significant, there are still many controversial and insufficiently studied issues, such as the formation of effective mechanisms and the choice of technological transfer tools, the dynamics of innovative and technological development of countries, the use of various institutional structures in the regulation of innovative processes, etc.

The formation of an organizational and economic mechanism focused on ensuring the process of obtaining, implementing and disseminating the results of scientific research and development with the involvement of all participants in the innovation process will contribute to increasing the efficiency of technology transfer. The formation of a system based on the integration of state institutions, the scientific and educational sphere and business in the BRICS member countries will contribute to the creation of competitive products, improving the status of countries in the global technology market.

References

1. R. Aoyama. J. of contemp. East Asia Studies, **5**, 3 (2016)
2. China Development Zones Association (2020) www.cadz.org.cn/
3. J. Edler, J. Fagerberg. Oxford Review of Economic Policy, **33**, 2 (2017)
4. Russian Statistical Yearbook (2020) https://rosstat.gov.ru/
5. Russia in numbers (2020) https://rosstat.gov.ru/
6. L. Shkvarya, O. Grigorenko, A. Strygin, V. Rusakovich, S. Shilina. Centralnaya Aziya i Kavkaz, **19**, 119 (2016)
7. L. Shkvarya, H. Yu. *Modern Global Economic System*, 1209 (2021)
8. Yu. Solovieva, M. Chernyaev, A. Korenevskaya. J. of Appl. Econ. Sci., **12,** 1473 (2017)
9. Yu.V. Solovieva. *Analysis of Intern. Relations*, 78 (2018)
10. Yu. Solovieva, Economic journal, **4,** 85 (2015)
11. The Global Innovation Index (2020) www.wipo.int/
12. World Intellectual Property Indicators (2020) www.wipo.int/

© The Authors, published by EDP Sciences. This is an open access article distributed under the terms of the Creative Commons Attribution License 4.0 (http://creativecommons.org/licenses/by/4.0/).

New trends and competition in the market of higher education services in the context of digitalization

Galina V. Astratova[1,2,*], *Tatyana K.* Rutkauskas[1], *Natalya V.* Parushina[3], and *Natalya A.* Suchkova[3]

[1]Ural Federal University named after the first President of Russia B. N. Yeltsin, Yekaterinburg, Russia
[2]Ural Juridical Institute of MIA of Russia, Yekaterinburg, Russia
[3]Orel State University of Economics and Trade, Orel, Russia

Abstract. The aim of the study was to analyze new trends, competition and structural changes in higher education in the context of digitalization. The main directions of higher education development in the XXI century are identified. It is shown that by 2020, new trends in the market of higher education services have appeared and fully manifested themselves, due to three key factors: 1) increased global competition; 2) the active introduction of digital technologies in higher education; 3) the impact of the global COVID pandemic on educational technologies. It is shown that the coronavirus pandemic was a powerful incentive for the transition to digital technologies in higher education. Thanks to digitalization, the structure of the market has changed, and the number of players has increased significantly, and the market consists of private and public institutions, ministries of education and government agencies, educational, consulting and test (rating) companies, etc. Moreover, the global competition of universities for quality is becoming more organized and technically equipped; it increasingly relies on impersonal methods of assessing quality using digital technologies. At the same time, the authors draw a number of conclusions about the negative consequences of distance education. It is shown that competitiveness and accessibility of educational services for different categories of citizens in the difficult conditions of global challenges and threats, the wave pandemic, is ensured through high-quality interaction of all participants in the process, the activity of scientists and practitioners around the world, and state support for teaching initiatives in all regions of the country.

1 Introduction

The outbreak of the coronavirus pandemic and its rapid spread around the world speaks of the vulnerability of human civilization to previously unknown, new diseases. It is also important that the pandemic occurred in the context of the global trend of digitalization of

* Corresponding author: galina_28@mail.ru

© The Authors, published by EDP Sciences. This is an open access article distributed under the terms of the Creative Commons Attribution License 4.0 (http://creativecommons.org/licenses/by/4.0/).

the economy. Accordingly, the response measures taken by the governments of the countries have led to serious changes in the socio-economic development of the regions, and the acceleration of the processes of digitalization of the economy has occurred mainly in developed countries.

The higher education system was in a special position in these conditions.

Indeed, the processes of globalization and digitalization, the ubiquitous spread of information and computer technologies, have led to changes in both the higher school itself (the "networking" of universities) and the learning process. For example, today, for universities that position themselves not only locally, but also internationally, the presence of an "online presence" is mandatory. In the new conditions, the methods of teaching and attracting new students, the nature of the work of the teaching staff, etc., are changing [9, 32].

Finally, the COVID-19 virus pandemic has illustrated the global need for online competencies for the majority of workers in the labor market, as well as the fact that in modern realities, it is almost impossible to implement higher education services without access to the Internet [2, 21, 34].

At the same time, as we noted earlier [2], in terms of the speed, nature, models of digital technologies and the results of the transition to online education, Russia is significantly behind both the pace and nature of the development of online education in the United States, the EU, China, and other countries, and the real needs of our country's economic development.

In addition, the process of developing a strategy for the development of higher education in Russia in the context of new global trends and structural changes in the context of digitalization is a very controversial and debatable event, which includes not only the coordination of various interests of the subjects of the education market, but also economic and mathematical modeling, assessment of the effectiveness of resource management, and a number of other problems.

In this regard, the purpose of the study was to analyze new trends, competition and structural changes in higher education in the context of digitalization

2 Materials and methods of research

The main research tools used in this work are system analysis. For the study of theoretical and practical material, the methods of desk research, the principles of formal logic, synthesis, statistical, logical methods of analysis, as well as methods of comparison, generalization of the experience of foreign and domestic higher education, structuring, based on the processing of a large volume of factual material, are used.

3 Digitalization and new trends in the development of higher education

One of the priority directions of Russia's development [8] and most of the world's economic leaders (USA, UK, Germany, China, etc.) [13] is the formation of the digital economy. The foundation of the digital economy is modern digital technologies, which have turned from a separate technological industry into a new way of life that affects all types of human activities.

The most obvious consequence of access to the Internet has been the explosive growth in the production and consumption of information around the world. So, according to the World Bank Report [4, 38], "a typical day on the Internet" represents $ 207 billion e-mail

© The Authors, published by EDP Sciences. This is an open access article distributed under the terms of the Creative Commons Attribution License 4.0 (http://creativecommons.org/licenses/by/4.0/).

messages; 8.8 billion video views on YouTube; 4.2 billion search queries in Google; 36 million purchases in Amazon, etc.

The new, digital way of life is represented today by at least ten key areas of activity:

1. 5G Internet;
2. Electric cars and self-driving cars;
3. 3D-printing;
4. Artificial intelligence and Big Data;
5. Virtual (VR) and artificial (AR) reality;
6. Cloud data storage;
7. Internet of Things;
8. Quantum computing;
9. Blockchain;
10. Brain implants [15].

However, as noted in the World Bank Report [4, 38], the benefits of digital transformation of the economy are inferior to the pace of the spread of digital technologies (especially the Internet, mobile technologies and other devices for collecting, storing, processing, analyzing and exchanging information in digital form). To date, digitalization from the simple introduction of technologies in the business models of individual companies has transformed into a global process of transition to the digital economy, where the processing and analysis of digital data not only increases the efficiency of business processes, but also becomes a key factor in the development of society.

A special place in the digital economy is given to education in general and higher education in particular. This is due to the fact that in the XXI century in developed countries, higher education, together with the R&D (Research and Development) sector, is becoming the largest sector of the global services market. Thus, according to the WTO [39], in 2019, global trade in services amounted to $ 13.3 trillion. At the same time, only from 2005 to 2017, the world trade in commercial services increased significantly, broken down by sector [39, p. 25]:

- telecommunications, computer and audio-visual services by 13.2%;
- Research and Development (R&D) by 1.4%;
- educational services by 0.8%.

At the same time, one of the fastest growing sectors of the global education market (+23% annually in the period from 2012 to 2017) is digital education, although its share in the total volume of the educational services sector before the coronacrisis was not large: less than 3% [12].

Moreover, in 2008, the global market for higher education services was $ 100 billion [1]. According to experts, its value may grow to $ 600-800 billion by 2030 [36]. The United States, Great Britain, Canada, Australia and other developed countries are leading in this market, where higher education services are developing mainly as an export industry [6, 14]. It should be noted that the total income from the Russian export of educational services is also growing. For example, only from 2007 to 2015 the growth from sales of educational services abroad increased by 1.64 times (in US$) and 3.28 times (in rubles) [7]. At the same time, the share of the Russian Federation is, according to various estimates, no more than 7% of the world market of higher education services (if we exclude representatives of the CIS from the total number of foreign students studying in Russian universities, then Russia's share in the global market of higher education services will not exceed 2%).

We consider it necessary to emphasize that education has traditionally been considered a conservative industry, subject to slow and smooth changes over a long time. At the same time, scientific and technological progress, changes in the productive forces and industrial relations in the XX century, and especially in the first quarter of the XXI century, led to

© The Authors, published by EDP Sciences. This is an open access article distributed under the terms of the Creative Commons Attribution License 4.0 (http://creativecommons.org/licenses/by/4.0/).

changes in society and, accordingly, influenced the fact that the market for higher education services has changed several mainstream trends only in the last few decades.

First, by the end of the last century, as noted in our earlier studies [3, pp. 22-23] and according to the results of research by other scientists and practitioners [1, 6, 14, et al.], the following main changes occurred:

1) Rapid growth of the scale of the educational services market:

2) Increasing mass market of educational services market and ensuring the right to education for everyone (the opportunity and equal chances for everyone to get an education in an educational institution of any type, regardless of national or racial affiliation);

3) Diversification of the territorial structure of the educational services market and the emergence of new "players" ;

4) Globalization and internationalization of higher education, based on the comprehensive nature of knowledge and the mobilization of the collective efforts of the international educational and scientific community;

5) Universalization of the content of all types of education, which cannot be stopped in the era of the information revolution and the spread of the Internet;

6) Democratization of the education system and the accessibility of education to the entire population of the country, the continuity of its stages and levels, the provision of autonomy and independence to educational institutions;

7) Increasing the range of educational and organizational activities aimed at both satisfying diverse interests and developing students ' abilities;

8) Turning education into a priority object of financing in the developed countries of the world;

9) Cross-cultural integration, tolerance to different cultures, traditions and customs; etc.

In Russia, at the same time, there is an educational reform, during which the following factors have a significant impact on the market of higher education services: 1) introduction of the Unified State Exam (USE), 2) transition to a multi-level system of higher education, 3) introduction of multi-channel financing of education, 4) reduction of the teaching staff, 5) consolidation of universities. These questions are discussed in more detail in our previous study [3, pp. 22-23].

Secondly, by 2010-2019, the emphasis on the market of higher education services is shifting somewhat, and the key trends, according to the results of our earlier studies [2] and conclusions from the results of research by other scientists and practitioners [10, 26, 37, etc.], are the following phenomena:

1. Changing global demographics due to increased life expectancy and declining birth rates (Africa alone is the only region where the youth population will grow until 2025) [37];
2. Increasing access to education for everyone, although there are still problems of gender and class inequality;
3. Development of national strategies for the internationalization of education, including, inter alia, the involvement of foreign students for training, as well as highly qualified migrants for conducting educational programs in universities;
4. Strengthening multi-sector cooperation, which implies not only the development of cooperation between universities and industrial enterprises, but also cooperation in the field of international higher education and science, R&D (for example, virtual international internships on the EU-VIP, PROVIP platforms. Pathway, etc.);
5. Use of new educational technologies. There are two main tools of educational technologies in universities: a) online learning and b) technologies that affect teaching, learning, and administration. As technologies (Internet of Things, blockchain, 3D printing, adaptive technologies, mobile technologies, etc.) [43] continue to integrate into the real world, changing and transforming it, higher education should also change over time and

© The Authors, published by EDP Sciences. This is an open access article distributed under the terms of the Creative Commons Attribution License 4.0 (http://creativecommons.org/licenses/by/4.0/).

integrate educational technologies that contribute to the development of international higher education;

6. Growing demand for specific competencies and skills, due to the fact that with the future automation of workplaces, there will also be a radical transformation of the labor market and the market of educational services. Accordingly, universities will have to change the educational programs in such a way as to guarantee graduates the appropriate competencies and skills;

7. Strengthening the influence of the English language. Although many persons consider English as a means of international business communication, but English as a way to attract foreign students for international higher education plays a much greater role, and this role is only growing every yea.

Third, by 2020, new trends in the market of higher education services have emerged and fully manifested themselves. These trends are related to three key factors, in our opinion: 1) increased global competition; 2) the active introduction of digital technologies in higher education; 3) the impact of the global COVID pandemic on educational technologies. We considered it appropriate to focus on the problems of global competition in this article.

4 Strengthening of global competition in the market of higher education services

Speaking about the problem of increasing global competition, it should be noted that competition in the education market is very high. First, thanks to digitalization, the market structure has changed, and the number of players has increased significantly. For example, only in Russia can we count at least nine key players in the education services market, including: schoolchildren, students, language platforms, aggregators, programmers, etc. [12].

Secondly, the structure of the higher education market has also changed dramatically over the past two decades.

As noted in the report of Transparency Market Research [28], the number of players in the global market for higher education services is increasing. Now it consists of private and public institutions, ministries of education and government agencies, educational, consulting and test (rating) companies, etc. Here is a list of key players in today's global higher education market: Smart Technologies, Inc.; Xerox Corporation; Panasonic Corporation; Oracle Corporation; EduComp Solutions; Dell Inc.; Cisco Systems Inc.; Three River Systems; IBM; Blackboard Inc.; Adobe Corporation; and others.

Accordingly, educational institutions are becoming increasingly receptive to the introduction of digital technologies, which play an important role in education, allowing students and teachers to interact and take advantage of the new learning opportunities that open up. It should be noted that in the market of higher education services, competition is growing not only between public and private universities. In order to attract students and teachers from all over the world, universities must, in addition to participating in international rankings, attract business partners, associations and foundations, both to conduct research and to find sources of funding [28].

Today, higher education institutions are characterized by economic competition in a tough global market environment, where governments are no longer key players [18, 28]. Thousands of universities from one and a half hundred countries offer their services, although the main struggle for the consumer is among the United States, Canada, Great Britain, Germany, France, Australia, and Japan, where more than 80% of all international students study [4, 22, 23].

As many researchers note [18, 19, 20, 27, 31, 33], over the past twenty to thirty years, the global system of higher education has dramatically changed not only the forms, types

© The Authors, published by EDP Sciences. This is an open access article distributed under the terms of the Creative Commons Attribution License 4.0 (http://creativecommons.org/licenses/by/4.0/).

and patterns of competition (from the struggle for the number of students to the struggle for the amount of budgets and the number of best professors), but also the intensity of competition. This eventually led to the emergence of new forms of competition-especially where these changes appeared, developed and were strongest, i.e. in research universities in the United States and Europe. It is argued that competition in higher education is no longer just between individuals, universities, and countries. Competition has become institutional and global, which has led to a multi-level form of competition and the transformation of universities into full-fledged business competitors.

In particular, since the 60s of the last century, in the United States, the editors of US News decided to invest in educational ratings as a way to distinguish themselves from competitors, offering society "news that can be used" in marketing strategies [24]. Since that time, the rankings have been published annually. In particular, at the moment the most well-known lists are the following: THE, QS, ARWU, etc. For many years, the top ten universities in the world are universities from the United States and Great Britain (Table 1).

Table 1. The "Top ten" of the best universities in the world and the "top ten" of the best Russian universities in the world university ranking in 2021, according to THE (Times Higher Education) [40]

Rating	Name, country, region	Number of students, total	Number of students per 1 teaching staff	Number of international students	Ratio of women to men
1.	University of Oxford, UK	20 774	11,1	41%	46: 54
2.	Stanford University, USA	16 223	7,4	23%	44: 56
3.	Harvard University, USA	21 261	9,3	25%	49: 51
4.	California Institute of Technology, USA (	2 238	6.3	33%	36: 64
5.	Massachusetts Institute of Technology, USA	11 276	8,4	34%	39: 61
6.	University of Cambridge, UK	19 370	11,0	38%	47: 53
7.	University of California, Berkeley, USA	39 918	19,8	17%	51: 49
8.	Yale University, USA	12 910	6,0	20%	50: 50
9.	Princeton University, USA	8 091	8,0	23%	46: 54
10.	University of Chicago, USA	4 292	5.9	31%	46: 54
...					
174.	Moscow State University, Russia	28 525	8,4	32%	52: 48
201-250	Moscow State University of Physics and Technology, Russia	5736	11,4	18%	25: 75
201-250	Ural Federal University, Russia	26 807	12,9	13%	48: 52
251-300	Higher School of Economics, Russia	25 268	14,8	11%	н / д
301-350	Peter the Great Saint Petersburg Polytechnic University, Russia	16 768	9,2	31%	39: 61
401-500	National Research Nuclear University, Russia	5797	10,5	26%	34: 66
501-600	Saint Petersburg State University of Information Technologies, Mechanics and Optics, Russia	10 094	14,5	17%	36: 64
601-800	National University of Science and Technology, Russia	7 168	10,8	26%	40: 60
601-800	Novosibirsk State University, Russia	7 043	6,2	20%	н / д
601-800	Kazan Federal University, Russia	25 940	9,7	25%	68: 32

As follows from the data of THE, no more than 50 Russian universities are included in this rating and are far from being in the leading positions. The annual leader in this list is

© The Authors, published by EDP Sciences. This is an open access article distributed under the terms of the Creative Commons Attribution License 4.0 (http://creativecommons.org/licenses/by/4.0/).

only Lomonosov Moscow State University, which is included only in the top 200 ranking of the best universities in the world.

Global competition between universities takes place in the form of a competition for quality, where the main indicators are the following [40, 41]: forms of education, international cooperation, multi-channel funding (primarily extra – budgetary), R&D volumes, citation indexes, etc.

In fairness, it should be noted that different models of cultures and traditions of higher education still exist in different countries. However, in reality, universities around the world take a hybrid approach, where all the well-known university traditions interact with each other. At the same time, there are some universities that still retain the reputation and traditions of a certain model, and to some extent this tradition distinguishes them from all others [30]. However, it is not the purpose of our study to analyze these trends.

Thus, competition for quality among universities (and especially research universities) turns from competition for resources (budget, students and staff) into competition for status (image, reputation, citation), which ultimately affects resources as well [19, 33].

For example, competition to attract talent from all over the world is growing very quickly, as more and more countries offer additional positions to graduates and postgraduates for non-residents of the country and, as a rule, with higher wages compared to citizens of the country [17, 30]. Moreover, fast-growing economies like China and Singapore are investing heavily in developing their higher education systems and making them friendlier to talented people from all over the world. In addition, the advent of new technologies has changed the traditional model of higher education, where physical presence is no longer a requirement [30, 42].

Competition for attracting finance is also growing. Although the distinctive feature of the Russian online education market is the initial focus on paid models, while in the United States both paid and free online university courses existed and developed in parallel, but in general, Russia is very much inferior in terms of the share of education spending in GDP. So, in 2016, public and private spending on educational institutions, as a % of GDP, was the following:

- In OECD countries – 5,0%;
- In USA – 6,0%;
- In Russian Federation – 2,4% [12].

In this regard, within the framework of the national project of the digital economy of Russia [8], experts of the HSE conducted experimental calculations of the internal costs amount for the digital economy development, based on available statistical information. As a result, the composition and amount of costs for the digital economy development were identified, and proposals for changes in the federal statistical observation were developed. It is shown that spending on the digital economy development (including in education) will grow exponentially in the near future [13].

Accordingly, the global competition of universities for quality is becoming more organized and technically equipped; it increasingly relies on impersonal methods of assessing quality using digital technologies, such as [11, 16, 25, 26, 29, 37]: Internet of Things, blockchain, cybersecurity, augmented and virtual reality, Big Data, artificial intelligence and machine learning, universal accessibility, etc. New digital technologies are used to carry out the following operations and procedures in the fight for quality: various forms of accreditation and rating of universities, competitive selection of teaching staff, scientometric assessments of publications, awarding of scientific grants, etc. For example, the association of the university name (or name of the university professor) with a high-status grant (for example, with the grants of the European Research Council – ERC in Europe, or TÜBITAK in Turkey, or NSF in the USA, etc.) increases the personal status, as

© The Authors, published by EDP Sciences. This is an open access article distributed under the terms of the Creative Commons Attribution License 4.0 (http://creativecommons.org/licenses/by/4.0/).

well as the status of the educational institution, since the above grants are now an indicator of the academic quality of the university and/or professor.

Studies by various authors have shown that this growing form of competition has given rise to new classifications and categories. Moreover, at the same time, there are unions of participants belonging to the same category (for example, leagues, guilds, or university associations), whose cooperative behavior allows for various forms of benefits and, accordingly, contributes to the sustainability of these alliances [33].

Ultimately, the increasing competition in modern conditions also leads to the complication of the university management process. This also leads to the need to use various digital technologies [11, 16, 25, 26, 29, 37, etc.] in order to automate the processes of collecting, processing, storing and analyzing heterogeneous external and internal information, as well as to translate communication between divisions (institutes and departments), teaching staff and employees, partners, contractors and customers into electronic form.

It should be noted that the coronavirus pandemic has undoubtedly been a powerful impetus for the transition to digital technologies in education. As noted above, all subjects of the educational process had to urgently master distance learning.

At the same time, our educational practice shows that there are a number of negative consequences of distance learning during the pandemic: 1) digital technologies free the student from the main type of his activity – mental work; 2) analysis of independent work of students indicates that screen addiction contributes to illiteracy; 3) typing more on the computer, there is a loss of writing skills, which contributes to the loss of creative abilities; 4) communication with the computer slows down the development of students, there are problems with the expression of thoughts both orally and in writing; et al.

The problems that arose in the educational process at the beginning of the pandemic, such as technical, questions of the organization of the educational process, psycho-physiological, and a year later did not lose their relevance, but their order and content changed. Due to innovative equipment and technologies, problems of a psycho-physiological nature have come to the fore, technical and methodological problems have moved to the background. At the same time, there is a need for third-party services to enhance interactivity, which requires further restructuring of the distance learning process.

Digital technologies open up the horizons of large-scale changes in higher education, both in terms of opportunities for improving the educational system, and in terms of the need for fundamental modernization of business processes, organizational design, and the development of the potential of teachers. Many educational services will be digitalized, and related technologies that regulate and optimize the educational process will be developed.

5 Conclusion

In the context of the global processes of the post-industrial way of life, the digital economy, the productive forces and production relations in all markets are radically changing, but especially noticeably in the higher education system.

Classes at universities take place in a remote format, the latest achievements of digital technologies are used and new educational technologies are emerging. Meanwhile, today the situation in the market of higher education services has worsened due to the challenge of Covid-19 on a global scale, which has affected all areas of socio-economic development of all regions of the world, and, above all, the Russian Federation.

The widespread use of remote technologies in the educational sphere with the simultaneous involvement of the population of different ages in obtaining educational services dictates the need to include in the educational programs courses that would be aimed at mastering modern technologies, internet outlets, information access portals,

© The Authors, published by EDP Sciences. This is an open access article distributed under the terms of the Creative Commons Attribution License 4.0 (http://creativecommons.org/licenses/by/4.0/).

electronic platforms for communication interaction. We are sure, in order for Russia to become a world leader in the development of digital technologies, the availability of information services should be free and online for all universities in the country. Moreover, it is important that educational information courses include both the basic development of practical skills to use the Internet, and professional skills to work with specialized resources and programs.

The development of the labor market, the threat of a second wave of the COVID-epidemic, and the systematic monitoring of the ranking of the most popular professions on Russian and global recruitment portals confirm the need to expand the flexibility of higher educational organizations in offering professional programs. The individual educational trajectory should be built not only at the stage of choosing a profession or in the process of mastering it at the training stages, but also in the possibility of offering universities a set of professional programs of additional education. These programs should be available to students and trainees, both on a paid and free basis, to our mind. Moreover, the availability of programs can be realized through grant support, joint global national projects and funds from the national projects "Demography", "Education", "Science", "Small and Medium-sized enterprises". In the context of digitalization of education, programs are implemented on digital educational platforms known to a wide range of students, or in cloud services.

The emphasis on practical training in the direction of the development of professional education should be placed on the exit of students and listeners outside the classroom. This is not only the expansion and equipment of practice-oriented laboratories and youth laboratories of production startups on the university campus, but also the constructive interaction of young people with mentors and specialists in production laboratories that are created at enterprises. A virtual, augmented reality and digital business model will allow you to join the production process and the value chain of the product on-line under the supervision and advice of a mentor. Practical training, thus, will meet its intended purpose and will allow getting a direct economic and educational effect from new studies in the digital educational space.

The competitiveness and accessibility of educational services for different categories of citizens in the difficult conditions of global challenges and threats, the wave pandemic, is ensured through high-quality interaction of all participants in the process, the activity of scientists and practitioners around the world, and state support for teaching initiatives in all regions.

References

1. I.A. Aydrus, V.M. Filippov, *World market of educational services,* 194 (2008)
2. G.V. Astratova, World of Science. Pedagogy and Psychology, **3**, 17 (2020)
3. G.V. Astratova, Science of Science, **4**, 27 (2016)
4. World Bank. 2016 year. World Development Report 2016 Digital Dividends. Overview. World Bank, Washington DC. License: Creative Commons Attribution CC BY 3.0 IGO, https://openknowledge.worldbank.org/.
5. World Education Report. Comparison of world statistics in the field of education (2007) http://www.ifap.ru/
6. V.A. Galichin, *International market of educational services: main characteristics and development trends* (2015)
7. I.G. Korotkov, V.A. Zubenko, World of new economy, **3**, 14 (2017)
8. Program "Digital Economy of the Russian Federation". Approved by the order of the Government of the Russian Federation dated July 28, 2017 No. 1632-r.

© The Authors, published by EDP Sciences. This is an open access article distributed under the terms of the Creative Commons Attribution License 4.0 (http://creativecommons.org/licenses/by/4.0/).

9. Russian education - 2020: a model of education for a knowledge-based economy: to the IX Intern. scientific. conf. "Modernization of the economy and globalization", Moscow, April 1-3, 2008, ed. J. Kuzminova, I. Frumin; State un-t - Higher School of Economics. - M .: Ed. House of the State University Higher School of Economics, 39 (2008)
10. T.V. Semenova, K.A. Vilkova, I.A. Shcheglova, Education Issues, **2** (2018)
11. M. A. Sklyar, K.V. Kudryavtseva, Economic revival of Russia, **3(61)**, 103 (2019)
12. Trends in the development of the online education market (2017), https://www.marketing.spb.ru/
13. G. I. Abdrakhmanova, K. O. Vishnevsky, L. M. Gokhberg, Trends, competencies, measurement: reports. to XX Apr. international scientific. conf. on the problems of economic and social development, Moscow, April 9–12 (2019)
14. E.N. Shcherbak, *World market for higher education* (2009)
15. 10 key technologies of the last decade (2020), https://habr.com/
16. J. Ainslee, Digitization of Education in the 21st Century (2018) https://elearningindustry.com/
17. M. Barber, K. Donnelly, S. Rizvi, L. Summers, *An avalanche is coming: Higher education and the revolution ahead* (2013)
18. J, Brennan, Ten Years on: changing education in a changing world. Center for Higher Education Research and Information, Open University Press, Maidenhead, UK, 23 (2004)
19. N. Brunsson, L. Wedlin, Competition for Status: The Example of Universities, Communication at the Competition(s) Workshop at the Copenhagen Business School, for the 'Performances of Value: Competition and Competitions Inside and Outside Markets' project, supported by the Leverhulme Trust and led by David Stark (2016)
20. N. Brunsson, K. Sahlin-Andersson, Organization Studies, **21**, 721 (2020)
21. Ceylan Rahmiye Figen , Ozkan Burhan, Mulazimogullari Esra, The European Journal of Health Economics (2020)
22. Education at a Glance 2016. OECD indicators, OECD publishing, http://www.keepeek.com/
23. Education at a Glance 2019. OECD indicators, OECD publishing, https://www.oecd-ilibrary.org/
24. W. Espeland, *The World of Indicators: The Making of Governmental Knowledge through Quantification*, 56 (2015)
25. D. Fedirko, 8 Top Trends Of Digital Transformation In Higher Education. 24 March. (2019) https://elearningindustry.com/
26. Gilbert Nestor, 19 Higher Education Trends for 2020/2021: Latest Forecasts To Watch Out For, https://financesonline.com/
27. R. Hasse, G. Krücken, Sociologia Internationalis, **51**, 181 (2013)
28. Higher Education Market – Global Industry Analysis, Size, Share, Trends, Analysis, Growth and Forecast 2016 – 2023, https://www.transparencymarketresearch.com/
29. S. Kamsker, G. Janschitz, S. Monitzer, International Journal for Business Education, 160, 22 (2020)
30. T. Kromydas, Palgrave Commun, **3(1)**, 1 (2017)
31. G. Krücken, Beiträge zur Hoschschulforschung, **39**, 10 (2017)

© The Authors, published by EDP Sciences. This is an open access article distributed under the terms of the Creative Commons Attribution License 4.0 (http://creativecommons.org/licenses/by/4.0/).

32. Viola A. Larionova, Alexander V. Tolmachev, Evgeniy V. Sinitsyn, Tatiana Y. Bystrova, Paper Materials of the 1st China-Russian Scientific Readings «Urbanization Level, Rural Labor Transfer and Economic Growth in the XXI-st Century: Economic Models, New Technologies and Management Practices», dedicated to the 60th anniversary of Lanzhou University of Finance and Economics professor Galina V. Astratova and 20th anniversary of her scientific school "Theory, methodology and practice of marketing and marketing research" , 155 (2019)
33. Musselin Christine, Socio-Economic Review, **16(3)**, 657 (2018)
34. Evgeny V. Sinitsyn, Alexander V. Tolmachev, Alexander S. Ovchinnikov, Electronic Scientific Economic Journal, **6(3)** (2020)
35. The Condition of Education 2020. Education Expenditures by Country. 2016. Chapter: 4. International Comparisons Section: Finances, https://nces.ed.gov/
36. Universities 2030: Learning from the Past to Anticipate the Future, Edited by Adam R. Nelson & Nicholas M. Strohl (University of Wisconsin-Madison) (2014) https://globalhighered.wordpress.com/
37. Weimer Leasa. 10 trends changing global higher education. Policy & Strategy. 26 July 2017, https://www.eaie.org/
38. World Development Report 2016: "Digital Dividends", https://unctad.org/
39. World Trade Report 2019. The future of services trade, https://www.wto.org/
40. World University Rankings 2021, https://www.timeshighereducation.com/
41. World University Rankings 2015-2016, https://www.timeshighereducation.com/
42. L. Yuan, S. Powell, J. Cetis, MOOCs and open education: Implications for higher education, JISC CETIS (2013)
43. 2017 Horizon Report (2017) https://library.educause.edu/

© The Authors, published by EDP Sciences. This is an open access article distributed under the terms of the Creative Commons Attribution License 4.0 (http://creativecommons.org/licenses/by/4.0/).

Peculiarities of the EEU countries interaction in pharmaceutical value chains in pandemic period

Elena Frolova [1,*], *Irina* Turgel[1], *Alexander* Ishukov [1,2], and *Zulparuza* Abdurahmanova [3]

[1]Ural Federal University named after the First President of Russia B.N. Yeltsin, Mira Str., 19, 620002 Ekaterinburg, Russia
[2]Institute of Economics, Ural Branch of the Russian Academy of Sciences, Moskovskaya Str., 29, 620014 Ekaterinburg, Russia
[3]Karaganda University of Kazpotrebsouz, Academic Str., 9, 100009 Karaganda, Kazakhstan

Abstract. Taking into account that new trends are emerging in the field of international economic integration in pandemic world the authors aimed to study the role of the country's participation in global value chains (GVCs) in the development of the Eurasian Economic Union (EEU or EAEU) national pharmaceutical industry. The article reveals the contradictory essence of modern concepts of interaction between the countries of integration associations and GVCs including its specificity for the EEU. An analysis of the pharmaceutical industry in EEU before and in the pandemic is carried out, and the features of such interaction are revealed in more detail in Kazakhstan case. The results showed the presence of low volumes of pharmaceutical products in EEU intraregional trade and minimum presence in GVCs. The hypothesis was partially confirmed that the countries localized the GVC "production" stage on their territory received gain as the increase in domestic production in pandemic period, despite the fact that this stage corresponds to low added value. The novelty is in the development of a reproductive approach to the study of integration groups. Recommendations are given to increase the share of economic entities of the EEU countries in the process of forming regional value chains.

1 Introduction

In the last decade, the structure of the world economy has been determined by the activity of added value chains (VC). Depending on their geographic coverage, they are called global (GVC) or regional (as a rule, in the regional integration associations). Many countries are involved in VC, therefore, there are several models for the national actors participation in them, depending on the priority of the goals, as follows [1, 2, 3]: from one side, the GVC-theory mentions that a country must locate on its territory the links with the maximum added value, from the other side, in the pandemic period, when export is restricted due to quarantine, countries that have localized on their territory the "production" link, which

* Corresponding author: frol-uved@yandex.ru

© The Authors, published by EDP Sciences. This is an open access article distributed under the terms of the Creative Commons Attribution License 4.0 (http://creativecommons.org/licenses/by/4.0/).

corresponds to the minimum added value, win (because, in this case the final pharmaceuticals are produced in their own country and do not need to be imported).

The main GVCs operate within the framework of TNCs, whose headquarters are located in the USA, Europe. For the EEU (as EAEU in different some sources) countries, there is a rather acute problem of increasing their role in the process of forming VC in the world [1]. First of all, it is necessary to study the internal factors that are holding back the development of the EEU through the prism of VC, including the low volume of intraregional trade, insufficient use of imported components for GVC (today, the final pharmaceutical product produced on their basis is sent not for export, but for domestic consumption) [2, 4].

Countries, including EEU ones, are particularly interested in non-resource industries, including the pharmaceutical. Despite the fact that the share of the EEU in the world market of pharmaceuticals is only 2.6%, but an increase in trade turnover in mutual trade of the union countries, a decrease volumes of imports from non-member countries are evidence of the sustainable development of the EEU pharmaceutical industry [6]. However, modern pharmaceuticals production is a complex process: to develop an effective pharmaceutical product, huge investment and scientific resources are needed. Nevertheless, scientists name the main priority in the development of networks as the intensification of supplies with the neighboring regions, the search for mutually beneficial cooperation with leading companies located in the same region. This enhances the relevance of the study in terms of searching for new opportunities for the formation of pharmaceutical GVCs in groups of countries, i.e. regional integration associations.

Despite the fact that the number of publications devoted to the interaction between the EEU and the GVC has recently increased [1,2,3], some of them reflect specific issues only, for example, the formation of the concept of a roadmap for the development of a single pharmaceutical market of the EEU in the context of pharmaceuticals safety, import substitution [4, 5]. But the new characteristics of the world economic environment require an in-depth industry study of this problem using GVC analysis, i.e. through the prism of international reproduction process.

Therefore, the authors made an attempt to identify the significance and features that make a specific model of participation of the EEU countries in global (regional) pharmaceutical value chains, as well as their role in the development of the national pharmaceutical industry. Since the material on the leading countries is informative, the study was carried out on Russian and Kazakh markets.

2 Theoretical aspects of the value chains functioning of within the EEU framework

Based on the analysis of academic literature we can formulate some of the most important conceptual statements.

First. GVC is an international reproduction process with some features: the process is fragmented strongly (from raw materials to the final product), its stages are located in the territories of several countries, trade of components is dominated within its frame [1, 2, 3]. According to the opinion of T. Meshkova, E. Moiseichev [3], G, Grereffy [7] it is very important also to underline that stages have different level of added value (in pharmaceutical industry the links with the maximum level of added value are R&D and the brand). Depending on the geographic coverage, they are global (GVC) or regional (they operate within a region, a regional integration association, or "when most of the added value of the AVC is created in regional blocks") [4, p. 76].

© The Authors, published by EDP Sciences. This is an open access article distributed under the terms of the Creative Commons Attribution License 4.0 (http://creativecommons.org/licenses/by/4.0/).

In the case of the essence of GVCs there is practically no disagreement among scientists, but as for using them for the national economy development there are some models [1, 4, 7]:

- One group of scientists, primarily Western, promotes a monopolar configuration of the world economy: supranational interests dominate in it, links of the chain are recommended for localization, along which the country has competitive advantages; for developing countries, this can lead to de-industrialization, and for developed countries, the placement of links with a high level of added value on their territory (R&D, brand management);
- Another group, primarily Russian, proposes a model that is aimed at developing the real sector of the national economy: national interests dominate in it, participation in GVCs (including through localization) is recommended only if the state is able to protect the national industry from dependence on external factors (on terms of cooperation).

Despite the possible risks of being in a GVC (for example, transformation the country into a "global" factory with bad ecological environment, etc.), regional value chains (if used skillfully) promise great benefits. Since the leading GVCs operate within TNCs (mainly through a joint-stock with foreign direct investment or contract form), these benefits are well known: access to new technologies, new markets, etc. But whether the GVC participants actually receive these benefits is a big question to be investigated. For example, V. V. Perskaya [1] and I. A. Manuilov [2] emphasize the importance of preserving and dominating national priorities in the field of economic integration when involved in GVCs, especially for the EEU.

The second provision concerns the essence of such concepts as "participation in the GVC", "presence in the GVC", "introducing GVC onto the territory of the EEU countries" which characterize the involvement in the global production fragmentation process. On the one hand, researchers distinguish between "forward participation" and "backward participation", while GVC analysis distinguishes the national and international parts of the value added within the chain [2, 3]. On the other hand, the form of "introducing GVCs" into the territory of countries is the localization of GVC links, i.e. location of factories, assembly shops, scientific laboratories, decision-making centers, etc. on the territory of the country. At the same time, the chain is looking for favorable conditions for the localization of its links, for example, the presence in the territory of a set of factors that form the low cost of transactions, including competitive local national structural units, infrastructure for effective business communication and logistics, etc., this is the case of creating a stimulating environment for GVCs [1, 2, 7]. And the national economy, in turn, should get its own benefits: jobs, FDI, joint development of high-tech links. The EEU countries are capable, in cooperation with each other, to form such conditions on mutually beneficial base. We want to note one more contradiction also:

- On the one hand, stages with high added value are subject to localization in order to increase the national part of the world added value (for the pharmaceutical chain, this is R&D link, which is located mainly in the developed countries);
- On the other hand, the developed countries (including USA and China), localize on their territory the "production" stage with low added value in order to avoid the risks of international trade and solving economic problems.

In other words, when choosing a form of participation in GVCs, it cannot be guided by one principle only - the placement of links with high added value on the territory of their country. The interests of TNCs, within the GVCs framework, should be balanced by the solution of the national development tasks (jobs for reduction unemployment).

Third. Some factors ensure the successful development for the countries of integrations through the tool of GVC:

© The Authors, published by EDP Sciences. This is an open access article distributed under the terms of the Creative Commons Attribution License 4.0 (http://creativecommons.org/licenses/by/4.0/).

- A high level of reproductive openness, measured by the "percentage of the sum of the country's exports and imports to the volume of its GDP" [1, p. 30], which shows the degree of national economy involvement into the world ties system;
- Significant volume of intraregional trade, because of the intensive trade in components within GVCs [2, 4];
- Effective regional trade agreements (RTA) in the context of foreign trade regulation: since within the framework of the GVC, components are traded, and within the framework of integration associations, regional trade agreements are in force, then RTA intensifies trade in component (reduction of customs duties, non-tariff barriers, etc.) [2];
- Favorable business environment: in addition to harmonizing standards, an infrastructure is needed to enhance mutual trade, namely a developed customs and transport infrastructure, the pandemic has raised the issue of digital platforms.

All of the above reveals the essence of the reproductive approach to the study of integration associations, while the literature focuses on the institutional component.

3 Research objective and methodology

In order to study and assess the interaction of regional VCs and the integrations (incl. EEU) (this is the research goal), it is necessary to use such indicators as the volume of intraregional trade as an indicator of the development of GVCs, the number of leading companies for cooperation, the level of manageability of the introduced GVCs (through a share in the authorized capital), specialization of the country, quality of the business environment. To achieve the goal, the following tasks were set: to identify the presence and features of regional pharmaceutical VC operating in the EEU countries, to compare the results of the development of the pharmaceutical industry before and in the pandemic (using the example of the leading states) based on case study and GVC-analysis. GVC-analysis means an analysis throw the prism of the stages of GVCs. The study was carried out in accordance with the following author's algorithm.

Stage 1. Analysis of the main indicators of the state and dynamics of development of the pharmaceutical market of the EEU member states in order to identify the volumes of trade in pharmaceutical products in the EEU before and in a pandemic, as well as the volumes of mutual (intraregional) trade in pharmaceuticals as an indicator of the development of GVCs, the ratio with the volume of trade with non-member countries, as well as assessing the attractiveness of the business environment for the introduction of pharmaceutical VC to the territory of the EEU countries.

Stage 2. Analysis of the pharmaceutical industry and the domestic market in each of the participating countries (the result for the leading country is also informative) to identify features and development trends.

Step 2.1. Compilation of the characteristics of the pharmaceutical industry and the pharmaceutical market (using the example of one of the leading countries) before the pandemic and during the pandemic according to such indicators as: the total volume of production and trade of pharmaceutical products, the rate of their changes, the dynamics of export-import volumes, as well as the identification of national industry characteristics and business -environments.

Step 2.2. GVC-analysis of the country's pharmaceuticals, i.e. analysis through the prism of the stages of pharmaceutical VC, including the identification of localization features of the "production" link.

Step 2.3. Identification of leading pharmaceutical companies, including from among the EEU member states acting as global or regional VCs, with the localization of their links in the selected country. The purpose of the stage is to identify the presence of leading

© The Authors, published by EDP Sciences. This is an open access article distributed under the terms of the Creative Commons Attribution License 4.0 (http://creativecommons.org/licenses/by/4.0/).

companies for cooperation, as well as the level of their manageability (through the national share in the authorized capital).

Step 2.4. Revealing the form of interaction between economic entities of leading world or regional pharmaceutical companies with the localization of stages in the selected country. The goal is to test the hypothesis of introducing pharmaceutical GVCs into the territory of the EEU countries, identify best practices, develop recommendations for the improvement of the EEU pharmaceutical industry on the basis of using the capabilities of GVCs.

4 Analysis of the interaction of the EEU countries in pharmaceutical AVC (on the example of Russia and Republic of Kazakhstan (RK): results and discussion

We obtained the following results at stage 1. *First result - in the pharmaceuticals production*: starting in 2015, when the volume of produced pharmaceutical products in the EEU countries amounted to $ 5372.3 mln., the volume of production grew steadily and pharmaceutical products worth $ 9243 million were produced in the Union in 2018. Growth in comparison with 2014 amounted to 36% and this gives the experts reason to state that the pharmaceutical industry in the EEU shows stable growth (on average by 2-4% per year). The Eurasian economic commission considers this sector a priority for the industrial cooperation. Among countries in terms of volume, Russia is the leader 90.8% of the total amount of produced products, followed by Belarus (6.5%), Kazakhstan (2.4%), Armenia (0.2%) [6, 8, 9].

Second result - in sales: the pharmaceutical market of the EEU countries in aggregate is one of the most dynamic and fastest growing world markets, the cumulative growth of which before the pandemic was about 10%. Among the member states, the leaders with an average annual increase in volumes before the pandemic over 10% are RK and Russia. The main partners of the Union countries are the USA, Germany, Switzerland, and France (for example, in RK, the volume of foreign trade with the EU countries is 65.1%) [6, 9]. Another feature is that the EEU pharma market is import-dependent: the main final products and its components are produced mainly abroad (local pharmaceuticals in the countries, except Russia, provide their own market by less than 50% (in Belarus - 7%, in RK - 2%) [8].

Despite the fact that the issues of creating a common market and integration within the EEU industry markets have been worked out since 2014, the volume of intraregional trade in pharmaceutical products is still minimal (Fig. 1).

We see in diagram that the volume of pharmaceuticals intraregional trade of the EEU member states is increasing. But its share in the total international trade with all countries is very low. We can calculate it not more than 4,5 %. The reason is huge import (final products and components) from European countries.

Third result - recent changes in the business environment, including institutional and legal, indicate the possibility of increasing the volume of mutual trade. Firstly, there are improvements in terms of the formation of a common information base, taking into account the peculiarities of the digital economy, harmonization of requirements for the quality of products and the introduction of quality management systems based on the unified international GMP standard, the adoption of an Agreement on common principles and rules of circulation of medicinal products in EEU [8]. Secondly, the activities of the EEU Association of Pharmaceutical Manufacturers are also aimed at strengthening the single pharmaceutical market of the Eurasian Economic Union, the purpose of which is not only to strength the domestic pharmaceutical industry competitiveness but intellectual

© The Authors, published by EDP Sciences. This is an open access article distributed under the terms of the Creative Commons Attribution License 4.0 (http://creativecommons.org/licenses/by/4.0/).

interaction also [10]. For this, the member states of the Association have major centers of competence, including research laboratories, their own know-how and other intellectual property in the development of pharmaceutical technologies, especially in Russia [10]. Thirdly, in the future, in 2021, it is planned to introduce product labeling to track their movement by regulatory authorities, which will make the passage of pharmaceuticals between borders more free and transparent [8]. The COVID-19 pandemic has made its own changes in developing new mechanisms, for example, the EEU has intensified the transition to distance format for pharmaceuticals production inspection, including EU framework.

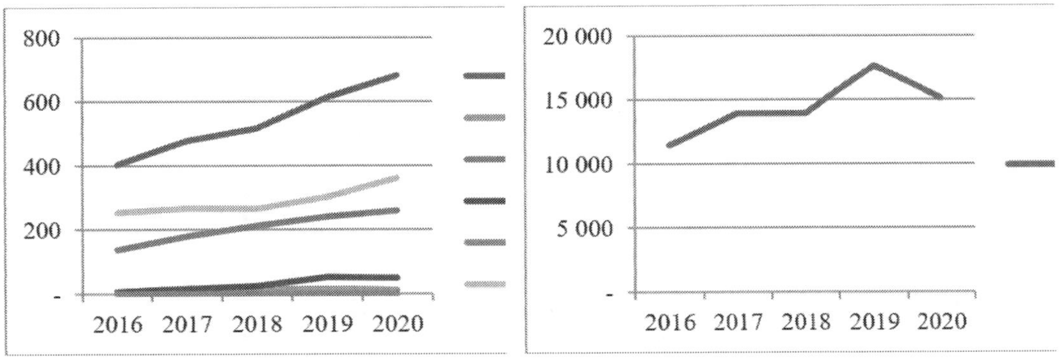

Fig. 1. Volumes of internal and external pharmaceuticals trade of the EEU member states 2016-2020, million dollars [9]

To analyze at stage 2 the pharmaceutical industry and the domestic pharmaceutical market of one of the EEU countries, the RK was selected as one of the EEU leaders. Obtained in step 2.1 results are summarized in Table 1.

Table 1. Comparative characteristics of the pharmaceutical industry and market development in Kazakhstan [6, 8, 9, 11, 12]

Indicator	Pre-pandemic level and trends	Pandemic level and trends
Pharmaceutical production volume	Volume of pharmaceutical production in the period from 2010 to 2018 increased fourfold: from 19.9 to 78.8 bln tenge or $ 228.6 mln; for the pre-pandemic (2019) volume increased by 18% and amounted to 92 bln tenge	For eight months of 2020, domestic enterprises, providing a high demand for pharmaceutical products, achieved an increase in production by 34.1 %, or 81.5 bln tenge [6]. By the end of 2020, the volume of pharmaceutical production increased by 47%
Pharmaceutical market volume	Market volume has grown steadily since 2016 at the end of 2019 amounted to $ 1.6 bln	Market size in 2020 continued to grow and amounted to $ 1.8 bln
Export volume of pharmaceutical products	Exports increased from $ 22.3 mln in 2016 to $ 34.9 mln in 2018. In 2019, the volume increased by 69% and amounted to $ 59 mln, $ 19 mln in four months 2019	Volume of exports after a 15% decrease in two months and in general for four months of 2020 amounted to $ 20.4 mln
Import volume of pharmaceutical products	Imports increased from $ 968 mln in 2016 to $ 1,178 mln in 2018 and $ 1,136 mln in 2019. In four months 2019 amounted to 395 mln doll	Imports for four months of 2020 amounted to $ 462.3 mln, which exceeds the same indicator for the corresponding period of the previous year by 17%.

Analysis of the table data demonstrates growth in all indicators. It is certainly driven by the demand for pharmaceuticals during the pandemic. However, some national peculiarities are noteworthy. Firstly, it is the high import dependence of the RK pharmaceutical market.

© The Authors, published by EDP Sciences. This is an open access article distributed under the terms of the Creative Commons Attribution License 4.0 (http://creativecommons.org/licenses/by/4.0/).

It is confirmed not only by large volumes of imports, but also by a small number of national manufacturers of leading companies [5]. This level of import dependence of the pharmaceutical market is also typical for other EEU countries (for example, 64% in Belarus, 97% in Kyrgyzstan), local pharmaceuticals provide their own market for less than half [5, 8]. However, in a pandemic year, the growth of production volumes in RK, according to the Ministry of Industry and Infrastructure Development, was provided by national enterprises [12].

The next step 2.2 through GVC-analysis of the national pharmaceutical industry will allow to understand how RK managed to create its own pharmaceutical production in such a short time (in 2014 there was not a single national enterprise-leader in RK) [5]. Despite the fact that the "production" stage of GVC has the minimum added value and is not recommended for localization, nevertheless, RK, realizing the importance of the domestic industry, seeks to progress the process of producing its own pharmaceuticals, bringing their level in government orders above 50% [5].

But having left this decision from import dependence in terms of ready-made pharmaceuticals, the national enterprises of the EEU countries did not leave it in terms of pharmaceutical substances for pharmaceuticals, since purchase them abroad [8]. For example, in RK, leading factories with modern technologies have been built, but they are only engaged in "assembly": imported substances are supplied to the enterprises, where pharmaceuticals, mainly generic are then bottled [8]. Analysis of the structure of imports for the 1st quarter of the pandemic year 2020 confirms this feature: the import of substances in RK amounted to 52.9%. So, it is necessary to create the raw material base in EEU.

Another feature is that production in RK is represented by two modifications: filling imported syringes with imported suspension and filling local syringes with it. For RK, a significant increase in the volume of domestic syringes production should be noted: in the first 5 months of the pandemic 2020, it increased more than 2.5 times and amounted to 64.6 mln pieces compared to the same period in 2019 with a volume of 23 mln units [13]. Organization of this production we consider as a transition to a new stage in pharmaceutical GVC, and the pandemic has accelerated it.

The analysis of economic entities at step 2.3 showed that among the Kazakhstan enterprises that made the largest contribution to the growth of production volumes in the pandemic 2020, experts name the following [12, 13, 14]:

- JV "Abdi İbrahim Global Pharm" LLP: produces more than 70 pharmaceuticals, currently sells its products through distribution networks and exports to countries such as Azerbaijan, Uzbekistan, Tajikistan, Kyrgyzstan, and in the near future it is planned to expand the geography of exports to Russia, Belarus, Turkmenistan, Moldova and Armenia (the company's goal is to obtain a leading position in the export of pharmaceuticals among the CIS countries);
- JSC "Khimfarm": the volume of manufactured products and the export has doubled over the past 2 years, the company's pharmaceutical products are in demand in the markets of Central Asia and the EEU, are exported to Kyrgyzstan, Tajikistan, Turkmenistan, Mongolia and Russia, it is planned to enter the Uzbekistan market, but the central task is to enter the pharmaceutical market of the European Union (by 2023, it is planned to increase the volume of exports to 16 billion tenge or 2.79 billion rubles);
- "NOBEL Almaty Pharmaceutical Factory": the company is one of the three leading companies in the Kazakhstan market, in the last 2-3 years it has demonstrated a good increase in production and sales; the share of exported products from the total volume of products produced at this enterprise is 25%

These companies operate on the basis of imported components. During the pandemic, the situation was as follows [14]: many suppliers of pharmaceutical substances stopped or

© The Authors, published by EDP Sciences. This is an open access article distributed under the terms of the Creative Commons Attribution License 4.0 (http://creativecommons.org/licenses/by/4.0/).

restricted their work, and prices for the most popular raw materials increased; the world's main suppliers of some substances even imposed export embargoes. The pandemic has demonstrated that the production of pharmaceutical substances of some drugs must be established within the country, because countries can stop production, or close borders. Therefore, Kazakh enterprises plan to establish the production of several pharmaceutical substances on the territory of Kazakhstan.

Experts associate the aforementioned fourfold increase in production volumes in the pharmaceutical industry of the Kazakhstan in the period from 2010 to 2018 with attracting foreign investment: such large producers as Polpharma (has a share in the authorized capital of JSC "Khimfarm"), Nobel (has a share in the authorized capital of Nobel AFF), Pharmstandard, Kelun and a number of others. Currently, the issue of attracting investments from TNCs among the world's leading pharmaceutical producers Big Pharma - Top 50: Pfizer, Boston Biopharma, Avantgen and Servier is being considered [8]. The analysis demonstrates that almost all of these companies operate as TNCs with a registration outside the EEU: mainly the United States (Pfizer, Boston Biopharma, Avantgen), Turkey (Nobel, Abdi Ibrahim), Poland (Polpharma), China (Kelun). Among EEU countries, only Russia (Pharmstandard) is presented, and if we turn to the Association of Pharmaceutical Manufacturers of the Eurasian Economic Union, in addition to Pharmstandard, there is a Russian full-cycle producer "Biocad" in Kazakhstan (for comparison: In Belarus, 3 out of 5 largest companies have European "citizenship") [5, 10].

The form of interaction between these companies is the localization of the GVC links "secondary packaging" of the finished substance from Russia (investments from Biocad) [15], "production" in the form of a spill of the incoming substance into syringes (investments from Pharmstandard in "Karaganda pharmaceutical complex" [16]), and with the subsequent consumption of pharmaceuticals in Kazakhstan, and not export to non-member countries, which provides for participation in the GVC.

To enhance the development of the pharmaceutical industry (through the prism of GVCs) Kazakhstan plans [8]:

- On the "raw materials" link - to reduce the rates of import customs duties for raw materials and components; to develop measures of state support in terms of reimbursing the costs of purchasing expensive imported raw materials,
- For the R&D link - to increase the amount of funding for scientific research; to develop a separate regulatory framework for the development of Kazakhstan laboratories that meet GLP standards, as well as measures of state support in terms of reimbursing the costs of conducting clinical trials of pharmaceuticals.

In our opinion, separately, none of the companies of the EEU countries, except for Russia, will be able to fundamentally finance the formation of pharmaceutical AVC, since their development is associated with a huge volume of R&D and corresponding investment resources. Therefore, in addition to the already existing AVC from Russian Pharmstandard, a potential phytochain from Kazakhstan's Phytokhimia, we recommend the EEU countries to design the formation of a strategic regional TNC (consortium) in order to concentrate investment resources on R&D stage on the basis of cooperation and specialization by links of all EEU member states.

5 Conclusions

The following features of the EEU countries industry, trade and interaction in pharmaceutical GVCs are identified:

- Despite the pandemic period, the pharmaceutical industry of all EEU countries shows growth, but the volume of intraregional trade in pharmaceuticals compared to the

© The Authors, published by EDP Sciences. This is an open access article distributed under the terms of the Creative Commons Attribution License 4.0 (http://creativecommons.org/licenses/by/4.0/).

total volume of the country's foreign trade is insignificant, which indicates the weak development of the regional VC;

- Russia is the leader among the EEU countries (it accounts for 90.8% of the total EEU pharmaceutics volume) than Kazakhstan, so they have to accumulate investment and initiate together the R&D stage generation in regional GVC and than as a final to create the common strategic regional corporation with all EEU countries participation ;
- The measures taken in the EEU have improved and are improving the business environment, which allows us to speak about the intensification of interaction and cooperation between the countries members of this integration association.

The Kazakhstan case as a leader of EEU revealed features of its participation in the pharmaceutical GVCs:

- this country interacts with regional VCs in such forms as a share in the authorized capital, localization of "production" and "packaging" stage (Russian companies from the EEU countries are present only), but this cannot be considered as a full-fledged participation in the GVC, because, firstly, there is no export to third countries, secondly, the benefits promised by the participation in GVC in the form of access to new markets are not realized;
- all pharmaceutical products produced from the imported component in the frame of value chains are consumed within the country, but in the case of organizing their export to third countries, participation in the GVCs can be full-fledged;
- RK managed to develop domestic pharmaceutical production by attracting foreign investments in the framework of Turkish ("NOBEL", "AbdiIbrahim"), Polish ("Polpharm"), Chinese ("Kelun"), Russian ("Pharmstandart") GVC;
- despite the low level of added value in the "production" stage of GVCs a new modern factories with a large number of jobs were built; despite the high share of foreign investors in the authorized capital of Kazakh companies, the country managed to find a balance between national and supranational interests due to a favorable business environment.

Acknowledgements

The research has been prepared with the support of the Russian Foundation for Basic Research, the Project № 20-014-00011 "Comparative analysis of the integration of national companies in the GVC: the case of Russia, Belarus, Kazakhstan", within the framework of the state task of the Institute of Economics, Ural Branch of the RAS (2021-2023)

References

1. V. V. Perskaya, Economic strategies, **5** (2015)
2. O. V. Biryukova, I. A. Manuilov, ENSR, **1(76)** (2017)
3. T. Meshkova, E. Moiseichev, Foresight and STI Governance, **1** (2016)
4. Y. M. Kukushkina, International trade and trade policy, **4(8)** (2016)
5. P. A. Nalimov, D. Y. Rudenko, Bulletin of the Tyumen State University. Socio-economic and legal research, **1** (2018)
6. *EAEU pharmaceutical industry shows stable growth* (2019) https://news.rambler.ru/
7. G. Gereffi, K. Fernandez-Stark, Global Value Chain Analysis: A PRIMER (2016) https://www.researchgate.net/

© The Authors, published by EDP Sciences. This is an open access article distributed under the terms of the Creative Commons Attribution License 4.0 (http://creativecommons.org/licenses/by/4.0/).

8. A. Ivanilova, The pandemic played into the hands of the pharmaceutical market of Kazakhstan (2020) https://mk-kz.kz/
9. EAEU statistics (2020) http://www.eurasiancommission.org/
10. Association of Pharmaceutical Manufacturers of the EAEU (2020) https://www.gosrf.ru/
11. Statistical Yearbook of Kazakhstan. Key indicators by type of economic activity. Manufacturing of basic pharmaceutical products. Committee on Statistics of the Republic of Kazakhstan (2021) https://stat.gov.kz
12. At the end of 2020, the volume of production in the pharmaceutical industry of the Republic of Kazakhstan increased by 47% (2021) https://www.trend.az/
13. The volume of production of pharmaceutical products in the Republic of Kazakhstan in January-May 2020 increased by 21.2% (2021) https://pharm.reviews/
14. Kazakhstan pharmaceutical industry and COVID-19. Impact and forecasts for the future (2020) https://yandex.ru/
15. Biocad (2021) https://biocadglobal.com/
16. Karaganda pharmaceutical complex (2020) http://kphk.kz/

© The Authors, published by EDP Sciences. This is an open access article distributed under the terms of the Creative Commons Attribution License 4.0 (http://creativecommons.org/licenses/by/4.0/).

State intervention in the domestic market restructuring: an example of the Russian sunflower market

Sergey U. Chernikov, and *Alexander* Zobov*

Peoples' Friendship University of Russia (RUDN University), Miklukho-Maklaya Street, 6, 117198 Moscow, Russian Federation

Abstract. In recent years, the Russian market of sunflower oil has shown steady growth, while the key trend is the increase and deepening of exports. In 2020, the market showed an increase in exports to China and India, reaching a value of 35% of the total export volume. At the moment, Russia is considering a formula for an export duty on sunflower oil, as it struggles with food price inflation. It is assumed that the export duty on sunflower oil will be structured similarly to that used in the wheat market. The government brokered the deal to mitigate the impact of high international prices on a key domestic product, and was forced to lock in retail sunflower oil prices after domestic sunflower and sunflower oil prices began to rise following last summer's low sunflower harvest in the Black Sea region. It is expected that the introduction of a new export duty will be able to stabilize domestic food inflation.

1 Introduction

The global economic situation is gradually deteriorating, and global financial institutions say that the recovery will be long and uneven. Governments in the world's leading economies are pouring unprecedented amounts of liquidity into national banking and fiscal systems, but the economic impact of these measures is still negligible at best. It is obvious that the global depression will last for several years and will cause many significant changes in the economic, social, and possibly political landscape of the world.

In the current conditions of the turbulence of the world economy and the reformatting of traditional trade relations, the trends towards import substitution and self-sufficiency in large countries are becoming more pronounced. With increasing inflationary trends in world markets, States seek to protect domestic markets and jobs by closing the production circuits of their most important products within controlled economic territories - preferably their own. These transformational aspects of modern globalization have primarily affected one of the most sensitive areas of the international economy - food production. The world population continues to grow by about 80 million people a year, which directly affects the level of food security of countries. To meet the growing demand for food, many countries have focused their efforts on developing new technologies for high-tech agricultural chains,

* Corresponding author: chernikov-syu@rudn.ru

© The Authors, published by EDP Sciences. This is an open access article distributed under the terms of the Creative Commons Attribution License 4.0 (http://creativecommons.org/licenses/by/4.0/).

which can significantly increase labor productivity. Typically, this is the adaptation of the latest and advanced technologies of industry 4.0 - fast-growing growth methods, growing crops in precise conditions of temperature and humidity, promoting soil fertility and balanced nutrition of crops, managing logistics chains and transport losses of crops, value added and post-harvest management, etc.

It is assumed that the application of these approaches leads to a significant transformation of the traditional format of the work of agriculture and the food industry. New agro-industrial enterprises are much less dependent on the use of water, fertilizers and pesticides in whole fields. Instead, farmers will use minimal quantities or even eliminate them completely from the supply chain. Agricultural technology startups in the world as a whole are showing steady growth, and entrepreneurs and investors are showing growing interest in this sector. Also, in the current conditions, the role of national producers of agricultural products is increasing, as opposed to large multinational corporations.

However, large monetary and non-monetary investments are necessary to implement all these technologies, and governments today seem to prioritize the exceptionally problematic issue of securing the agricultural stability at the home market. Russia has seriously turned to the ideas of self-sufficiency and import substitution in 2014, after the Ukraine crisis. The efforts led to particularly good results in agriculture and military-focused industries. In many agricultural markets Russia switched from importing to exporting activitis. However, it seems that the current global inflation issue in world economy creates a requirement in further prioritising the internal comsumption, rather that mere partial import subsititution.

This hypothesis made up the primary goal of the current article, which can be formulated as evaluating the possible consequences of further state intervention to restructuring the sunflower oil domestic market and industry. This particular market segment is an exceptionally good example due to its obvious export orientation and strong positions in the global agriculture industry.

2 Materials and Methods

The paper was composed on the content-analysis basis of a range of statistical resources and publications relevant to the research topic. The statistical data derived from both specialized national and international commodity databases and information portals, while the discussion points concentrate on topical journals.

3 Results and Discussion

3.1 Global sunflower market outlook

In the world market of sunflower oil, Ukraine and Russia are the permanent leaders, they hold the leading positions in international supplies of sunflower oil. In 2019, the share of exports from Russia and Ukraine covered almost 80% of the market, despite the unfavorable price situation and low trade activity. Figure 1 shows data on the world's main sunflower oil exporting countries in 2019 and changes compared to 2018.

© The Authors, published by EDP Sciences. This is an open access article distributed under the terms of the Creative Commons Attribution License 4.0 (http://creativecommons.org/licenses/by/4.0/).

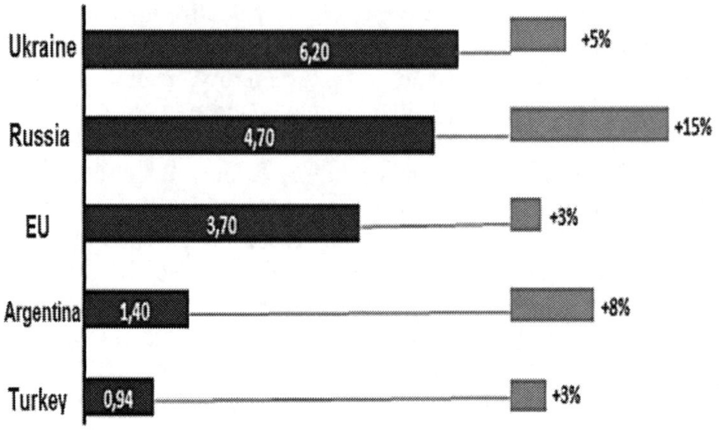

Fig. 1. Majour exporters of sunflower oil in 2019 ([1]).

The top-5 sunflower producing countries according to the FAO report include Ukraine, Russia, Argentina, Romania and China (see Figure 2).

The Asia-Pacific region is the fastest growing sunflower market. This region dominates the sunflower market in terms of consumption after Europe, occupying about 30% of the global market in 2019. India and China occupy the leading positions in exports at the country level, this is primarily due to the high marginality of exports. The market in this region is determined by consumer growth, and over the past 15 years, consumption has approximately increased fivefold.

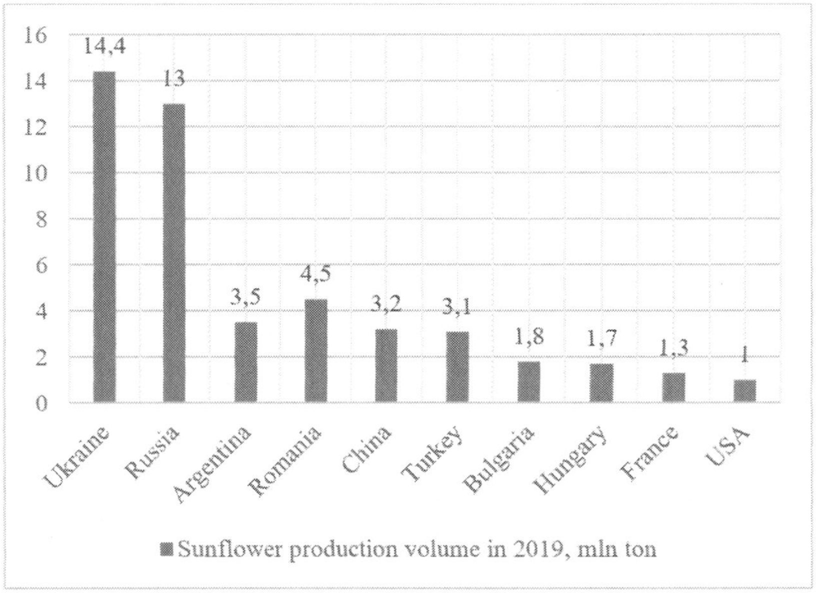

Fig. 2. Top-10 countries in terms of sunflower production in the world in 2019 [2].

Constantly growing stocks of sunflower oil in the context of increasing global consumption contributed to the expansion of supplies to the foreign market [3]. Figure 3 shows the world leaders in sunflower acreage in 2019. At the same time, the total area of sunflower crops in 2019 amounted to 26 million hectares and 69% of all crops fall on the five world leaders.

© The Authors, published by EDP Sciences. This is an open access article distributed under the terms of the Creative Commons Attribution License 4.0 (http://creativecommons.org/licenses/by/4.0/).

It should be noted that, despite quite significant differences in the export capacities of the main producers, the pace of global competition remains quite high and continues to grow every year.

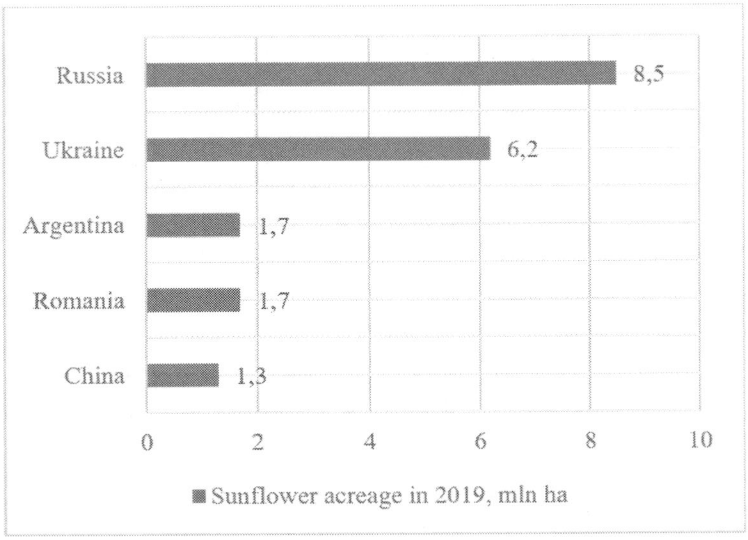

Fig. 3. Top 5 countries by sunflower acreage in 2019 [2].

Russian sunflower oil suppliers compete with Ukraine for demand from major sunflower importers in China [4]. Figure 13 shows the largest importing countries of Russian sunflower oil. Thus, the main consumers of Russian sunflower oil, according to the presented schedule, are Iran, Turkey, China and Egypt [5].

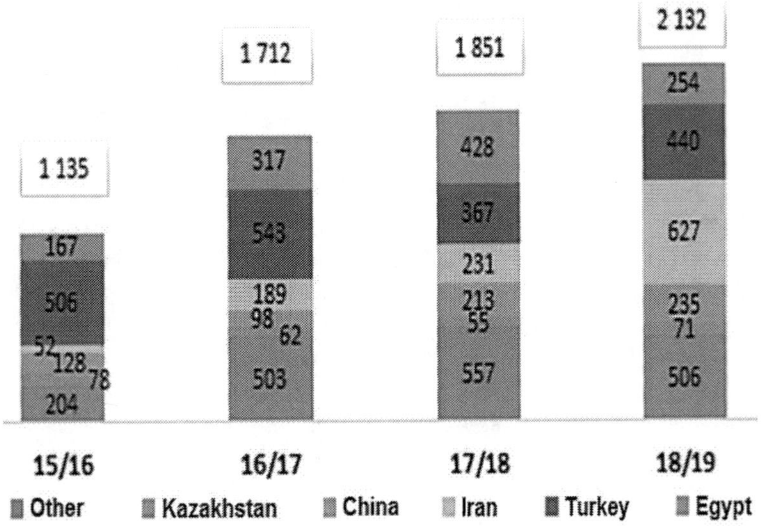

Fig. 4. Importing countries of Russian sunflower oil, thousand tons [6].

It is important to note the interrupted long-term trend of increasing sunflower production in the world [7]. The maximum volumes of oilseed harvesting stimulated the companies to increase their capacity and load the existing lines. However, according to the results of the 2020/2021 season, there was a sharp decline in production, caused by an unexpected crop failure in the leading countries of sunflower production.

© The Authors, published by EDP Sciences. This is an open access article distributed under the terms of the Creative Commons Attribution License 4.0 (http://creativecommons.org/licenses/by/4.0/).

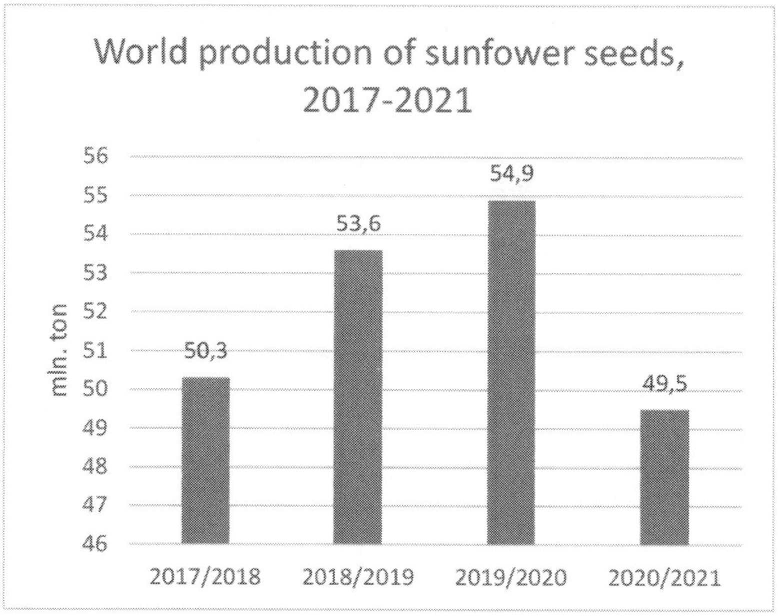

Source: composed by author from Worldwide oilseed production by type [8].

Fig. 5. Forecast of sunflower production in the world in 2017-2021.

Considering the situation on the global sunflower market, it is important to note the following trends that will continue over the next five years [9, 10]:

— stable growth in sunflower production and consumption, by an average of 4% per year;
— increase in the share of industrial sunflower consumption;
— market instability due to price dynamics;
— reduced yield;
— сокращение reduced acreage and reduced harvest in 2020 due to the pandemic;
— sunflower seed export growth;
— рост increased production of sunflower cake and meal;
— difficulties with logistics;
— rising prices for sunflower seeds and their processed products, caused by consistently high global demand, rising global prices for vegetable oils, and a reduction in the volume of sunflower seed collections.

The increase in world prices that began in the summer of 2020 was caused by unexpectedly very low production levels in Ukraine, Russia and Argentina, following the loss of a crop estimated at about 6 million tons (see Figure 6).

© The Authors, published by EDP Sciences. This is an open access article distributed under the terms of the Creative Commons Attribution License 4.0 (http://creativecommons.org/licenses/by/4.0/).

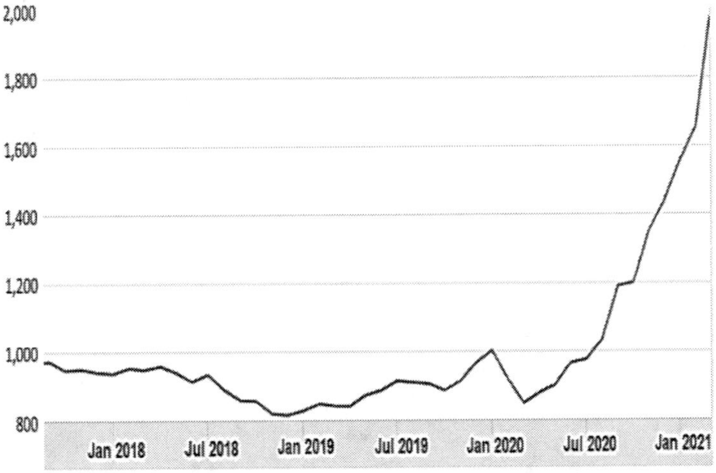

Source: fred.stlouisfed.org [11].

Fig. 6. Average price of sunflower oil on the world market per ton, USD, 2018-2021.

3.2 Controversy of Russian sunflower market

In Russia, the increase in sunflower prices, in addition to these factors, is also caused by the devaluation of the national currency [12]. As mentioned above, Russia is the world's second largest producer and exporter of sunflower oil after Ukraine. With the increase in refining capacity in Russia, the country is also increasing its share in the global market. Russian sunflower oil production is growing steadily, while domestic consumption will remain virtually unchanged (see Figure 7).

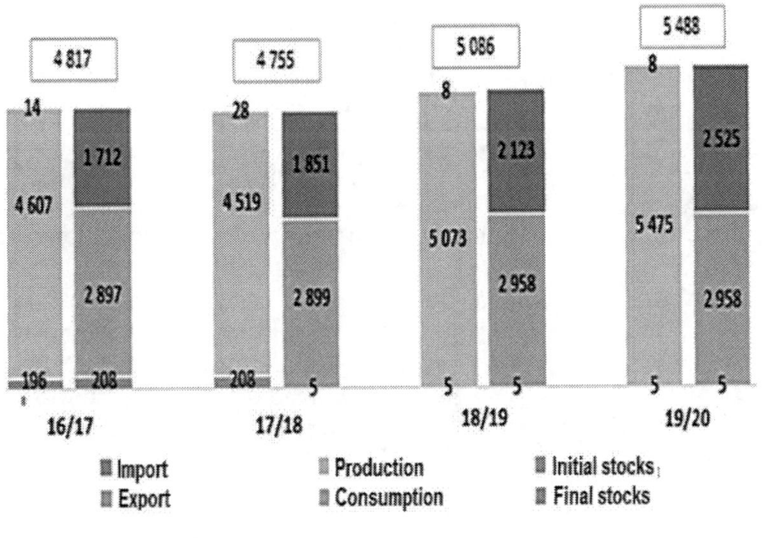

Fig. 7. Supply and demand balance of sunflower oil in Russia, thousand tons [6].

According to the FAO "Food Outlook" report, it was expected that by the end of the 2020 season, the supply of sunflower oil from Russia would reach a new level due to competitive prices and high production volume. Against this background, an increase in the supply of sunflower meal and cake from Russia was also expected. However, the forecast for the 2020/2021 season showed a drop in sunflower seed production in the Russian

© The Authors, published by EDP Sciences. This is an open access article distributed under the terms of the Creative Commons Attribution License 4.0 (http://creativecommons.org/licenses/by/4.0/).

Federation against the background of a return to the average yield level. As of 2021, Russia is the world's second-largest exporter of sunflower oil after Ukraine. From Russia, deliveries are made to more than 40 countries around the world, and this number is growing from year to year. The increase in the forecast of world production and the decline in consumer demand were the main reasons for the development of large-scale rolling stocks of sunflower oil, which, in turn, affected prices and slowed down the pace of trading in the foreign market [14]. The consumption of sunflower oil in Russia is quite stable, for 2020 it is about 80% of the total consumption of vegetable oil, while 58% is accounted for by unrefined sunflower oil . Mainly, sunflower oil is used in cooking. Over the past five years, the volume of sunflower acreage in Russia has grown by more than 1 million hectares (see Figure 8).

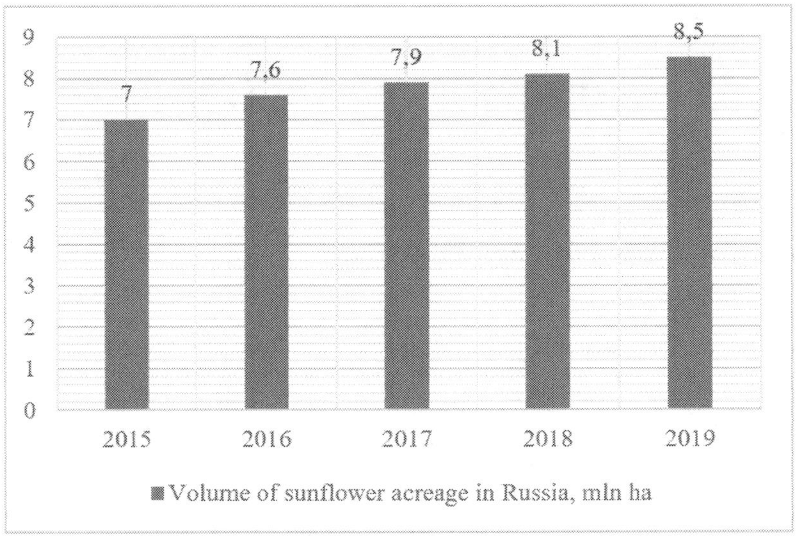

Source: composed by author from ab-centre.ru [15].

Fig. 8. The volume of sunflower acreage in Russia in 2015-2019, million hectares.

The main share of imports for sunflower production in Russia are the shipments of sunflower seeds (see Figure 9). Fertilizer and equipment used, to the contrary, are predominantly of domestic origin. This is an important notice, showing that unlike other agricultural sectors, the sulflower oil prices dependace on exchange rate is moderate at best.

Deliveries of sunflower seeds for sowing in Russia in 2019 amounted to 29,321. 2 tons, which is 5.6% (1,555. 0 tons) more than in 2018. For five years, the volume increased by 33.9% (by 7,416. 6 tons), for ten years - by 229.5% (by 20,422.5 tons), by 2001 - by 2,530. 1% (by 28,206. 4 tons). In January-June 2020, the volume of imports of sunflower seeds for sowing in Russia was at the level of 20,301. 2 tons, which is 10.1% (2,290. 3 tons) less than in the same period of 2019.

© The Authors, published by EDP Sciences. This is an open access article distributed under the terms of the Creative Commons Attribution License 4.0 (http://creativecommons.org/licenses/by/4.0/).

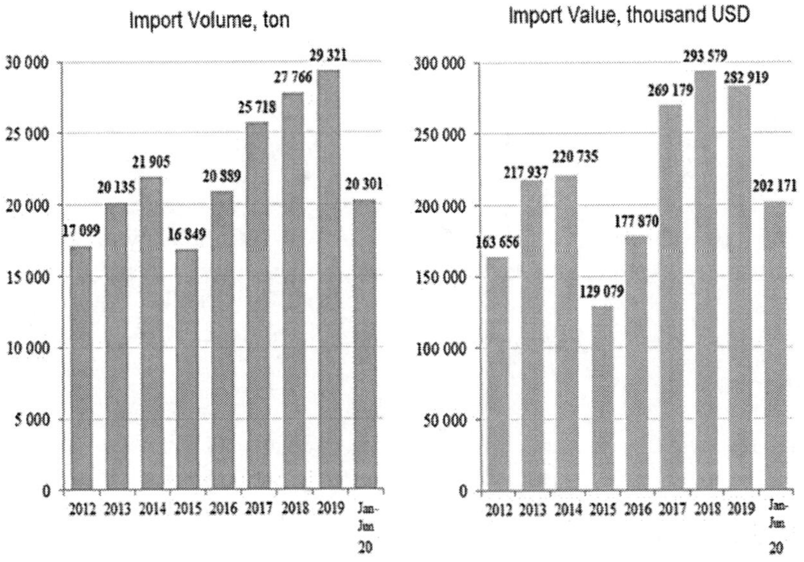

Fig. 9. The volume and cost of import of sunflower seeds for sowing in Russia in 2019 [16].

The largest market segment of sunflower is the production of sunflower oil. Oil, in turn, is used in the food industry as an ingredient for the production of semi-finished products or final food products, and is also bottled as a final product. About 90% of sunflower oil is used in cooking, less than 10% goes to the production of margarine. It is important to notice that the consumption of sunflower oil (both refined and non-refined) is deeply embedded into daily household routines. The consumption of soya or palm oils for the purpose of preparing meals is extremely low, thus making the market in question a crucial pillar of social stability and economy [17].

On the average, Russians consume approximately 30 kg of sunflower oil per capita, which is higher than almost anywhere else in the world. At the same time, the preceding analysis shows a clear export pattern in the market construction. While the home market provides the "bulk" sales to the main producers, the "profit" part of commerce lays beyond the national border. Operating in the market economy, over several decades the sunflower oil producers were balancing between the higher profits and domestic market share, and comparing the opportunity costs of selling internally and externally. The increase in the world sunflower oil prices with simultaneous drop in domestic target audience revenues provoked a rise in the product cost on the shelves of Russian stores. Since Russia exports almost 40% of the sunflower oil produced, companies sought to keep the pre-crisis margin and even saw the opportunity to earn extra profits from the situation.

In December 2020, the Ministry of Industry and Trade, the Ministry of Agriculture, the largest Russian retailers and food producers signed agreements to stabilize prices for sunflower oil in order to contain inflation after a sharp increase in prices caused by a low harvest in the Black Sea region. According to the agreements, the selling prices at the manufacturing plants should be no higher than 36 rubles for sugar, and no higher than 95 rubles for sunflower oil. In retail, prices should be no more than 46 rubles for sugar, and no more than 110 rubles for butter. On March 30, the Russian government extended the agreements on the stabilization of sunflower oil prices until October 1, 2021 [18].

The Russian government also offered a subsidy of 10 rubles per liter of sunflower oil to maintain a stable supply of sunflower oil to the domestic market. The subsidy is expected to last until September 2021 to ensure that the industry can adhere to the fixed price agreements for retail sunflower oil that the government introduced earlier. The quota will

© The Authors, published by EDP Sciences. This is an open access article distributed under the terms of the Creative Commons Attribution License 4.0 (http://creativecommons.org/licenses/by/4.0/).

be distributed among all market participants who supply bottled sunflower oil to the domestic market. According to experts, the costs of the industry can reach up to 19 billion rubles (256 million dollars). Currently, the Ministry of Agriculture of Russia is implementing measures aimed at improving the competitiveness of the industry. Among them are subsidies that stimulate the production of oilseeds, and concessional lending at a rate of 1% to 5% [19]

Finally, after long preparation and research, the government has announced Russia's new export duty on sunflower of 50%, up from 30% currently, with a minimum of $320/tonne. A formula-based export tax on sunflower oil will run for one year starting 1st of Sept 2021.

4 Conclusions

The measures taken by the government drew strong criticism from industry representatives. The latter stated that in the conditions of the existing "export disparity" in the market, that is, with the divergence of prices in the domestic and foreign markets, the lost income of sunflower oil producers from sales in the domestic market is 30-40 thousand rubles per 1 ton. The freezing of prices for sunflower oil has led to large losses for producers, which are estimated at 15 billion rubles. At the same time, by the end of 2021, manufacturers may lose more than 50 billion rubles.

Critics also stressed that the strengthening of state regulation in the domestic market and the introduction of export duties will have negative consequences in the long term, since none of the major processors will build serious plans for the development of exports, since the situation in the domestic market can cancel them out. And this, in turn, significantly limits the planning horizon of producers: no one will invest in processing in the target markets, no one will seriously invest in quality.

Despite the fact that this feedback from the expert community should be taken carefully, the sequence of decisive actions of the government in relation to one of the most profitable and export-oriented branches of agriculture in Russia allows us to draw important conclusions. First, these actions showed that the government's priority is to meet the needs of the domestic market, and not the profitability of producers or export revenues. The needs of globalization and participation in the world economy are clearly declared secondary to the problems of the national economy, which is an extremely serious change from the previous 30 years of the evolution of the Russian economy and the "orientation towards foreign investment". Secondly, the nature of the measures introduced is obviously more favorable to large companies that have significant lobbying and lending opportunities to meet these new conditions. This is a sign of the rapid consolidation of the industry and the displacement of small and even medium-sized processing companies from it, which generally indicates a further movement towards the formation of state capitalism in the Chinese format. Given the importance of the industry for domestic consumption, with the next jumps in world prices for sunflower oil, it is quite possible to expect the creation of another state corporation, following the example of Rostec, Rosneft, etc., responsible for the umbrella regulation of this strategically important industry for the country.

References

1. Ukraine and Russia on the global market of sunflower oil — strong leadership positions, https://www.apk-inform.com
2. Food Outlook 2019, http://www.fao.org
3. N. Shevtsova, A. Kogteva, Scientific result Economic research, **6**, 2 (2020)

© The Authors, published by EDP Sciences. This is an open access article distributed under the terms of the Creative Commons Attribution License 4.0 (http://creativecommons.org/licenses/by/4.0/).

4. A. Baryshpolets, S. Devadoss, European Review Of Agricultural Economics, jbab017 (2021)
5. A. Koocheki, M. Nassiri, S. Gliessman, A. Zarea, J. of Sustainable Agriculture, **32(1)**, 95 (2008)
6. Russia gradually increases its presence on the global market of sunflower oil, , https://www.apk-inform.com
7. F. Santeramo, L. Di Gioia, E. Lamonaca, Land Use Policy, **101**, 105102 (2020)
8. Worldwide oilseed production by type, https://www.statista.com
9. J. Cui, J. Martin, Energy Economics, **65**, 148 (2017)
10. A. T. Rahoveanu, M. Rahoveanu, R. Ion, Land Use Policy, **71**, 261 (2018)
11. Economic data, Global price of Sunflower Oil, https://fred.stlouisfed.org
12. A. Cheptea, C. Gaigné, European Review of Agricultural Economics, **2**, 684 (2020)
13. Russia gradually increases its presence on the global market of sunflower oil, https://www.apk-inform.com
14. B. Xiong, F.Yi, Y. Li, China Agricultural Economic Review, **2**, 303 (2017)
15. Sown areas of sunflower in Russia, Results of 2019, https://ab-centre.ru
16. Analysis of sunflower seed import for sowing in Russia, https://www.oilworld.ru
17. K. Bentsen, P. E. Pedersen, British Food J., **3**, 1083 (2021)
18. D. Ternovsky, N. Shagaida, Economic Development of Russia, **1**, 24 (2021)
19. M. Mironova, Modern economy: problems, solutions, prospects: collection of scientific papers, 347 (2021)

© The Authors, published by EDP Sciences. This is an open access article distributed under the terms of the Creative Commons Attribution License 4.0 (http://creativecommons.org/licenses/by/4.0/).

Technology transfer of the military-industrial complex as a factor in increasing the science intensity of the civilian industry

Daria Akimkina[1], *Evgenii* Khrustalev[1], *Nina* Baranova[2,*], and *Daria* Loginova[1,2]

[1]Laboratory of simulation modeling of the interaction of economic objects, Central Economics and Mathematics Institute RAS (CEMI RAS), 47, Nakhimovsky prospect, 117418 Moscow, Russia
[2]Faculty of Economics, Department of Economic and Mathematical modeling, Peoples' Friendship University of Russia (RUDN University), 6, Miklukho-Maklaya Str, 117198 Moscow, Russia

Abstract. The high innovation potential of the Russian military-industrial complex can become a source of technology for the civilian industry. In the context of the diversification of the military-industrial complex, the development of an effective technology transfer can become one of the key elements in building a competitive Russian economy. The article contains a diagram of the mechanism for the innovative technology development and transfer between the military and civilian sectors. On the basis of this method, it is possible to build effective schemes for technology transfer and exchange and joint work of the military and civilian industries. The emergence of a technology broker, as an intermediary between the military-industrial and civilian complexes, will accelerate the development of active information and scientific exchange between these sectors, will make it possible to bring technologies to its implementation faster, and industries become more knowledge-intensive. Integration of the military and civilian sectors will help to reduce the technological gap within the country, lagging behind the world level.

1 Introduction

The rapidly growing globalization directly affects all areas of activity of any country: economic, political, social, cultural. On the one hand, close international cooperation leads to an increase in the communication speed and method, a rise in the number of partnership approaches, the international capital flow, migration, etc. On the other hand, this aggravates a number of problems: economic instability, the gap between wealth and poverty, technological gap, national safety, etc.

In the pre-crisis conditions, Russia pinned great hopes on such a form of cooperation as technology borrowing in order to reduce the technological gap. In the context of anti-Russian sanctions and access restriction to foreign technologies, an alternative source of advanced technologies must be found [1].

* Corresponding author: baranova_nm@pfur.ru

© The Authors, published by EDP Sciences. This is an open access article distributed under the terms of the Creative Commons Attribution License 4.0 (http://creativecommons.org/licenses/by/4.0/).

The R&D development and introduction requires high expenditures, long lapse of time and availability of highly qualified staff. This process is always associated with risks of various sorts: economic, scientific and technical, engineering and production, that not all of enterprises are ready and able to bear.

The military-industrial complex (MIC) can become a launch pad for the breakthrough technologies development in both the military-industrial and civilian industries. It is this complex that has recently been well funded by the state, is engaged in research and development and has highly qualified staff. Therefore, a study of the process of the advanced technologies transfer from the military-industrial complex to the civilian must be conducted and vice versa, to study its shortcomings.

2 Materials and Methods

The paper studied the scientific works of Russian and foreign scientists, devoted to various aspects of the process of technologies transfer and renewal. In Russia, such studies were carried out by the scientists Khrustalev E. Yu., Batkovsky A.M., Semenova E.G., Fomina A.V., Khrustalev O.E. [2-3] and others. The researches of the scientists Batkovsky A.M. [3], Slavyanov A.S., Khrustalev E. Yu. [4] Anischenko T.V. [5] and others are devoted to the military-industrial complex diversification and the possibility of using military technologies. The proceedings of Ignatova A.I. and others [6] are devoted to the study of the technology transfer system in Russia. The proceedings of such scientists as Zharova E.N., Gribovsky A.V. [7] and others focused on the assessment of the technology transfer efficiency, demand assessment and search for market outlets.

The foreign scientists have also investigated various components of the technology transfer process from the military to the civilian area – Bukkvoll T., Malmlöfb T., Makienko K. [8], etc. considered the conditions that would facilitate this process. Van Norman, Eisenkot R. [9], etc. analyzed the technology transfer process from a legal standpoint. Schmid J. [10] et al. examined the impact of government funding on R&D. The researches of Anischenko T.V., Mendi A.F., Nacar M.A., Solmaz M. [11], Bozeman B. [12] and others are devoted to the issues of functioning and building a business model within the framework of technology transfer.

The technology transfer is discussed in the articles by Li Yang, Hu Wang [13], Yuan Z. [14], Bakulina A.A., Sokolova E.S. [15], etc. The integration of the military and civil sectors of China is studied by Ng J. C. Y., Song K. K. W., Tan Q [16].

The issue of the impact of technology transfer on the science intensity growth and the technological level of civilian industry is not sufficiently covered in the economics literature, despite a large number of studies on the technology transfer from the military-industrial complex to the civilian. The synthesis method and comparative analysis are used to study this issue. The comparative analysis method is needed to analyze the experience of technology transfer from the MIC for the development of civilian products in Russia and abroad. The synthesis method delivers to construct a diagram of the development and technology transfer stages and highlight the main directions for further improving this process.

3 Results

The MIC is a tech-intensive sector that can be a scientific and technological base for innovation-driven growth. Many technologies came to the civilian sector from the military sphere during the period of hostilities [17]. The military technologies transfer contributed to

© The Authors, published by EDP Sciences. This is an open access article distributed under the terms of the Creative Commons Attribution License 4.0 (http://creativecommons.org/licenses/by/4.0/).

the emergence of the Internet, GLONASS, GPS, satellite communication and television, jet passenger aviation, etc. [4]

The diversification of high-tech industries, including the MIC, determined the fundamental directions of Russian industrial policy [15]. The State Program “Scientific and Technological Development of the Russian Federation” (2019-2030) identified a problem hampering scientific and technological development. It is a lack of technology and knowledge transfer between the military and civil sectors.

The technology transfer [12] is defined as the process of transition of technologies and the corresponding rights to them, knowledge, skills, production methods from the transferring party to the receiving party for the purpose of their subsequent implementation and use [18]. The diagram of technology development and transfer between the military and civilian sectors and their relationship was developed for illustration purposes (Fig. 1).

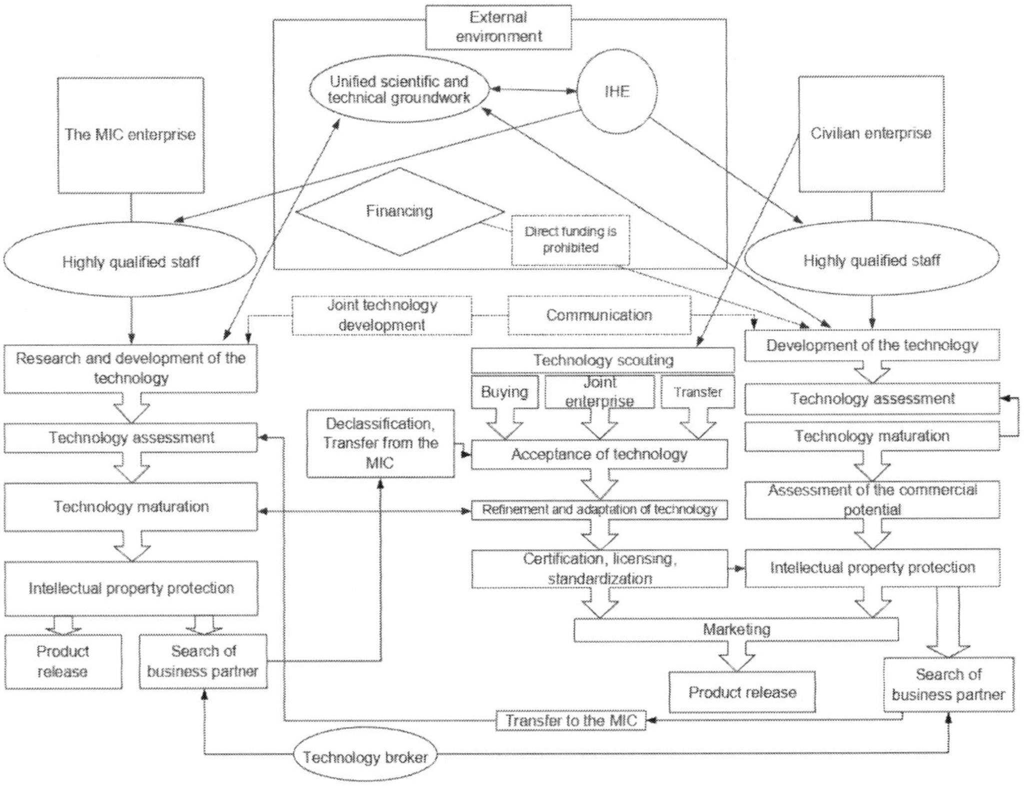

Fig. 1. The technology development and transfer stages. *Source:* prepared by the authors [19].

The creation and development of R&D for military and civilian enterprises requires highly qualified staff, a scientific and technological base, and funding [7].

The stages of technology development at military and civilian enterprises are basically the same (Fig. 1). At a military enterprise, this process has the stages as follows:

1) research and development of technologies;
2) technology assessment;
3) technology maturation (development of technologies, prototype production of products);
4) intellectual property protection;
5) product release.

A civil enterprise innovation process needs to include two stages: assessment of the commercial potential of innovation and marketing. The competitive capacity of an

© The Authors, published by EDP Sciences. This is an open access article distributed under the terms of the Creative Commons Attribution License 4.0 (http://creativecommons.org/licenses/by/4.0/).

enterprise directly depends not only on the innovation development, but also on its successful implementation. This should offset the costs of developing these technologies.

If a civil enterprise is short of resources to develop technologies on its own, it decides to borrow them. In this case, technologies can be obtained in several ways: acquisition of use right, foundation of a joint venture, or transfer. Moreover, the transfer has not always commercial nature. Since there is a ban on direct financing of commercial products, the military technology transfer can be a good alternative, since this does not apply to military developments [20]. This method of civil development support is widely used abroad. But there is a limitation – the transfer of military technologies to the civilian area is possible only after they go through the declassification procedure.

When the enabling technologies and the budget contributor are found, agreements are concluded, the technologies are accepted in compliance with all legal norms. Then the technologies should be further refined and adapted for civilian use through R&D, manufacture a prototype of products, eliminate shortcomings, and improve technologies. Then the procedures of certification, licensing, standardization of products, marketing research, commercialization of products (service release), and product launch are carried out.

In the process of technological transfer, two-way relationships are established between the transmitting and receiving parties. And these relations do not end after the settlement of transaction, but are strengthened through the information and knowledge exchange, joint technology development, and military-civil integration. Such integration will reduce production costs, enter the civilian market and gain competitive experience for defense enterprises, and for civilian enterprises, it will accelerate the creation of innovative science-intensive products with new consumer performance of civilian or dual-use. The long-term cooperation can help resolve any issues with production capacity.

When the integration scheme is implemented, the problem, as follows, on more than one occasion arises: military and civilian manufacturers may not know about each other or may not know how to establish a mutually beneficial relationship. That issue should be tackled through an intermediary required in the technology transfer, or a technology broker (Fig. 1). Its tasks include the search and selection of business partners, the establishment of relationships with them, assistance in the refinement and adaptation of technologies, legal protection, marketing research, the innovative environment formation [5, 8]. The National Association for Technology Transfer (NATT) is an example of such a broker. The NATT carries out expert, consulting and educational activities in the technology transfer field, helps to take technologies to implementation faster, and to increase the science intensity of the industry.

4 Discussion

The technology transfer could become an effective mechanism for enhancing innovation processes in the civilian science-intensive industry. Defense technology can be used as a basis for civilian developments, reducing development and innovation costs and time. Civilian companies scarcely ever have the resources to innovations, and defense enterprises are better funded by the government. Therefore, the technology transfer can be regarded as a reallocation of funds, which will optimize the cost of budgetary funds to maintain scientific and technical potential.

The technology transfer from the civilian sector to the military is less frequent, since military technologies are usually more advanced, and civilian technologies, as a rule, are for commercial purposes [13]. However, there is a practice abroad of attracting private companies in the interests of production high-tech military products, for example, in the IT

© The Authors, published by EDP Sciences. This is an open access article distributed under the terms of the Creative Commons Attribution License 4.0 (http://creativecommons.org/licenses/by/4.0/).

industry. For example, the US Department of Defense is involved in attracting small businesses to solve technological problems [15].

Integration and mutual enrichment of military and civilian technologies has a multiplier effect, which leads to increased flexibility in creating an innovative science-intensive product. The cooperation of military enterprises, joint project organization with private high-tech enterprises will permit defense enterprises to enter the market environment, learn how to manage risks, improve management efficiency, master cost-saving method, improve marketing and project management [21-22].

5 Conclusions

To sum up, the accumulated scientific and technical experience of the military sector can be effectively used in the process of scientific and technological modernization of mechanical engineering, increasing the level of its science intensity through the technology transfer. This will contribute to a reduction of the dependence on imports, foreign technologies and suppliers. Working with enterprises manufacturing components will contribute to the formation of a base of high-quality, reliable components in Russia, which is essential for the stable enterprise operation.

Medium-tech companies and industries can also accept technology assistance from the defense industry. This will contribute not only to the technological advancement of the industry, an increase in the high-tech products share, but also the transformation of the industry from the medium-tech to high-tech and science-intensive.

Integration of the military and civil areas would bring together efforts for technological advancement, reduce the technological gap within the country, and lag behind the world level. This will contribute to the development of the national competitiveness, sustainable economic growth, and national security.

Acknowledgements

The article was prepared with financial support by the Russian Science Foundation, grant № 21-78-20001 «Development of the theory and model tools for optimizing the management of the diversification of defense production in the context of the economic crisis and growing threats to Russia's national security».

References

1. N. M. Baranova, N. A. Shevtsova, E. G. Dmitrieva, *Global Challenges and Prospects of the Modern Economic Development*, 1, 1877 (2019)
2. E. Yu. Khrustalev, National Interests: Priorities and Security, **35**, 61 (2011)
3. A. M. Batkovsky, E. G. Semenova, A. V. Fomina, E. Yu. Khrustalev, O. E. Khrustalev, Indian J. Sci. & Technol., **9**, 1 (2016)
4. A. S. Slavyanov, E. Yu. Khrustalev, Scientific J. of KubSAU, **6**, 115 (2020)
5. T. V. Anischenko, *Management of the economy, systems, processes*, 1, 26 (2019)
6. A. I. Ignatova, http://nauchkor.ru/pubs (2016)
7. E. N. Zharova, A. V. Gribovsky, VUiT J. of Economics, **4**, 1 (2017)
8. T. Bukkvoll, T. Malmlöf, & K. Makienko, Post-Communist Economies, **29**, 232 (2017)
9. G. A. Van Norman, R. Eisenkot, JACC: Basic to Translational Science, **2**, 85 (2017)

© The Authors, published by EDP Sciences. This is an open access article distributed under the terms of the Creative Commons Attribution License 4.0 (http://creativecommons.org/licenses/by/4.0/).

10. J. Schmid, Defence and Peace Economics, **29**, 595 (2018)
11. A. F. Mendi, M. A. Nacar, M. Solmaz, ICTE, **1**. (2020)
12. B. Bozeman, Research Policy, **29**, 627 (2000)
13. Y. Li, P. Yang, J. Hu, *Development of Marine Electronics and Communication Navigation Technology* (2011)
14. Z. Yuan, Sci. and Technol. Management Research, **28**, 487 (2008)
15. A. A. Bakulina, E. S. Sokolova, National security / nota bene, **5**, 1 (2019)
16. J. C. Y. Ng, K. K. W. Song, Q. Tan, Int. J. of Innovation Studies, **2**, 33 (2018)
17. WIPO2012, https://www.wipo.int (2012)
18. GOST R 57194.1-2016 (2016)
19. A. Yu. Pronin, VectorEconomy **12**, 58 (2020)
20. A. B. Slivitsky, Russia: Trends and Development Prospects, **12**, 421 (2017)
21. O. V. Samovarova, S. V. Zhurkina, E. A. Gorin, Innovation, **9**, 9 (2019)
22. A. V. Kirsanova, Innovation and investment, **2**, 232 (2018)

© The Authors, published by EDP Sciences. This is an open access article distributed under the terms of the Creative Commons Attribution License 4.0 (http://creativecommons.org/licenses/by/4.0/).

The analytic study of the Russian experience in the context of corporate social responsibility (CSR): the lessons for Iran and developing countries

Ismael Rezaeinejad*, *Amol* Khaniwadekar, and *Karan* Singh Rana

Peoples' Friendship University of Russia (RUDN University), Miklukho-Maklaya Street, 6, 117198 Moscow, Russian Federation

Abstract. There is no doubt that large Russian firms are aware of the traditional symbols and concepts of CSR, and, more importantly, they admit that it benefits them to include these concepts in their corporate strategies. Large companies adopted internal and external policies to improve their employees' well-being in terms of health, housing, and a wide variety of educational and cultural programs. The emphasis on CSR is primarily on domestic issues. So far, little attention has been paid to global warming, environmental protection, fair trade, or business activity's carbon effects. The use of CSR by Russian companies in various fields is mainly from the companies operating in the leading industry, where foreigners have significant investments. However, there is only a balanced level of attention to national standards, and therefore governmental reporting is limited according to standards of international organizations in Russia. There has been a corporate social responsibility in Russia in response to its history and policy. This pattern followed a very different way from other non - socialist countries because Russia's market economy is still very new, increasing attention to CSR, such as corporate governance, is still not grown up, but that is mainly occurring in large firms with an international profile or under the State's influence to participate in CSR activities. This paper attempts to review the Russian country's experiences due to its similarity in the former Soviet times with the Iranian economy's current circumstances, analyze and compare and solutions that a government must consider to be transferred from a government-run to a market economy.

1 Introduction

The development and expansion of financial firms, although it has direct public and private benefits for individuals and society, simultaneously with the services or loss to the community, which in the literature of economics refer to the economy's external consequences. In most modern countries, one of the government's main tasks is to enter and intervene in such activities through appropriate mechanisms to regulate the market. The

* Corresponding author: mailto:1032205254@pfur.ru

© The Authors, published by EDP Sciences. This is an open access article distributed under the terms of the Creative Commons Attribution License 4.0 (http://creativecommons.org/licenses/by/4.0/).

government has minimized the development of regulations, or by the receipt of taxes and duties, social losses arising from firms' activity. In contrast, by paying subsidies or legal incentives, social benefits arise from the action of investors who have positive externalities. Although this mechanism has become commonplace in developed countries, multinationals' movement in other countries is not necessarily moderated by host governments' actions. Therefore, CSR affected by their experiences or pre - assumptions and normative, cultural, and social patterns decide to compensate for their country and host region activities. Such voluntary activities add to legal duties that the host nation may impose on multinationals. Today, what has become a common practice for multinational corporations in countries and host communities is CSR, under the name of compensating for environmental or social impacts[2].

In contrast, in developing countries, small and large firms have different behavior regarding their activity. Indeed, large firms in these countries have not only reached the maturity level of voluntary participation in compensation of the damage imposed to the host region, but in the absence of appropriate legal duties, they are also exempt from direct social and environmental damages.[3]

In Iran too, corporations and large public and private firms in the country, with government support and under the pretext of participation in job creation and economic growth, rarely have to compensate for their external consequences or voluntary payments to compensate for their social and environmental damage to the surrounding environment and host areas. In particular, due to the weakness of organizational and technological capabilities available in such economic units, the damages caused by external consequences of their activities are usually more extensive than similar companies' actions in multinational companies.

The new approach to solving adverse external effects such as environmental pollution is the category of CSR. Corporate Social Responsibility involves several different concepts and definitions [2]. It seems that the description is given by (Verheugen G. (2007), 'CSR and competitiveness: a view from the European Commission), summarizes the underlying points of the concept in a nutshell[12]. According to the report, the social responsibility of large companies: Social and environmental concerns in business processes and engagement with stakeholders are voluntary Social responsibility is not only to meet legal expectations but also to go beyond obedience and to invest more in human capital, the environment, and relationships with society. The European Commission's vice president says: "Businesses in all sizes to consider their role in today's society when taking strategic and operational decisions" [12].

Corporate social responsibility covers all aspects of corporate governance. It is about how businesses conduct their businesses morally and considers the effects of human rights' economic, environmental, and social factors. Social responsibility can improve social life quality in social dimensions. It also increases corporate social responsibility to maximize profit. In this case, attention to stakeholders' interests and the interests of society is also considered.

2 Methodology

Carroll (1991), in the article "Four facets of corporate citizenship," writes that a company is expected to show responsibility, just as it is expected from other citizens. Social responsibility problems become relevant not only for companies but also for the States[4]. Several studies show that decision-making within the framework of corporate social responsibility has a positive impact on the effectiveness and efficiency of a business. How do Russian companies try to change corporate behavior to meet prevailing social norms, values, and expectations?

© The Authors, published by EDP Sciences. This is an open access article distributed under the terms of the Creative Commons Attribution License 4.0 (http://creativecommons.org/licenses/by/4.0/).

https://doi.org/10.1051/shsconf/202111401028

CSR in Russia has specific features that make it different from other foreign countries' practices. What are the reasons, objectives, directions, and expected results of CSR in Russia? Are there any specific features of CSR in Russia?

This paper attempts to review the Russian country's experiences due to its similarity in the former Soviet times with the Iranian economy's current circumstances, analyze and compare and solutions that a government must consider to be transferred from a government-run to a market economy.

The analysis and synthesis of fragmented data on Russian companies' activity in CSR are used as research methods. Comparative evaluation to identify CSR's causes and objectives in the Russian market is based on comparing the results of the content analysis of reports placed by Russian companies and external studies' results.

3 Results and Discussion

3.1 CSR in Russia

Crane and Matten (2007) wrote in their famous book, Corporate Social Responsibility: in the relevant section of emerging transitional economies: while there are many different approaches to these countries, it is argued that in some respects, Russia and China are first and pioneer in CSR. On the other hand, Russia has experienced privatization and a return to capitalism when the state institutions so weak and corrupt. It has reached a place where the "cowboy economy " is said[2]. Therefore, it is no wonder that it is still a much-unknown concept in Russia, and it brings a strong resemblance to a communist time for many Russian business owners [3].

According to the definition, transition economies experience dynamic changes. In Russia's case, the latest changes - 2008 - have strengthened and concentrated many activities in different government agencies. Simultaneously, there has been considerable growth in the private sector of the economy and the rapid development of companies, and the adoption of modern private corporate governance practices.

From 2004 to 2006, the corporate management center at the " School of Economics " held more than five seminars in four different cities across Russia about the various aspects of corporate governance, which was also part of its social responsibility. Since then, the demand for teaching and writing about CSR has become so high that the corporate management center has taught itself a wide range of recommendations in the significant commitment to corporate management. The search for knowledge about the appropriate corporate management practices by the private sector has increased exponentially. Due to its adoption, the adoption of a variety of traditional corporate social responsibility practices has increased.

Other evidence, but all of the interest in Russia, is that in November 2008, if the term "Russian Corporate Social Responsibility" entered Google's search engine, over half a million cases- exactly 559000 - were found. While this number is not as large as other countries (Brazil, Russia, India, and China; Brazil with over 2 million cases), or European countries (754000 for the UK) or the United States (with close to 1120000), it is still a lot. It states that in Russia, something is more than an "unknown concept."

Of course, many cases do not relate to understanding or performing in Russia. A simple vector case tells us that most of the issues come from private organizations operating around the world up to now. These organizations use the "WorldWideWeb " to explain their position in exchange for use - and it has always been used in many countries that operate (including Russia). It seems that about half of all cases are of this type[7]. While these are not much more than a conscious guess - the content of all these blogs has not been investigated yet - yet it means that in the past year, about 600 websites are entirely devoted to CSR in Russia.

© The Authors, published by EDP Sciences. This is an open access article distributed under the terms of the Creative Commons Attribution License 4.0 (http://creativecommons.org/licenses/by/4.0/).

While this type of available evidence does not show strong or sufficient interest from the Russian companies, it is large enough that CSR's minimum conclusion is an " understood " concept in Russia[7].

On the contrary, we argue that CSR was always in a large part of Russia and has been a more critical element in the strategies and operations of the growing number of firms in the Russian business community.

3.2 The unique characteristics of corporate social responsibility in Russia

One of the big problems in accessing and comparing CSR between different countries is that the idea itself has a wide variety of meanings. Tried to deal with the proposal that it meant significant amounts to the person defining it - government, business communities, corporations, non-governmental associations, or corporations - and notes several methods in these categories. Suppose this lack of definition between different groups is not enough to make it challenging to assess CSR. In that case, CSR's implications also vary widely from country to country and from time to time. As a result, all definitions of CSR should be considered carefully. Social, cultural, and historical developments of nations are so vast that there is no set of CSR principles with everyone, or it is virtually inconvenient.

Indeed, the extent of CSR in Russia in the 21-st century is the product of its unique geographical, historical and political conditions.

3.2.1 Summary of CSR history in Russia

3.2.1.1 Soviet era

One of the reasons for the early development of CSR in Russia is the vast area. Most of the Siberia cities and the Far East are far from each other and, of course, the community's socialist organization in Soviet times, the emphasis was on resource development. This action's primary method was through an extensive planning program and large industrial enterprises built around them. This kind of development's consequences was that the industrial factories and the cities blended and intensified themselves. Government agencies produced products from factories and made deep integration in all community activities by providing heat and electricity to homes, building roads, supporting schools, etc.

They also supported the sports teams, financed and built the stadium and the swimming pool, organized sports leagues, and conducted monetary supply for the orchestra and the dance. Workers were dependent on these companies for housing, transportation, leisure and holidays, health care, and shelter. Depending on the person's place in the Russian Federation, the company or institution that people worked for was responsible for any services from the cradle to the grave. These cities were literally parts of the factory, embedded in social infrastructure.

Indeed, it can be confirmed that Russian state enterprises started implementing CSR earlier and more sharply than most of the activities that were later found as a significant part of CSR in many European and North American countries.

3.2.1.2 Transitional years

In the mid-1980s and Perestroika time, Russia remained at the time of struggling to move to the market economy from a revolution, perhaps even after the Bolshevik revolution; this was a really tough time when the entire structure of the economy changed due to the enormous change in ownership of the firms. Many immense Russian industries became extinct in the

© The Authors, published by EDP Sciences. This is an open access article distributed under the terms of the Creative Commons Attribution License 4.0 (http://creativecommons.org/licenses/by/4.0/).

hands of eminent and influential landowners. Thousands of medium-sized companies have been privatized and are owned and managed by groups with insufficient knowledge about how the private market works and the changes that need to be successful in their business. Finally, many organizations remained in the government's hands and practiced by people who tried to implement a management system in a market economy used under the communist regime. The result was chaos and, as many managers did not know how to work - disruptions in their supply resources, limited demand for their goods, and a low commitment to contracts. The government was corrupt, and bribery became widespread; businesses began to seek ways to reduce their cost; only staying upright was the primary goal of most people.

What is not sufficiently understood in most analyses of this period is the transitional period, which included transferring property to the private sector; the privatized companies inherited a wide range of social services performed by companies in the Soviet era. Necessary activities such as transportation, heating, water, and sewage disposal in some cities suddenly fell to these companies and their new owners, most of whom did not care to preserve them. In places such as Norilsk, Tyumen, and other cities dominated by corporations, large companies put their traditional role as social service providers to maximize profits. These cities were converted into little-equipped local governments, and governments responsible for providing the infrastructure necessary to sustain community activity. As a result, in many regions, cities, and villages, all sorts of infrastructure, including cinema, theatres, transport, and communications, started to collapse.

In addition to reducing the original service, the relationship between workers and employers also varied. The Soviet era's cornerstone was to provide employers with benefits such as housing, health care, vacation, and other services; during privatization, the services were cut off, and wages were converted into money with much less purchase power. As a result, a whole generation remained without monetary resources to purchase homes and other services that were usually prepared by the companies that worked for them[5].

As is well known, without adequate government in any of the levels, the situation went out of control, such that in 1998, almost every form of the product was deficient, the shops became vacant, the banking system collapsed, and the central government was unable to pay its debts. Finally, Russia went bankrupt.

3.2.1.3 Start of the Stability

As the 21st century approached, there was a better understanding among the managers and business owners of the Russian business about how the market functioned, the more stable government emerged and began to grow due to a rapid rise in prices and demand for natural gas, petroleum, precious metals, and agricultural products from Europe and the Middle East. During the privatization period, many of the most powerful corporations, mainly in Boris Yeltsin's time, adopted modern management methods and started to operate efficiently and profitably. Since the big companies flourished, thousands of suppliers and services required by giant companies also grew. The boom of the industries into other fields of the economy and the rise in Russia's growth rates between 1998 and 2004 were among the fastest-growing developments experienced by a modern country[5].

Under such circumstances, it is not surprising that the first decade of the 21-st century was a significant change period for Russia. During the Soviet and transitional periods, the Russians did not travel much, were exposed to Western culture, and very little foreign investment. After Perestroika's end, the Russians began to travel widely across all society levels, fascinated by Western television and literature. The exchange of ideas was typical among the researchers and business owners. The companies increased the international trade volume to a large proportion, and the government joined international organizations such as the Organization for Economic Co-operation and Development(OECD). Once again, Russia

© The Authors, published by EDP Sciences. This is an open access article distributed under the terms of the Creative Commons Attribution License 4.0 (http://creativecommons.org/licenses/by/4.0/).

became an active country in the global community. During this period, other countries and international agencies' development and some of the country's donor states affected the Russian companies and governments' method of applying CSR[6].

3.3 Infrastructure restoration and CSR

The most crucial issue in Russia in the first years of the 21st century was who should pay to rebuild the economic infrastructure, which was essential to its health and prosperity. Besides, who should pay for the social service expense, which is crucial in developing and maintaining an effective and efficient working force?

The period's economic growth shared in the government's stability and provided a foundation for taxes and raising government funds to rebuild infrastructure. However, business entities paid much of the infrastructure costs and varied cultural, educational, and other services that people enjoyed under the Soviet system in the past.

In short, CSR's issue was at the center of the relationship between the relatively strong private sector and a strong central government.

Undoubtedly, the issue that the central government should expect from the private sector would have a significant role to play with CSR issues. In fact, at the annual session of the " Congress of Industry Owners " - a large business forum from private enterprises - in November 2004, Vladimir Putin announced that the Russian government expects the Russian business to increase its investments in social, science, education, and development so-called " human factors[8].

It was a significant announcement because, unlike the mid-1990s, when the government was weak, and the business dominated most aspects of Russia, the government regained its place and became its leading political player. Private sector leaders were well aware that a new economy driven by the Russian market was expected from them.

Besides, the situation made the private sector more involved in many activities that generally collapsed in some parts of the country. Unfortunately, over the years, when local and regional governments were under control, specific government services were weakened to become unreliable and, in some cases, almost nonexistent. As a result, with a boom, not necessarily because they wanted to, but because of their survival, a large number of companies began operating in urban services in different cities once again .these were linked with the local government to make the road, draw gas lines (not only to their plants but to residential consumers) and to develop electricity and sewage disposal facilities. The fact is that Russian companies attracted much of society's essential cost, but they also managed to cooperate with local authorities. In fact, they accepted almost everything in soviet times. However, there was a significant difference with Soviet times. Now, most of the local governments still found themselves as partners of new private companies. Generally, while the participants were recognized and prepared, they divided social services through taxes.

Nevertheless, when they wanted to expand or make changes, they could only deal with certain people. Otherwise, it is unlikely even to gain the services to advance their business and grow their business. At the time, these fees are known as "the cost of voting rights."For many businesses, bribery was accepted as the cost of doing business and maintaining order in civil society.However, many companies' desire to take responsibility for such a large number of services and bribes was more than taking services to work for efficient action on the agenda. The introduction of such a new concept as private property, even to its most incredible supporters, was uncomfortable for many people - especially the older ones. Suppose the transformation from an ordered economy to a market economy is to be successfully implemented. In that case, companies need all the support they can get from the society in which they are based. As a result, rather than reducing their support for local activities, they tried to maintain and increase it.

© The Authors, published by EDP Sciences. This is an open access article distributed under the terms of the Creative Commons Attribution License 4.0 (http://creativecommons.org/licenses/by/4.0/).

At the same time, market-based businesses did not want to return to the former Soviet system, where companies offered all services. As a result, the discussion of the correct amount of CSR in Russia is not so different from that in other countries. How is the proper distribution of responsibility for various social activities in the private sectors?

3.4 The impact of international trade and investment on CSR

One of the results of the transformation from the transition economy to the market economy was the growing number of prominent Russian companies. As these companies grow, with a tremendous global demand for resources, their need for capital has been developed.

Under such circumstances, it is not surprising that large Russian companies, mainly controlled by oligarchs, started listing their shares in London and New York stock exchanges. Once they did, they had to meet all the standards required to be on the list. While most of these standards were related to government requests - transparency of functions and completeness of reporting, and so on, but stock owners also expected the listed companies to operate under the social responsibility style. It meaning Russian companies admitted that they are responsible for socially responsible behavior and started to proclaim their declarations in annual reports and other literature they had[7].

Meanwhile, foreign companies in Russia invested money and increased their activities in Russia; many of these companies have been connected to industry - mining, energy, forestry -which in many countries, according to strict standards regarding safety, disposal of waste, environmental protection, and sustainability-When they arrived in Russia, they brought with themselves approaches to issues such as safety, environmental protection, etc.

CSR guidelines in other countries partially fulfilled their activity, affecting Russian and Russian partners' work in the same style. For example, according to various reports of Severstal (a British- Russian company) annually, it provides a regional and national investment policy and an action plan to provide housing for disadvantaged social families, treat hepatitis, and support medical institutions' financial shortage. It also supports sports teams and organizations in their areas under their activity and published a social report of their actions. At least, foreign companies provide standards based on which internal functions are measured.

3.5 The effect of privatization on management and CSR

However, it is not just foreign investors and professional consultants' feet that make large enterprises participate in CSR's diverse aspects. New Russian companies were born under international management from used managerial practices that developed in Western firms over a century have been implemented in a decade. Today's management in many Russian companies is very modern.

After deregulation and following customers' intentions in the supply chain and support from different government levels, the Russian managers found that CSR's acceptable practices as part of its strategic planning constitute a more effective performance. For example, Severstal reported in his company documents in 2007 that the social policy that has been inherited from the past and the active part of its new strategy makes it actively shaping its employees ' quality of life[6].

It also reports: The social responsibility of Severstal comes under the following: protecting the health and improving the living standards of company employees and charitable work [and] in 2006, the company allocated $ 50 million for social and philanthropic programs [10].

In recent years, while in Russia, as in Western countries, CSR has not prevailed, but some Russian companies have been very active in supporting different sports and cultural events.

© The Authors, published by EDP Sciences. This is an open access article distributed under the terms of the Creative Commons Attribution License 4.0 (http://creativecommons.org/licenses/by/4.0/).

They do not only act as a marketing tool but as part of the more extensive CSR programs. Although such an activity is not considerable compared to this kind of financial support in many other countries, for example, Coca - Cola who supports the Olympics, is the financial support of a fast-growing element in many countries' strategy.

It is worth mentioning that some Russian companies support higher education's financial institutions by paying tuition to talented students and supporting college members and young specialists' research.

By examining the annual reports and other documents, it is determined that huge companies are committed to applying Value and ethics in everything they do - how to perform their businesses to interact with employees to how to influence communities and regions. The implementation of its obligation is subject to review and reporting, and such reporting is not as widespread in Russia as in many other countries. However, this statement's expression can be incorrect that the highest management level in large enterprises is unaware of CSR's significant issues and, at least in their strategic business planning, is not considered part of them. The reason is not only because of the desire accepted among the big ones but also because of history and the willingness to engage with government targets to participate in CSR's long-term profitable partnerships.

3.6 Humanitarian action impact on CSR

There is a very long and lasting tradition of private philanthropy in Russia. Early in the 17th and 18th centuries, merchants and families of Russian-born nobility built dance halls and poetry halls to introduce artists, dancers, sculptors, poets, and writers. In the 19th century, Russian art supporters, such as those in Europe, were famous throughout Europe for their unique artworks' significant collections[1]. In soviet times, these collections, including many others, were nationalized and displayed at the national art museum.

During the communist era, there was no such thing as corporate or corporate philanthropy. However, under every leadership, there was a fixed friendship - strong support for cultural activities, especially music and dance.

The Russian Opera Company and Russian Opera became both organizations of global credit. The ballet symbolized the Russian culture. The support of these organizations and cultural organizations is always strong by the Russian people and their governments.

In recent years, private philanthropy significantly increased. Many wealthy Russians mimic the role of American Capitalists. Those had filled their mansion with pictures and other European art types at the end of the nineteenth century. The extent of participation of large Russian companies in various humanitarian activities, particularly in CSR, is still so large that Russian collectors are now a dominant factor in the global art community. In fact, in recent years, Russian oligarchs have been one of the principal patrons of modern artists[11].

Nevertheless, the situation is changing, mainly due to foreign companies working in Russian agencies or establishing Moscow offices. Such charity is part of their public performances. They support all kinds of philanthropic and humanitarian activities in other countries, so this is just part of their regular work supporting nonprofit organizations that work in health, education, and public welfare in Russia. After accepting some corporate social responsibility, the usual practice of specialized foreign companies in Russia, most of its clients mimicked their behavior.

3.7 The impact of national and international organizations on CSR

Undoubtedly, the concept of CSR has entered the vocabulary of Russian business culture. The major commercial enterprises of Russia, such as the " Russian Union of Industrialists

© The Authors, published by EDP Sciences. This is an open access article distributed under the terms of the Creative Commons Attribution License 4.0 (http://creativecommons.org/licenses/by/4.0/).

and Entrepreneurs, " Association of managers, and Chamber of Commerce and Industry of the Russian Federation," followed the direction of Western colleagues, focusing their attention on the establishment of guidelines for the role of businesses in the community. Russian Union of Industrialists and Entrepreneurs (also known as the "red " group of directors) used the "social prism" that draws their business role in society and urges its members to become an active part of the communities in which they are stationed. In addition to institutions and business associations' efforts, NGOs facilitate CSR's entry to Russia, with financial support from more than 30 seminars and conferences on CSR since 2004 and translating international standards on CSR reporting in Russian[9].

The United Nations achieved through the Russian business and government leaders to be part of the UN program to engage companies in resolving social and environmental issues. The United Nations publishes a journal that reports about developments related to CSR in the majority world, as follows:

CSR has been stable at the beginning of 2007; part of the business plans more than 40 Russian companies. Fifteen of them presented such reports as separate sections of their annual reporting, while eighteen of them produced individual social reports. Eight were reported on sustainable development, and the remaining four companies presented environmental reports. Thirteen of the documents were based on the basis and indicators of international standards.

The United Nations has also supported information, research, and advisory activity to develop Russia's CSR practices. While most Russian businesses and business organizations have dealt with social and political risks, only a few pioneers of Russian enterprises have begun to address CSR's environmental issues. However, the change was on the way, "the Russian Association of Directors" added an environmental component to CSR's definition. It has been reported that the application of environmental sustainability policies and a mechanism for reporting how to implement them in 43 firms are in the early stages.

The conservation of the environment in Russia remains mostly a matter that must be followed. The majority of Russian business leaders still believe that international environmental management standards are very costly. Therefore, major Russian companies have not focused on these issues and are not worried about participating in the rankings and international layouts. Unfortunately, only four Russian companies currently offer environmental reports.Some Russian national systems have been devised, such as Russia's environmental responsibility, the largest industrial enterprises, rating the "social-environmental union and an independent environmental rating agency value."

Besides, values are variable, but they are all designed to encourage Russian firms to apply social goals - in a broad definition - in their strategic planning.There is a difference in the definition between CSR in Russia and CSR in the Western world. CSR in a state of development, analysis, and judgment is that the State provided all citizens' services. It is not surprising that CSR is closely linked to Russia's political, social, and economic background towards it in any other modern industrial country. In most Western industrialized countries, large and giant CSR issues are related to the decision about what social issues are the companies' responsibility and what concerns the business must undertake the government. In Russia, the government indirectly represented all social benefits and benefits in the Soviet era, and that the private companies performed in the post-transition periods largely complicate the issue. Within this historical context, the current development of CSR in Russia is under implementation and development.

Since the financial crisis and the high price of goods, significant economic growth brings the expectation that major Russian companies, especially in oil, gas, metals, and mining and manufacturing, must use large amounts of the general philosophy of the soviet era. Those who clearly understand the nature of personalization recognize that the private company is now a completely different instrument from the soviet - Era Company and cannot offer all

© The Authors, published by EDP Sciences. This is an open access article distributed under the terms of the Creative Commons Attribution License 4.0 (http://creativecommons.org/licenses/by/4.0/).

expected benefits and remain competitive in the free market. Many Russian companies have been stranded in the past heritage. Workers and the government are looking forward to their return to their model when given money and offered almost all services. Of course, the companies will resist attempts to force them to accept these costs.

As a result, CSR's debate is about the voluntary acceptance of specific social, cultural, or economic responsibilities by businesses and the overall spectrum of social services. Recently, at all levels, the government has focused on the need for Russian enterprises to improve their benefits and social infrastructure. In response, firms are continually seeking a long-term planning process to manage public expectations regarding CSR activities' appropriate scope. Many companies also try to distance themselves from programs designed based on soviet collective policy to implement policies that get the group of target employees, such as young professionals and highly talented specialists.

While discussion of the distribution of social costs between businesses and governments is not unique in most countries, it has different nature in Russia. It is not about whether many communities-based activities undertaken in CSR's name are appropriate for Russia's businesses. Conversely, this situation is seen in other countries. More importantly, assumptions that emphasize the volume of literature on CSR may indicate the need for a unique set of strategies and tactics for fans of a practical approach to CSR in the new world free market in Russia.

4 Conclusions

All the factors mentioned above had affected the development of CSR in Russia, the need for restructuring infrastructure, the impact of trade and foreign investment, applying modern management methods, philanthropy development, and international and domestic organizations' functioning.CSR in Russia has a long history that backs to Soviet times. If CSR is defined in terms of the relation of business institutions to the society in which they are present, during that period, under the socialist philosophy, firms provided the "cradle to grave "services to their employees and cities headquartered in it. Nevertheless, during the Great Transformation from a government economy to a market economy, all CSR disappeared with every intention and purpose in the middle of the public turmoil. During the period of return at the end of the 20th century, due to the collapse of the usual infrastructure and social services during the transition period, many corporations with governments contributed to the activities considered in most industrialized countries only to governments. With the growth of the economy and companies' development, the discussion of the right balance between private support and government support for such services increased. There is no doubt that large Russian firms are aware of the traditional symbols and concepts of CSR, and, more importantly, they admit that it benefits them to include these concepts in their corporate strategies. Large companies adopted internal and external policies to improve their employees' well-being in terms of health, housing, and a wide variety of educational and cultural programs.

The emphasis of Russian companies on CSR is mainly on internal issues. So far, little attention has been given to global warming, environmental protection, fair trade, or commercial activities' carbon effects. The use of CSR by Russian companies in different fields has mainly been driven by companies operating in the leading industry, where foreigners have significant investments. However, there is only a balanced level of attention to national standards, and therefore governmental reporting is limited according to standards of international organizations in Russia. CSR's teaching and writing have accelerated to the management teams, and CSR for Russian managers is not a new concept.

There has been a corporate social responsibility in Russia in response to its history and policy. This pattern followed a very different way from other non - socialist countries.

© The Authors, published by EDP Sciences. This is an open access article distributed under the terms of the Creative Commons Attribution License 4.0 (http://creativecommons.org/licenses/by/4.0/).

Because Russia's market economy is still very new, increasing attention to CSR, such as corporate governance, is still not grown up, but that is mainly occurring in large firms with an international profile or under the State's influence to participate in CSR activities.

4.1 CSR lessons from Russia

1. If any society has reason to shift from a transition economy to a market economy, it is crucial to set up plans and maintain public services. The inability or not to do so will cause chaos.
2. Local governments that have not been involved in providing local infrastructure have sufficient readiness and equipment to undertake responsibilities during transitioning.
3. Privatized companies must be able to supply those state-owned companies in the socialist era to continue at the same level as before; otherwise, this situation led to the collapse. In 1998, this situation led to the collapse of the Russian economy.
4. A strong government is an essential factor in encouraging CSR.
5. The techniques and methods of modern management quickly and freely promote in the free communities.
6. The knowledge about CSR is spread in Russia and continues to grow.
7. The leading force in acquiring knowledge about CSR in Russia was large Russian corporations in foreign stock exchanges.
8. Investors and foreign advisors of the Russian companies have been the leading forces in CSR's arrival concerning CSR in Russia.
9. The educational institutions in Russia contributed to expanding the knowledge linked to CSR.
10. International agencies and international organizations, which are famous in Russia, have an impact on promoting CSR.
11. CSR is an essential element in large Russian companies and large Russian firms with foreign partners.
12. Even among larger firms, CSR focuses on internal issues rather than international issues.
13. Recently, CSR has become a factor in small and medium companies.
14. Philanthropy has become evident between major Russian companies and business leaders, especially in the field of art.
15. Undoubtedly, with firms' maturity in the relatively young Russian free market, other modes of industrialized countries are being adopted, and these firms welcome higher levels of CSR.
16. Indeed, it can be confirmed that Russian state enterprises started implementing CSR earlier and more sharply than most of the activities that were later found as a significant part of CSR in many European and North American countries.
17. CSR was at the center of the relationship between the relatively strong private sector and a strong central government.
18. When foreign companies entered Russia, they brought approaches to safety, environment preservation, CSR guidelines from other countries to adjusted their activity, affecting the Russian companies and their Russian partners to operate in the same style.
19. New Russian companies were born under international management ideas in the transition period to a market economy.
20. There is a long tradition of private philanthropy in Russia.
21. CSR has been stable at the beginning of 2007; more than 40 Russian companies took CSR in their strategy.
22. The United Nations has also supported information, research, and advisory activity to develop Russia's CSR practices.

© The Authors, published by EDP Sciences. This is an open access article distributed under the terms of the Creative Commons Attribution License 4.0 (http://creativecommons.org/licenses/by/4.0/).

23. There is a difference in the definition between CSR in Russia and CSR in the Western world.
24. CSR's discussion in Russia is about voluntary acceptance of specific social, cultural, or economic responsibilities by businesses and the broad spectrum of social services.

References

1. Art Newspaper: International Edition, **16(184)**, 1 (2008)
2. A. B. Carroll, Academy of Management Review, **4(4)**, 497 (1979)
3. A. Crane, D. Matten, L. Spence, *Corporate Social Responsibility: Readings and Cases* (2007)
4. A. Crane, D. Matten, A. McWilliams, J. Moon, D. Siegel, The Oxford Handbook of CSR (2007)
5. S. Dayman, Russia in 2008: ISO Corporate Social Responsibility in a Post- Socialist State, ISO Social Responsibility (2008)
6. J. Gillies, and O. Melitonyan, *The Canada/Russia Corporate Governance Program, 2000–2006* (2008)
7. S. Grafski, J. Moon (2004), *Comparative overview of Western and Russian CSR Models, Report on Social Investment in Russia* (2004)
8. D. Johnson, (2004), Social Responsibility of Russian Businesses, Johnson's Russia List, 8002
9. A. Kostin, Russia: the evolving corporate responsibility landscape (2007)
10. Severstal, Social Responsibility (2008)
11. D. Thompson, The 12 Million Dollar Stuffed Shark: The Curious Economics of Contemporary Art (2007)
12. G. Verheugen, CSR and competitiveness: a view from the European Commission, AccountAbility (2007)

© The Authors, published by EDP Sciences. This is an open access article distributed under the terms of the Creative Commons Attribution License 4.0 (http://creativecommons.org/licenses/by/4.0/).

The current state and potential of innovative developments in the service industry

Vasily Zharov*

High School of Service Federal State Budgetary Educational Institution of Higher Education RGUTIS, Russia

Abstract. The article analysis the actual state and examine the probable development course of the service industry in the light of the innovative scheme of the economic development. This work reviews the hypothesis that the service industry existing in the modern-day Russian circumstances where the market is at the stage of saturation with both the household and industrial goods becomes that branch of the economy which focuses not only on providing the functioning of the durable goods and satisfying the people's needs for various services but it is also one of the major economic sectors which is capable to adapt to the realities of the modern life while there is the transition to a new technological paradigm. The present work substantiates the necessity to work out the theoretical and methodological grounds and the recommendations for the management in the field of braches of the service industry in the face of development of the innovative digital economy in order to provide the formation of the new technological paradigm considering the current forecasts. The proposed methods for solving the tasks consider the evolutionary innovative processes in the service industry, the concept of serviceology, methods of the management study at various levels of the national economy as well as the researches results in the frame of the digital economy development while creating the new technological paradigm in the Russia Federation.

1 Introduction

An analysis of the works of a number of researches represents that the service industry becomes a major activity sector in the modern economy both from production and consumptions sides. The increasing role of the services also becomes a substantial part of the people's social life.

The author of this research article points out that it is the consumption of the services that starts to act as the dominant way of satisfying the human needs and the most important factor stimulating production, the main result and sphere of the labour application. In accordance with the author's opinion such activity can indicate the transition from the traditional economic paradigm to the innovative one – the service-digital economy.

This standpoint is substantiated by the research article [1] which contains the data analysis of the dynamics of the economical structure both the developing and developed countries,

*Corresponding author: basille@mail.ru

© The Authors, published by EDP Sciences. This is an open access article distributed under the terms of the Creative Commons Attribution License 4.0 (http://creativecommons.org/licenses/by/4.0/).

analyzes the architecture of employment and consumption in the society, and assesses the input of the services to the cost of the end product.

As it was said above the service activity either main or associated one is an integral part of the offered product and highly affects the relations between the business entities related to designing, manufacturing and providing various types of services. The activity in the service sector is a chain of the interconnected actions – processes for satisfying the consumers' needs in the corresponding services.

We can remark that the concept in which the service has to be provided in the most convenient, adaptive form for the customer, and sometimes it has to have an individual approach is being of a great importance. For that reason we can say that the service quality is one of the key element of the service itself, the poorly provided service can lead to the degradation not only of the properties of the goods sold along with the service but it also can have a sensible impact on the public image of all the participants involved in the corresponding production-service chain.

Based on the author's opinion it becomes not enough for the manufacturer to just produce a good quality product in the modern market environment because the quality of its sale service and the service support is also important in the battle for the client. Sometimes the high quality manufacturer service has the direct impact on choosing a product that is especially important while comparing similar types of goods by their characteristics. In this case according to many years of the author's observations the decisive difference will be the service and the maintenance starting from the purchasing stage till the end of the product's life cycle. For this reason the service and the maintenance of the goods can provide additional competitive advantages and ensure the income for both the manufacturer and the service company during the entire service life of the product specified by the manufacturer in the documentation which can be formulated in a life cycle contract.

Based on the above material we can say that in the modern market environment at the stage of the market saturation with both the household and industrial goods the service industry of the Russian Federation becomes that branch of the economy which focuses not only on providing the functioning of the durable goods and satisfying the people's needs for various services but it is also one of the major economic sectors which is capable to adapt to the practicalities of the modern life in the transition to a new technological paradigm, the service-digital economy,.

The 2020 COVID-19 pandemic demonstrated the demand for activities providing by the service industry, in particular, educational services, basic and supplementary, consulting services etc. and at this time a new effective connection was observed – the service industry and the digital technologies. During the entire period of lockdowns and self-imposed isolations it could be observed that the digital technologies are most in demand and are best adapted to the service industry. The author of this research work points out that the digital technologies actually establish new, innovative connections between the subjects which often allow achieving a more meaningful result of such interaction against the traditional connections. It is worth saying that it applies to the services that do not require direct person-to-person contact. This customer-focused service approach is significant for creating the contractor's competitive advantages when all other conditions being equal. As we can see the service and the maintenance are important for the economic relations between the manufacturers and the consumers. [2, 3]

Since the service is often the system of the person-to-person interaction when considering the actual service activity of a person – the service provider, conditions under which this service is being provided matters much as well as the culture of the service to which the system of the socially accepted norms and labour relations are generally referred to and which belong to a particular state and depend on the internal social organization that has been born in the course of its development. [4]

© The Authors, published by EDP Sciences. This is an open access article distributed under the terms of the Creative Commons Attribution License 4.0 (http://creativecommons.org/licenses/by/4.0/).

High quality level of the customer service is important not only for the service industry but for other economic sectors because it has a positive impact on the collective attitude of the customers both to the organizations providing the service and to the service industry as a whole. [5]

Today the processes in the economics of the most countries of the world are described in the publications as informational, innovative, global, network, knowledge, creative, digital ones etc. They are the consequence of the innovative technologies development in almost all economic sectors. Based on the information of the state economy development level and being focused on the certain innovative technologies this trend can be defined from the position of the predominance of a specific technology. [6, 7]

On the basis of the records [8, 9] the service industry is a component of the non-resource sector and its considerable importance is mentioned in the above materials. It should be stressed that in the future the innovations in the non-resource sector of the economy will be crucial in achieving the maturity and sustainability indicators of the national economy.

The author of this work concurs with the opinion of Karpova G.A. and Sushchinskaya M.D. that today these processes are the aspects of the systematic transformation of the economy as a complex of their interrelation together with mediating and stimulating the predominant role of the service industry and its development.

As the researchers remark [1] the economy servicization requires the indicators, the models to define the paths of its development, to define its value for the state economy and also to define the socio-economic suppositions and consequences in the transition to the service-digital economy.

That is why it is necessary to concentrate more on the research works which examine the processes in the economic systems impacting the servicization which is an integral component of the innovative processes of the transition to the service-digital economy described in the research articles as well. [10, 11]

2 Materials and methods

The author of this paper focuses on the necessity to carry out the researches in order to substantiate the theoretical and methodological grounds and develop the recommendations during the establishment of the innovative service-digital approach in the field of the management in the service industry within the frame of the digital economy development which is also highlighted in the document [11].

For that to happen it is necessary to review a set of the missions correlated with the study of the theoretical standpoints of the service industry management concept, the serviceology concept, and the methods applicable to the research of the management at various layers of the national economy. The examination of the innovative development potential of the digital economy should be carried with a view to determine the conceptual approaches to the service industry management.

So the management processes in the service industry become the object of the study and they are understood as the innovations in the service companies allowing to timely adapting to the changes appearing in the service market during the development of the digital economy and appearing of the new technological paradigm. [12]

A combination and interrelation of economical, technical, organizational, managerial fields of the service sector is a subject of this research article that is taking place in the formation of the innovative service-digital management approach of the service sector.

The research methodological basis is formed by using the systematic approach to investigating problems of the capacity assessment, the management factors as well as the research tools for determining the management processes in the branches of the service industry.

© The Authors, published by EDP Sciences. This is an open access article distributed under the terms of the Creative Commons Attribution License 4.0 (http://creativecommons.org/licenses/by/4.0/).

The research methodology presumes the usage of the quantitative and qualitative methods: the panel survey and the expert assessment, the focus groups and the projection methods, the statistical analysis, the economic and social forecasting, the observations, . [10, 13, 14]

The evidence base of the research is represented by the research works related to the problems of the service industry development, the formation of the service including the scientific works of many foreign and domestic scientists: Avanesova G.A., Bagdasaryan T.D., Burmenko T.D., Balaeva O.N., Danilenko N.N., Zvorykina T.I., Karnaukhova V.K., Krakovskaya T.A., Petrova A.N., Platonova N.A., Predvoditeleva M.D., Romanovich Zh. A., Tretyakova T.N., Turenko T.A., Khlebovich D.I., Khoreva I.V., Khristoforova I.V., Lovelock K., Haksever K., Russell R., Merdick R. and other.

The author of this article supposes that the mentioned above research methods are suitable for solving the most vital tasks of the service industry including: the survey of the service sector of the Russian Federation economy in a phase of the transition to the service-digital economy; the examination of the existing provisions and recommendations in the service industry management; the study of the optimization methods and criteria as well as development of the innovative management methods for the branches of the service industry; the determination of the priority tasks of the state regulation of the service industry; the development of the innovative theoretical management model for the branches of the service industry reflecting the informatization impact and the digitalization of the economy and society during the transition to the service-digital economy; the development of the assessment methodology of the service sector companies management effectiveness considering the innovations which allow to timely adapt to the changes arising in the service market in the context of the digital economy development and the formation of the new technological paradigm; the development of the practical guidelines for the management in the branches of the service industry accounting the innovations.

3 Results and discussion

After the USSR breakup the state participation in the service industry was mainly reduced to the tax control and suppressing the businessmen informal activities in this sector. [14, 15]

Concurrently, the lack of the control over the quality of the provided services by the state resulted in the fact that the market was filled with the unscrupulous businessmen from whose activities are suffering not only the quality of the providing the goods but the citizens' wellness and this arises the need to define the state participation as a subject of the service sector and as a controller. [5]

Recently in the Russian Federation it can be distinctly seen due to the increase of the opportunities provided by the digital technologies, the services are being carried out by 'the impersonal specialists', 'one-day specialists'. These activities resulted in the unreasonably high prices for the services, the fraudulent schemes are being used, and in fact there is no one to be accused of providing the poor service. It provokes the appearance of the task of the state regulation of the service industry related to the quality control of these services, the control over the qualification and expertise of the service providers and the cost of the provided service subject to taxation. [16]

At the same time the state of the economy as a whole will depend on what will be the policy of the state towards the processes happening in the economic and social spheres. It is impossible to ensure the regeneration of the labour force, increase the labour productivity, improve the educational level and qualification of the labour resources, create the innovative climate to provide the competitive scientific and technological progress, look toward improving the living standards of the citizens, provide appropriate spiritual and cultural life of the society without the involvement of the state instruments. [1, 17]

© The Authors, published by EDP Sciences. This is an open access article distributed under the terms of the Creative Commons Attribution License 4.0 (http://creativecommons.org/licenses/by/4.0/).

As part of this article theme we should mention two forms of the state participation in the service industry– as a manufacturer and as a controller. In this case the digitalization becomes the state's control tool and it facilitates access to the services, provides the transparency of the activities i.e. the self-regulation of this sector of the economy for the business entities and individuals.

The use of the digital technologies in the service industry allows ensuring the transparency of the providers, minimizing or totally eliminating the informal part of this economical sector, increase the tax collection. [10, 18]

The researchers in the research article [1] bring to the attention the statistically confirmed pattern that the processes of the economic servicization are the most typical for the developed economic systems. This is proven by the following indicators of the service industry share in GDP in comparison with GNI per capita determined by the methodology of the World Bank. The percent of the services in GPD for the high-income countries was 74%, for the average-income countries it was above the level of 57%, for the below average level income countries it was 52%, for the low-income countries it was 48% in 2004. [10, 19]

As it is stated by the researchers in the work [1] the increase number of employees in the service sector is the most important indicator of the transition to the service-digital economy.

By comparing the employment level by economic sectors for 2000 and 2012 the authors of the article [1] received the general characteristic of this process. According to the comparison within the above time intervals it was shown that 40,4% of all employed were employed in the agricultural sector and 39,1% in the service sector in 2000. As of similar indicators for 2012 there was a rise in the agricultural sector to 31,9%, in the industry - to 23%, the service sector demonstrated the greatest growth to 45% of the total number of the employed. In comparison with the similar indicators for 2000 and 2012 the developed countries also experience an increase in the number of the employed in the service industry from 67,3% to 73,9%, a reduction in the agricultural sector from 5,5% to 3,2%, a reduction in the industrial sector from 27,2% to 22,5%. [10, 20]

Karpova G.A. and Sushchinskaya M.D. observe the substantive difference in this indicator across the groups of the countries and for specific countries, however, the increase of the employment in the service sector is a general trend which shows the increase in the customers' demand for the services and is one of the indicators of the service sector growth.

As indicated earlier [9] there can be conservative or innovative scenarios of the economic development. With the decrease of the energy resources cost in the external market the innovative development scenario becomes more important which is shown, for instance, by the abrupt changes in those industries where the crisis situations in the economy open up a 'window of opportunity'.

So in 2020 during the pandemic and lockdowns the various services of the distance format became popular around the globe which did not happen earlier, the educational services using the service-digital technologies received the additional opportunities in training methods and the audience coverage. Services for delivery of medicines, food and goods became vital for many categories of the citizens. We can present more examples but they will have one thing in common – the increase in the employment in the service industry which is an important task in the modern economical conditions. The author thinks that these opportunities address and even emphasize the importance of the service industry as one of the most important sources of the employment for the people.

The above examples are applicable not only to a single state but to the majority of the European and Asian countries. This tendency describes the process of the transition to the innovative development scenario for the economics of the developing and developed countries – the transition to the service-digital economy. [10, 11]

The researchers Evmenov A.D., Bulochnikova I.Y. and Blagova state that the changes in the economy are the cause for appearing of the tasks in the service industry which require

© The Authors, published by EDP Sciences. This is an open access article distributed under the terms of the Creative Commons Attribution License 4.0 (http://creativecommons.org/licenses/by/4.0/).

solving of the methodological problems of the service industry management while reviewing the development innovativeness of this industry.

Joseph Stiglitz also speaks about the methodological problems in the service sector and notes that this is the topical problem especially for the countries where the service sector is the main one. [15]

As it was already brought up the transition to the innovative service-digital economy is characterized by the predominance of the service industry and services in the state economy. The researchers [7] note that the innovations in the service sector are the most in-demand especially taking into consideration the events associated with the 2020 pandemic.

In reliance on the data of the research article [9] it is emphasized that by reason of the growth of the industrial interdependences there will be conditions facilitating the increase of the investments from the foreign businesses. In such situation the share of the service sector funding will be one of the largest.

The authors of the article [7] state that the branches of the extractive sector were the top priority for the state for a long period of time and the service sector was not assumed as a significant part of the economy. This situation has led to the fact that the service sector as an important social tool in the state policy has practically no control which leads to the growth of the informal sector in this industry. The state tried to reduce the informal component of the service sector by allowing the citizens to register themselves as the self-employed persons.

The author of this research article offers an opinion that such measures are not enough to ensure the effective functioning of the Russian service industry and it is essential to progress methodology of the management processes for the service industry in the formation of the innovative environment during the digitalization of the service industry. The authors spoke out on the importance of the assigned problem in the works [10, 11].

This aspect is reflected in the document [8] where it is emphasized that it is not possible to create the innovative model suitable for all sectors of the Russian economy because of the fundamental differences in the approaches to the evaluation of the innovative activities. Besides it is required to estimate the impact of the internal and external factors on the study field in order to engage the appropriate resources when solving the assigned tasks.

To estimate the impact of the various indicators on the study field it is proposed to apply the methodology stipulated in the works [7, 10, 11]. The authors propose to use the indicators which describe the state of the environment of the innovation in the service industry in different time periods including the forecast indicators. This methodology allows assessing the significance of each of the chosen indicators in terms of the innovations.

The author of this research work draws attention to the problem of the informal sector of the service industry which can substantially distort the current state of the service sector; thus, it is suggested to apply the comparative analysis of the reporting indicators for the self-employed citizens in order to have a more accurate picture. [10, 11]

The analysis of a number of the works showed that it is rather difficult to evaluate the innovative potential. There are no clear indicators from the perspective of the resource approach. A number of issues related to the assessment of the innovative potential can be settled by involving the knowledge of the OECD experts (Organization for Economic Cooperation and Development). [11]

The author points out that it is a complex task to determine the innovative potential of the service industry and it is necessary to consider a wide range of both the current indicators and the indicators responsible for the prospects for further development of the industry as a whole.

© The Authors, published by EDP Sciences. This is an open access article distributed under the terms of the Creative Commons Attribution License 4.0 (http://creativecommons.org/licenses/by/4.0/).

4 Conclusions

The service industry is a diverse one and covers most sectors of the national economy. The mechanism is required to ensure the transparency and safety of the economic activities as well as facilitate the formation of the reasonable and justified pricing policy when setting the prices for the corresponding services.

The fall of the state participation in the service industry has recently led to the adverse effects – the tariff rates for the services in the various sectors are overpriced, there are the service providers who do not have the necessary industry-specific education and the corresponding qualification. Often the services of the low quality are being provided, there are problems with the warranty service of the goods, and the customer's degree of trust towards the service workers is decreasing.

According to the author of this research article this is happening due to the lack of the effective mechanisms for monitoring the carrying out of the activities in the service sector and non-compliance with the legislation of this industry. Here we can also speak about a large volume of the services provided in the "informal sector" and do not fall under taxation which, for sure, is the problem of the state oversight.

Since the respective individuals are the substantial part of the customers in the service industry the processes which are taking place have the impact on each of us, to say the least.

Apart from being the economic system the service industry is the social system as well and can effect on the people's attitude towards the state and this will depend on how the state mechanisms will look like in the field of the management or regulation of the service industry processes. The author of this research work offers an opinion that the consumer should feel protected by the state when interacting with the agents of the service sector which will make these relations more trusting.

We can make reference to the innovative approach as an illustration of such interaction between the state and the consumer – the digitalization in providing the state services on the corresponding innovative digital services. However, the excessive digitalization of the service delivery can become a positive factor for one group of the consumers and negative for another.

The positive result can be demonstrated by an example of providing the state services where the client receives almost the whole range of the services remotely – in the electronic format which has the same legal power as a real 'paper' document.

The negative factor for both the legal entities and the respective individuals can consist in the sensibility to the fraudulent activities and possible damage due to the insufficient protection of the personal data and data constituting commercial or medical secrets for almost all categories of the consumers. This situation signals the necessity of the state participation in the regulation of the service industry.

The clients – the legal entities obtain the services in accordance with the contracts between the business entities and the scope of their interaction is defined by the provisions of the contract which gives a more balanced quality evaluation of the provided services.

In this matter the innovative digital technologies give an additional advantage when managing the service companies remotely, ensure the progress and results of works by means of photo and video recording as well as the transparency of the accounting and other reporting which is especially important when cooperating with the state enterprises. [14, 17, 21]

It should be stated that the service industry for the legal entities can include both the private and state companies. In this case the provision of the services can be carried out in view of the various pricing mechanisms. As regards to the state organizations the price setting for the services can be carried out based on the auctions and tenders and in the case of the commercial companies it can be done on the basis of the commercial expediency.

© The Authors, published by EDP Sciences. This is an open access article distributed under the terms of the Creative Commons Attribution License 4.0 (http://creativecommons.org/licenses/by/4.0/).

This situation can lead to the imbalance of the tariff prices in the service industry and failure to cover the expenses for the providing services. This also shows that there is the need to form the methodology for the innovative environment during the digitalization of the service sector.

The service industry needs the criteria and methods that will ensure the combined functioning of the state control mechanisms and the market self-regulation mechanisms in this sector.

To transit to the service-digital economy in the branches of the service industry and to form the innovative environment methodology are vital issues in terms of the service industry digitalization and the development of the digital economy and the formation of the new technological paradigm.

The service activities cover almost all sectors of the national economy, therefore, the results of this work can be used not only to form the methodology of the innovative environment as part of the service industry digitalization but also in other sectors of the national economy.

References

1. G.A. Karpova, M.D. Sushchinskaya, *The service industry in the modern economy: monograph* (2016)
2. V.G. Zharov, N.V. Voronov, S.G. Kuzmishkin, Slavic forum, **2(24)**, 197 (2019)
3. K.V. Sukhoterina, *Business technologies in tourism and hospitality, Service technologies as a tool for increasing the efficiency and quality of the services*, 401 (2020)
4. V.G. Zharov, M.V. Runtova, Slavic forum, **1(27)**, 200 (2020)
5. L.V. Sumzina, A.V. Maksimov, V.G. Zharov, Industrial service, **4(69)**, 33 (2018)
6. A.B. Tulinov, A.B. Goncharov, A.A. Korneev, Theoretical and practical problems of service, **3(28)**, 50 (2008)
7. A.D. Evmenov, P.A. Bulochnikov, I.Y. Blagova, Journal of Legal and Economic Studies, **4**, 203 (2013)
8. Russia's innovative development strategy to 2020 (approved by the Decree of the Chairman of the Russian Government dated 08.12.2011 No. 2227-p, Rossiyskaya Gazeta, 4864 (2011)
9. The forecast of the long term socio-economic development of the Russian Federation for the period upto 2030, http://old.economy.gov.ru/
10. L. Gokhberg, K. Ditkovskiy, E. Evnevich et al., *Indicators of Innovation in the Russian Federation: 2020: Data Book*, National Research University Higher School of Economics. – Moscow: HSE (2020)
11. Oslo Manual, Guidelines for Collecting and Interpreting Innovation Data, 3rd Edition Manuel d'Oslo: Principes directeurs pour le recueil et l'interpretation des donnees sur l'innovation, 3e edition OECD/EQ (2005)
12. S.I. Kravchenko, V.M. Dolzhenkova, Scientific works of Donetsk National Technical University. Series: economic, **4**, 98 (2008)
13. I. Ansoff, *Strategic management* (1999)
14. N.V. Blinova, E.V. Gaganova, Human, socio-economic and social sciences, **2(11)**, 52 (2015)
15. J. Stiglitz, Economic science of modern Russia, **4**, 92 (2010)
16. N.M. Komarov, N.V. Ivanov, V.M. Safronov, V.G. Zharov, N.V. Duntsova, K.N. Komarov, Internet journal Naukovedeniye, **1(14)**, 30 (2013)
17. N.N. Teodorovich, A.I. Mokhov, Evolutionary processes of information technologies. Collection of works based on the materials of the 4th international scientific and technical conference, 97 (2019)

© The Authors, published by EDP Sciences. This is an open access article distributed under the terms of the Creative Commons Attribution License 4.0 (http://creativecommons.org/licenses/by/4.0/).

18. V. G. Zharov, Revista inclusions7, Número Especial (2020), http://revistainclusiones.com/
19. World Development Indicators: Structure of output, http://data.worldbank.org/
20. International labour organization. Global Employment Trends 2014, http://www.ilo.org/
21. A.V. Demenev, V.G. Zharov, R.N. Lopatko, In the collection of works: 2019 International Multi-Conference on Industrial Engineering and Modern Technologies, FarEastCon, 8934034 (2019)

© The Authors, published by EDP Sciences. This is an open access article distributed under the terms of the Creative Commons Attribution License 4.0 (http://creativecommons.org/licenses/by/4.0/).

The evolution of robotics in China: accumulated experience and prospects

Marina S. Reshetnikova*, *Irina A.* Pugacheva, and *Yulia D.* Lukina

RUDN University, Miklukho-Maklaya St, 6, 117198 Moscow, Russian Federation

Abstract. In the context of globalization and the fourth industrial revolution (Industry 4.0), the development of robotics and AI has become one of the most important factors of economic and technological growth. A leadership race has already unfolded between the developed countries. Today, China is the undisputed leader in this industry and the largest consumer of industrial robots, but this was not always the case. To achieve leadership in this sphere, the Chinese government supports robotics startups in automotive, electronics, and logistics. The support is provided in the form of tax incentives and subsidies for R&D. The purpose of the research was to forecast the industrial robots' market in China and answer whether it will keep its position in the future. The carried study showed that the market of industrial robots in China is increasing, significantly outpacing the dynamics of global indicators. The decline in the growth of the Chinese robotics market indicators can be explained by the consequences of the COVID-19 pandemic and trade tensions between the market leaders: China and the United States.

1 Introduction

In the context of globalization and the fourth industrial revolution (Industry 4.0), the development of robotics has become one of the most important factors of economic and technological growth [1]. As a result, a leadership race has already unfolded between the world's leading countries. Today, China is the undisputed leader in this industry and the largest consumer of industrial robots, but this was not always the case [2].

Until 2000, the use of industrial robots in China was negligible. In the 1970s, the country began research in industrial robots, samples, and the development of fundamental technologies. In 1990-2000 the development of prototypes of the initial models and their small-scale production began. In the late 1990s, the commercialization of domestic robot production began: industrial and research bases were created, and the first large national manufacturers appeared [3].

In 2001-2010, it was a stage for mass industrialization when Chinese companies began mass production of domestic industrial robots for general consumption. From 2011 to the present, a period of ultra-rapid development of the Chinese robotics industry has already started. As a result, in 2013, China has overtaken Japan to become the world's largest market for industrial robots.

* Corresponding author: reshetnikova-ms@rudn.ru

© The Authors, published by EDP Sciences. This is an open access article distributed under the terms of the Creative Commons Attribution License 4.0 (http://creativecommons.org/licenses/by/4.0/).

The accelerated growth of the robotics industry in China is associated with reducing labor costs and increasing labor productivity, ensuring product quality control. In addition, demographic factors played an important role: a shortage of skilled labor, an increase in its price, and an aging population. The Chinese leadership believes that the development of robotics is significant for national security and recovery of entire economic sectors after the COVID-19 pandemic [4].

To achieve its goals, the Chinese government supports robotics startups in automotive, electronics, IT, and logistics. The support is provided in the form of tax incentives and subsidies for R&D. The purpose of the research was to forecast the industrial robots' market in China and answer whether it will keep its position in the future.

2 Materials and methods

The authors used the theoretical and methodological foundations of scientific research as the general methodology. Empirical methods, methods of calculating indices, comparative analysis methods, and generalizations of statistical data were used to solve the problems. Together, these methods and scientific principles made it possible to ensure the study's reliability and the correctness of the conclusions.

The forecast of the industrial robots' market in China is performed using the linear regression method and an econometric model. The multiple linear regression equation will have the form:

$$Y^{\wedge} = b_0 * x_0 + b_1 * x_1 + b_2 * x_2 + b_n * x_n , \quad (1)$$

where: b0, b1, b2, bn – parameters of the equation calculated by the method of least squares (LS).

The resulting variable is the estimated value of the volume of the industrial robot market in the current year (Fig. 6).

The following variable names were entered for the model:

— VOLUME^ – the volume of the industrial robot market, billion USD (calculated values).

— VOLUME – the volume of the industrial robot market, billion USD.

— PROD – the volume of production of industrial robots, thousands of units.

— SALES – the volume of sales of industrial robots, billion USD.

— INVEST – the volume of investments, billion USD.

— DENSITY – the density of robotics, the number of robots per 10 thousand employed in the industry.

The estimated equation:

$$\text{VOLUME}^{\wedge} = 57.88 + 0.09 * \text{PROD} - 12.64 * \text{SALES} + 0.94 * \text{INVEST} + 0.26 * \text{DENSITY} \quad (2)$$

To identify the overall growth trend of the China robotics market and calculate the forecast values until 2025, an econometric model of the Lin-Lin type was constructed, which depends on several factors: the volume of production of industrial robots, the volume of sales of industrial robots, the volume of investments and the density of robotics, the number of robots per 10 thousand employed in the industry.

3 Results

Now Asia is the largest robotics market, which can be divided into three sectors: industrial robots, service robots, and specialized robots. In this article, the authors analyzed the

© The Authors, published by EDP Sciences. This is an open access article distributed under the terms of the Creative Commons Attribution License 4.0 (http://creativecommons.org/licenses/by/4.0/).

dynamics of indicators for industrial and service robots in the Chinese market. The immense market demand for robots in China is industrial robots – 71% as of 2018. On the other hand, service robots show promising potential in the Chinese market (21% in 2018), as many families use home robots. Furthermore, the rapid development of the healthcare sector creates prerequisites for the growth of this category in the future. (Fig. 1).

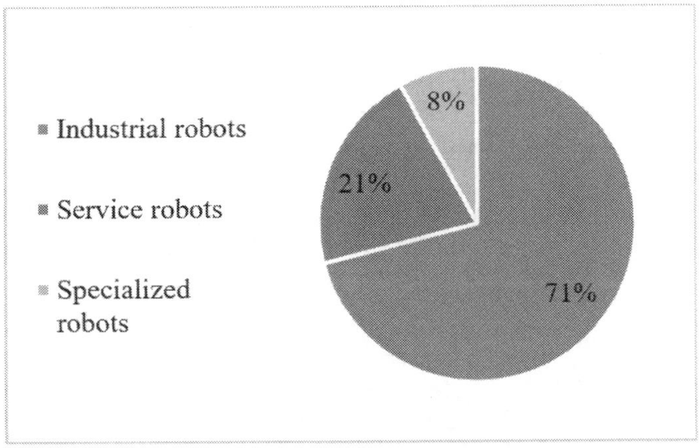

Source: Compiled by the authors, data collected from IFR reports: IFR (2019) [5], IFR (2020) [6].

Fig. 1. Types of robots. Market share in 2018.

The Chinese industrial robot market has maintained a positive trend throughout its development, occupying about a third of the global market. It is the world's largest market for industrial robot applications. According to IFR statistics, the density of industrial robots in China reached 187 units per 10 thousand employees in 2019, which is higher than the global average. It is expected that in 2021 this figure will reach a value of 232 robots, and by the end of 2025, this figure will increase by another 121 units (Figure 2).

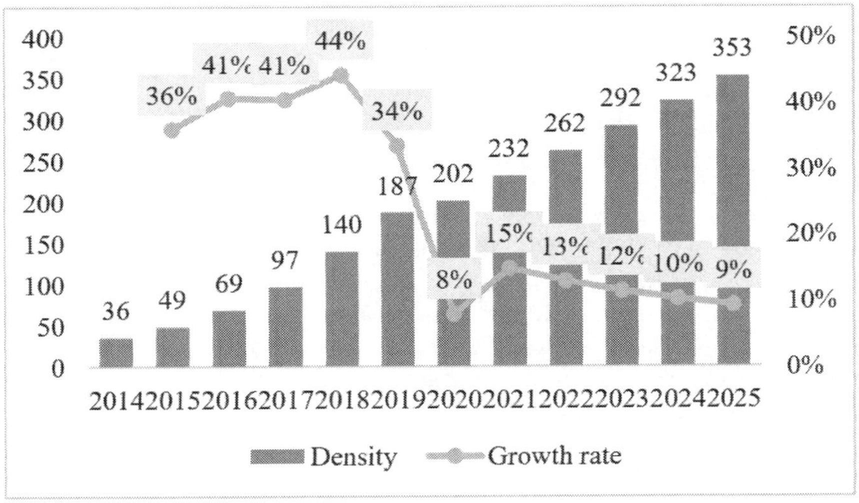

Source: Compiled by the authors, data collected from IFR reports: IFR (2019) [5], IFR (2020) [6].

Fig. 2. Dynamics of the density of robots per 10 thousand employees.

The Chinese robotics market is constantly growing. In 2019, the industrial robot market reached a value of $ 72.6 billion (+14.9% by 2018), while the service robot market reached

© The Authors, published by EDP Sciences. This is an open access article distributed under the terms of the Creative Commons Attribution License 4.0 (http://creativecommons.org/licenses/by/4.0/).

a value of $ 7.5 billion (+19.0% by 2018). However, it is worth noting that the growth rate of these types of robots is different: the growth of industrial robots fell sharply in 2015, but since 2017 began to increase rapidly, the growth of service robots since 2017, on the contrary, began to decrease (Figure 3). This is due to the continuous development of traditional industrial robots, simplifying their programming for complex tasks and the latest security systems [7].

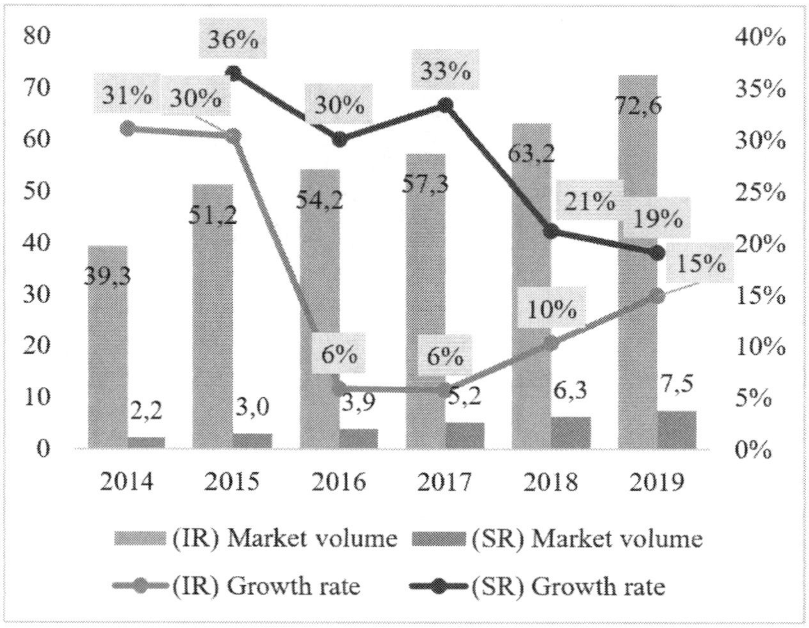

Source: Compiled by the authors, data collected from IFR reports: IFR (2019) [5], IFR (2020) [6].

Fig. 3. Dynamics of the market volume of industrial and service robots, billion dollars.

The service robots' market in China is rapidly expanding, becoming a bright application area of the robotics market. Due to the accelerating trend of population aging and the continued high demand for healthcare and education, China's service robots have huge market potential and development opportunities. As a result, autonomous mobile robots and automated vehicles used in logistics applications are expected to grow significantly, with sales exceeding $ 3 billion by 2022 [8].

Figure 4 shows the dynamics of manufactured industrial robots in China and their sales in physical terms from 2012 to 2019, with a forecast until 2025. At the end of 2019, 120 thousand robots were produced, and 128 thousand units were sold, which is 18,2% and 22,1% less than in 2018, respectively [9]. It can be noted that in 2019, the indicators decreased significantly, but in 2020 they began to grow again. Thus, positive dynamics are expected until the end of the forecast period. By the end of 2025, the volume of industrial robots produced in China may reach a value of 280.6 thousand. As a result, units and the number of sales will increase to 194,6 thousand units. In addition, starting in 2020, the gap between manufactured and sold robots may change: production will grow at a higher rate than sales.

© The Authors, published by EDP Sciences. This is an open access article distributed under the terms of the Creative Commons Attribution License 4.0 (http://creativecommons.org/licenses/by/4.0/).

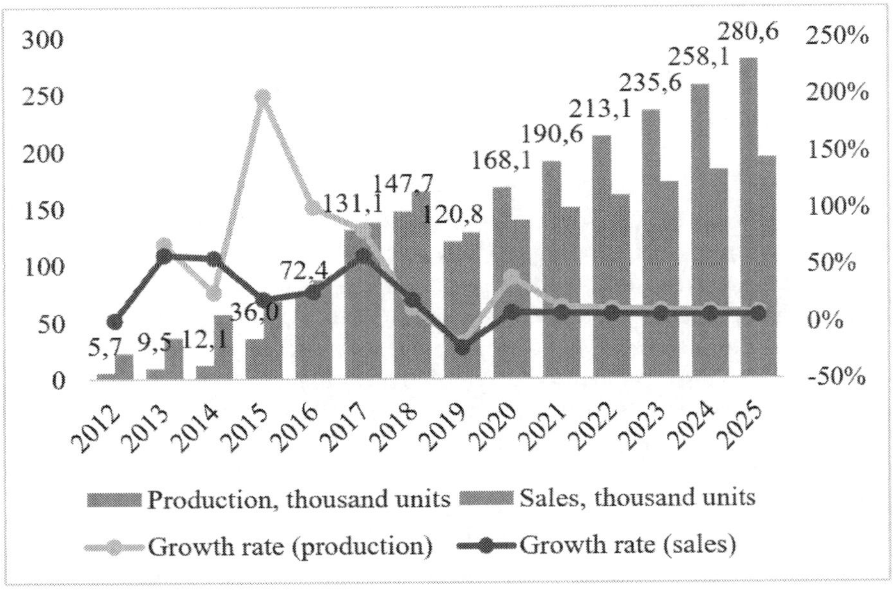

Source: Compiled by the authors, data collected from IFR reports: IFR (2019) [5], IFR (2020) [6].

Fig. 4. Dynamics of production and sales of industrial robots, thousand units.

China buys and manufactures more industrial robots than any other country, but its market is still dominated by Japanese companies, followed by manufacturers from Europe and South Korea [10].

In 2019, $ 54,2 billion was invested in robotics, which is 41% more than in the previous year. Investment is expected to reach $ 105,8 billion in 2025, but the drop in investment growth should be noted (Figure 5).

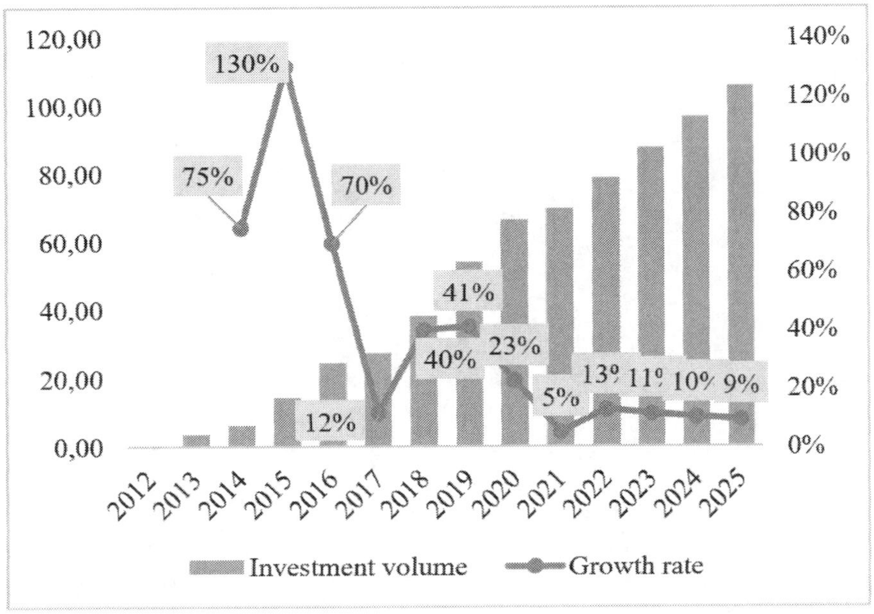

Source: Compiled by the authors, data collected from IFR reports: IFR (2019) [5], IFR (2020) [6].

Fig. 5. Dynamics of investment volume, billion dollars (2012-2025).

© The Authors, published by EDP Sciences. This is an open access article distributed under the terms of the Creative Commons Attribution License 4.0 (http://creativecommons.org/licenses/by/4.0/).

Often, investments in China are flown not in applied but basic research. Basic research focuses on developing manipulators and controllers, while applied research focuses on the component base and specific functions required by the end-user and industry. Because of this, the industrial R&D base lacks vital breakthrough technologies, which makes it less competitive [11].

According to the forecast of IDC Corporation, the robotics market in China will multiply and reach $ 80.5 billion by 2022, accounting for almost 40 % of the global total [12]. The Chinese government actively supports robotics production by providing subsidies, low-interest loans, tax incentives, and land lease benefits to manufacturers of robotic products and enterprises that automate production processes. Thanks to the strategic course aimed at developing the innovation potential and shifting the structural base of the national economy to more knowledge-intensive industries based on innovation, as well as the "Made in China" program, the country is constantly rising in the ranking of innovative economies according to the World Intellectual Property Organization. In 2008, the country was ranked 37th, and ten years later-17th [13].

In addition, the Chinese government considers international cooperation with Western countries in robotics as one of the most effective means of improving the technological level of products. Now, alliances and partner scientific and technical programs are being created with leading Western manufacturers and innovative foreign startups engaged in promising areas of robotics.

According to the forecast, in 2021, the growth of industrial robotics in China will be only 0,2% ($82,9 billion), after which it will increase at a slow rate of an average of 6.0% each year. By the end of 2025, the Chinese robotics market may reach more than $ 100 billion ($104,8 billion), which is 1,45 times more than 2019. As a result, the Compound annual growth rate (CAGR) for the entire study period from 2014 to 2025 will be 9,3%. Now, this indicator has a value of 13% (Figure 6).

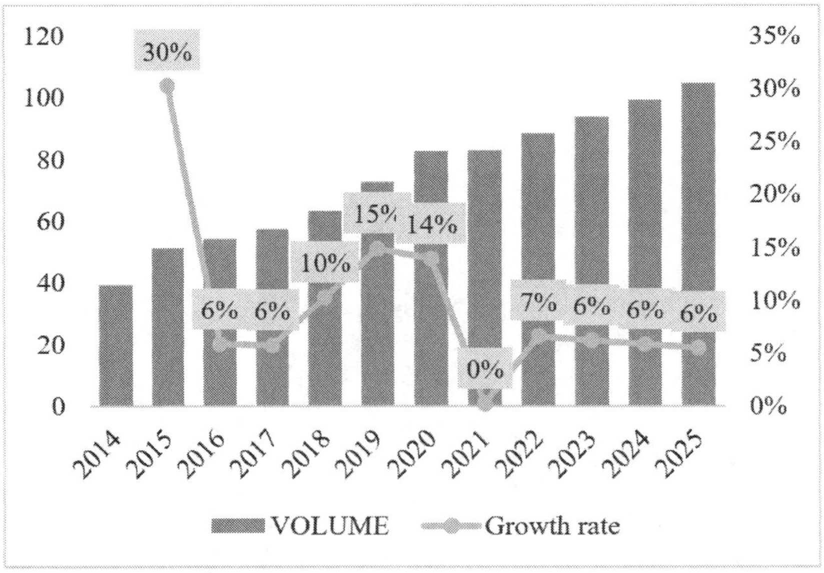

Source: Compiled by the authors, data collected from IFR reports: IFR (2019) [5], IFR (2020) [5].

Fig. 6. The forecast of China robotics market, billion dollars (2014-2025).

The decline in the growth of the Chinese robotics market is due to the consequences of the COVID-19 pandemic and trade tensions between the market leaders: China and the United States. Furthermore, problems in economic relations provoked the redistribution of

© The Authors, published by EDP Sciences. This is an open access article distributed under the terms of the Creative Commons Attribution License 4.0 (http://creativecommons.org/licenses/by/4.0/).

production value chains in 2019, which resulted in a reduction in the global fleet of industrial robots from 422 thousand to 421 thousand units. In addition, most Western countries have introduced strict protectionist measures to protect their strategically essential sectors of the economy, thereby significantly reducing the opportunities for mergers and acquisitions by Chinese companies in industrial innovation.

The Chinese market of industrial robotics can be safely called the largest in the world and, at the same time, rapidly growing. This is what attracts the world's leading manufacturers to open their offices and production sites here. Some foreign companies follow an exciting strategy: they sell their robots to China and then invest in the local industry. Such companies include the German company Kuka AG, the Japanese Yaskawa Electric Corporation, and the Swedish-Swiss multinational corporation ABB Ltd. The government's constant focus on developing the robotics industry encourages Chinese companies to establish close relationships with foreign organizations that possess the most advanced technologies [14].

Despite this, the Chinese market still has problems that hinder faster and better development because the country's potential in this industry is vast. First, it is worth noting the lack of young, highly qualified specialists in China, since today the government has poorly developed educational opportunities for training such personnel [15]. The most popular education system in the specialty "Robotics" is in the United States, and the number of Chinese applicants is growing. Students cannot get the necessary education at home due to the inability of the national education system to respond to market demands in the field of artificial intelligence and robotics flexibly [16].

China is still dependent on imports of industrial robots and their critical components from Japan, Sweden, Germany, Italy, and the United States. Until 2014, European companies imported more than 70% of industrial robots to China [17].

China is now lagging in key industrial robot control technologies. In particular, the international standards for programmable logic controllers (PLCs) and frequency change controls are not met. It is the poor quality of Chinese robotics, which reduces the service life of robotic systems, holding back the industry's economic growth.

One of the critical problems of the Chinese robotics market can also be called irrational competition among local manufacturers. Many production parks in the country compete to produce similar robots without studying the need and demand well. Since 2014, 40 new industrial parks have been built in the country, specializing in robotics. At the same time, it should be said that Chinese robots are not popular in China. This is quite a controversial fact, but the Chinese, investing in domestic production, still buy imported robots because of the higher quality and much lower cost.

Another critical problem is the dependence of Chinese enterprises on government funding. It is not difficult to notice a reverse side of the coin: many factories cannot afford innovative activities. Soon, about 80-90% of them may stop working without state subsidies. Now the Chinese government is trying to take measures to combat fraud and corruption, but this will still hinder the development of robotics production in the country in the coming years.

4 Conclusion

China is the primary winner in the implementation of the catch-up strategy. In a historically short period, from a backward country in which life expectancy did not exceed 40 years, it grew into the second-largest economy in the world while remaining in the shadow of innovative leaders.

The robotics industry is vital for the future of China's manufacturing industry and many other sectors of the economy [18]. Even though China started large-scale production of industrial robots late, it is now a major global consumer of these products.

© The Authors, published by EDP Sciences. This is an open access article distributed under the terms of the Creative Commons Attribution License 4.0 (http://creativecommons.org/licenses/by/4.0/).

The market of industrial robots in the country is increasing, significantly outpacing the dynamics of global indicators. China is accelerating the production of industrial robots and the introduction of related scientific advances. As a result, the large-scale robotization of the automotive, textile, chemical, logistics, electronics, and household appliances industries carried out by the country's leadership will significantly change national and global production in the future.

Now, this shadow weighs in China. It no longer wants to be a global factory but a "scientific and technological superpower." The worldwide dominance of AI and robotics will ensure this. By consistently implementing the state policy for a new generation of AI since 2017, Beijing has confidently moved into second place in development, behind only the United States.

References

1. I. Ermolov, Robotics: Industry 4.0 Issues & New Intelligent Control Paradigms, **272**
2. S. O'Meara, Nature, **572** (2019)
3. L. S. Dalenogare, G. B. Benitez, N. F. Ayala, A. G. Frank, Int. J. of Production Economics, **204** (2018)
4. B. Chen, S. Marvin, A. Marvin, Dialogues in Human Geography, **10** (2020)
5. World Robotics 2019 Preview (2019)
6. World Robotics 2020 Report (2020)
7. H. Fan, Y. Hu, L. Tang, J. of Economic Behavior & Organization (2020)
8. M. S. Reshetnikova, Modern Global Economic System: Evolutional Development vs. Revolutionary Leap, **198** (2021)
9. R. Bogue, IR, **48** (2020)
10. K. Rojko, IIASS, **13** (2020)
11. T. Zhou, E. Lyandres, K. Zhou, X. Cheng, SSRN J. (2020)
12. Worldwide Robotics and Drones Spending Guide (2020)
13. Technology Trends 2019: Artificial Intelligence (2021)
14. P. Aivaliotis, et al., Robotics and Computer-Integrated Manufacturing, **59** (2019)
15. D. Cyranoski. Nature, **553** (2018)
16. J. Stoff, China's Quest for Foreign Technology, 38 (2020)
17. M. Olszewski, PAR, **24** (2020)
18. H. Cheng, R. Jia, R. D. Li, H. Li, J. of Economic Perspectives, **33** (2019)

© The Authors, published by EDP Sciences. This is an open access article distributed under the terms of the Creative Commons Attribution License 4.0 (http://creativecommons.org/licenses/by/4.0/).

The impact of COVID – 19 on Collecting of Tax Revenues in Kosovo

Albina Sekiraqa*, *Edona* Sekiraça, *Verona* Sekiraqa, and *Skender* Ahmeti

University of Prishtina, Agim Ramadani str., 10000 Prishtina, Kosovo

Abstract. This paper aims to present the effect of the pandemic crisis COVID - 19 on the economy of Kosovo, namely tax revenues focusing on the impact of the pandemic situation on macroeconomic indicators such as Gross Domestic Product. The impact of the crisis on the collection of tax revenues, the impact of the crisis on tax evasion, the effect of the crisis on business turnover, and the ratio of tax revenues to GDP. The data that have been used in this paper are data from primary sources through questionnaires and data from secondary sources. The findings of this research show that the pandemic situation in 2020 has slowed down local economic growth, declined trend of tax revenues, declined trend of GDP compared to the previous periods, and other negative effects such as declined economic activity, business, and employment level.

1 Introduction

Throughout the history of humanity and its development, the world has faced various crises of an economic or even healthy nature. Each of them has left its negative effects in the states where they have been present. The latest crisis caused by the COVID - 19 virus first appeared in Wuhan, China, and then spread rapidly to all other continents. Like any crisis, the latter has had negative effects on health, politics, and the economy. The negative effects started from the individual level and in the households to continue in businesses and the state level. The situation forced policymakers through fiscal and monetary policy mechanisms to take measures to mitigate the impact of the crisis on the economy. However, the rapid spread of the crisis negatively affected many important microeconomic and macroeconomic indicators. As in other countries, in Kosovo employees began to lose their jobs due to declining business solvency to pay wages, businesses were unable to meet their obligations to creditors, suppliers, or tax authorities; on the other hand, their business activity began to decline due to declining demand for products and services and uncertainty in the market. Weaker business activity, reflected in the macroeconomic aspect, as business turnover decreased compared to previous non-pandemic periods, almost all domestic investments and foreign direct investment stagnated, tax revenues declined due to the declining in business turnover, which has directly reflected in the decline of the country's GDP; on the other hand, some government spendings were increased in order to finance the approved fiscal packages to help the local economy.

* Corresponding author: albinasekiraqaa@gmail.com

© The Authors, published by EDP Sciences. This is an open access article distributed under the terms of the Creative Commons Attribution License 4.0 (http://creativecommons.org/licenses/by/4.0/).

In this paper, through the primary data collected, we will analyze the determining factors in the turnover of businesses during the pandemic, and through the secondary data will be shown the effect of the pandemic on tax revenues of selected businesses, the impact of fiscal measures, especially for employees with tax revenues in Kosovo, the effect of the pandemic on tax revenues and analysis of the ratio of tax revenues to GDP during this period.

2 Materials Review

The global economy counts many phases of slowdown or decline, which are caused by financial crisis or "crisis" of a health nature such as viruses. The last virus that appeared in Wuhan, China in December 2019 caused the pandemic which affected the social and economic aspect in all the countries where it was present. At the initial of COVID-19 virus spread, countries mobilized through fiscal and monetary policies to mitigate the impact of the pandemic on the economic downturn. The closure of business activities caused the first effects on employees' incomes, therefore almost all fiscal policies in the financing programs of the affected business economies included supportive policies for employees.

In response to the pandemic situation, the International Labor Organization called for the protection of workers in the workplace, the stimulation of the economy, employment, and other financial and tax benefits, including micro, small, and medium-sized enterprises [5].

While COVID - 19 affected economies and businesses, the Tax Authorities in each country had to react quickly to change the way they functioned. For example, blocking measures required them to quickly introduce more digital services in order to protect the health of taxpayers and their employees [2]. Kosovo was also involved in the early reactions after the announcement of the first cases, where the implementation of Government decisions to support business entities through fiscal packages that provided tax relief began. Fiscal measures were supported by employees and businesses initially with a total financial amount of 180 million euros in order to neutralize the side effects on the economy during the emergency situation COVID - 19 [4].

The rapid spread of the virus in the countries of the region and beyond increased the need for the Tax Authorities to take the necessary measures to alleviate the economic situation of employees and businesses. According to a published report by Deloitte Albania [13] it is shown that governments in most countries in the world were taking measures by approving fiscal packages amounting to 10% of Gross Domestic Product. Whereas, Albania experienced this situation differently due to the earthquake of 2019 that caused a loss of about 1 billion euros. According to a survey conducted by Biznes Albania [14] for more than half of the surveyed enterprises have stated that the financial impact of the crisis is high, consequently, their revenues have fallen over 50%, which also reflects in tax revenues lower. According to another study in Albania [15] it is stated that 47% of enterprises completely stopped business activity during the pandemic and that the most common problems of enterprises were lack of customers, lack of liquidity, difficulties in paying salaries, and tax obligations. Measures to prevent the spread of COVID-19 have slowed down economic activity in northern Macedonia as well. According to a study conducted there [8] from March 2020, there was a rapid decline in tax revenues of about 11% compared to the same period last year; 17% less in VAT, 11% less in revenues from excises and 11% less in revenues from profit tax which have a negative impact on the budget.

A report published by the World Bank Group [17] shows that Western Balkan countries have adopted a wide range of social protection policy response measures aimed at supporting formal sector workers, the unemployed, the poor, and the elderly peopulation. Some of these measures were also wage subsidies which are valid in the short term but can

© The Authors, published by EDP Sciences. This is an open access article distributed under the terms of the Creative Commons Attribution License 4.0 (http://creativecommons.org/licenses/by/4.0/).

be costly in the long term. In northern Macedonia, for example, it is estimated that a comprehensive wage subsidy for 250,000 employees in the value of the minimum wage could cost more than 60 milion euros a month. In Montenegro, a two-month wage subsidy of up to 100% of the minimum wage for all firms in the affected sectors has been announced, as well as a six-month wage subsidy for newly declared employment in all sectors, to boost formalization of employment. In Serbia, in large companies, all employees whose contracts have expired due to the crisis have been paid 50 percent of the minimum wage. According to a report [6] in Serbia, measures have been taken through fiscal and monetary packages to mitigate the negative impact of the crisis, one of which included the payment of minimum wages and postponement of deadlines for payment of tax obligations. In this case, 65% of businesses have used this measure approved by the Serbian Government. Whereas a report by The OECD [7] shows that through emergency package measures, the Government's goal has been to mitigate the negative impact of COVID - 19, on the other hand, this situation from the crisis has caused the reduction of corporate income tax by 4% compared to the previous year. On the other hand, if we talk about countries in which the average income is relatively high, then subsidizing them is more challenging.

The German government in response to the crisis has introduced assistance to businesses through the provision of grants to assist their loans, guarantees, or tax deferrals for companies that are in a difficult economic situation [10]. Another research from Czech Republic [9] shows that in the framework of measures to mitigate the effect of COVID - 19 in the economy, the Czech Government applied the measure such as lending to SMEs with a 2-year repayment period, and it also has taken such measures such as postponement of deadlines for declaration and payment of taxes or non-calculation of interest in cases of delays.

The negative effects of the crisis were also in countries across the ocean. Based on a study [11] on the impact of North Carolina income tax it is estimated that the scenario at best is a 20 to 25% reduction in income tax for 2020, the scenario at worst is a 50% loss.

In Malaysia after the first cases with COVID - 1, the payment of taxes has been suspended for 3 months and the memoranda for payment of tax obligations has been reformulated in a softer way. In Panama because of the crisis, the tax administration decided to provide assistance to taxpayers in paying tax liabilities in terms of pending foreclosure proceedings and the sale of their assets. In Singapore in line with the automatic 3-month deferral of tax payments for companies and self-employed persons, all debt recovery actions against these qualifying groups have been suspended respectively until extended deadlines [16].

The financial difficulties of the crisis have been pronounced against another category of businesses. Everywhere in the world, tax administrations are faced with situations of finding businesses that are not registered or are registered but do not present accurate income declarations to the tax authority. Thus a study [12] shows that unregistered enterprises which are mainly self-employed and smaller businesses will not be able to access the financial support provided by the Government as they are unknown to the tax administrations. They will be completely excluded from access to financial support during the current crisis. On the other hand, registered businesses that do not declare all their transactions will be able to access support in accordance with their declared turnover level according to the number of employees declared. Coronavirus is pushing economies towards economic recession, at a time when all parameters and trends were in a positive trend [18].

The pandemic mainly affected the trade and services sector. Quarantine measures have a negative impact on the volume of exports and imports, which means that they have a negative impact on the economy [3]. According to a report on the Coronavirus and the economic situation in Eastern Europe [1], the extension of quarantine in many Eastern

© The Authors, published by EDP Sciences. This is an open access article distributed under the terms of the Creative Commons Attribution License 4.0 (http://creativecommons.org/licenses/by/4.0/).

European countries causes a recession in their economies, on the other hand, reduced funding will call for stricter fiscal policies, deepening the recession even further.

3 Research Methodology

In terms of composition, this paper includes the theoretical and practical parts. Whereas, in terms of implementation this study is divided into two parts which are significantly related to each other. In the first part we are focused on analyzing the impact of the pandemic crisis in business terms, by laying the econometric model and testing raised hypothesis. While in the second part we have analyzed the effect of the crisis directly on tax revenues by the Tax Administration of Kosovo.

The first part, used data provided from primary sources through questionnaires submitted to 200 businesses, while we received answers for 165 of them. The questionnaires were realized electronically and sent to businesses according to 4 different business categories. There are different ways of categorizing businesses as it is according to the number of employees, gross turnover, net income, and others. This study used categorization based on the gross annual turnover of the previous year according to the following table (the categorization is based on the Tax Legislation, where the businesses are categorized for the purpose of mandatory fines).

Table 1. Businesses Categorization.

Categories	The annual gross turnover
C 1	0-30,000
C 2	30,000 - 200,000
C 3	200,000 - 500,000
C 4	over 500,000

Table 2. Respondents' participation in the questionnaire.

Case Processing Summary						
	Cases					
	Included		Excluded		Total	
	N	Percent	N	Percent	N	Percent
Observations	165	82.5%	35	17.5%	200	100.0%

*Source: Authors Calculations.

As we can see, out of a total of 200 submitted observations, responses were returned by 165 of them.

Table 3. Participation in the questionnaire by categories.

		Frequency	Percent	Valid Percent	Cumulative Percentage
Valid	C 1	41	24.85	24.85	24.85
	C 2	37	22.42	22.42	47.27
	C 3	42	25.45	25.45	72.73
	C 4	45	27.27	27.27	100.00
	Total	165	100.00	100.00	

*Source: Authors Calculations.

© The Authors, published by EDP Sciences. This is an open access article distributed under the terms of the Creative Commons Attribution License 4.0 (http://creativecommons.org/licenses/by/4.0/).

The questionnaire is divided into three parts and the formulated questions have multiple answers in 5 categories: 1 (Not at all), 2 (Somewhere), 3 (Average), 4 (Good), 5 (Very good).

This paper analyzed the collected data exploiting SPSS software using the ORL model which helped us to test the hypotheses and analyze how the independent variables impact the dependent variable of the chosen model. The research questions that:

1. What is the most important factor in generating business revenue? and

2. What is the effect of the pandemic crisis on tax revenues? will find answers in the first and second part of this paper.

Questionnaires were realized with businesses to collect data for the determinants of their revenues, in order to assess the effect of the pandemic situation on their activity and indirectly their contribution to the state taxes.

We have raised the hypothesis assuming that:

3.There is a significant link between customer demand and business turnover

The hypothesis was tested by using the ORL model to test independently the impact of each standardized independent variable on the dependent variable and find the most influential factor. The achieved results will be presented through qualitative and quantitative evaluations as well as tabular forms. We have used a hierarchical regressive econometric model according to the following formula:

$$BT= \beta_0 + \beta_1 z_1 + \beta_1 z_2 + \beta_1 z_2 + \sigma^*$$

*Business Turnover

Where the dependent variable is Business Turnover, β0 is the constant which indicates how much is the value of BT if all of the other factors are unchanged (in most cases, it makes no economic sense), z_1 indicates the customer demand for the goods and services, z_2 represents the supply of goods for sales, the variable z_3 represents the financial ability to return liabilities (including employees salaries, rents, suppliers, etc.) and σ is a probabilistic component.

Table 4. Mean and variation compatibility of respondents on the impact of independent variables on the dependent variable.

Increase of Business Turnover	Mean	Median	St. Deviation	Variance	Observtions
Customers Demand	4.2	4	0.832	0.6922	165
Supply of Goods	3.7	3	1.020	1.040	165
Solvency	3.1	4	1.237	1.530	165

*Source: Authors Calculations.

In the table above we have presented the descriptive statistics for the collected data. According to the data presented we see that the variable which most clearly describes the central trend and the variation indicator is the variable "Customers Demand" and most of the respondents have a rating of 4.2 which means that they attribute the impact of customer demand on their turnover. The individual data of the respondents for this variable deviate from their average by 0.832 evaluation points.

© The Authors, published by EDP Sciences. This is an open access article distributed under the terms of the Creative Commons Attribution License 4.0 (http://creativecommons.org/licenses/by/4.0/).

Table 5. Parameter Estimates for hypothesis testing.

Increase of Business Turnover	Coefficient	Standard Error	df	t Value	p Value	95% Confidence Interval	
						Lower	Upper
Customers Demand	0.8642	0.1320	249	1.1876	0.0238	2.5693	2.56928
Supply of Goods	0.7582	0.1542	249	2.431	0.0381	1.1951	6.20488
Solvency	0.5391	0.1727	249	2.87	0.0413	0.6755	5.52452

*Source: Authors Calculations.

Based on the above estimated parameters, the independent variables have high coefficients during testing in the ORL model. The variable with the highest indication is "Customers Demand" with a value of 0.8642 which means that the selected econometric model is best explained by this independent variable, precisely in the turnover of businesses the highest impact has the demand of consumers, then the supply of goods and finally the ability to pay financial liabilities. This research has found that the pandemic crisis in Kosovo, has affected the customers to reduce the consumption of goods and services (elastic demand) due to insecurity, movement restrictions and the tendency to increase savings due to uncertainty of future and others. This directly reflects the decrease in business sales, which in a chain way has also affected the decrease of the supply of goods for the above reasons. This situation is reflected in the budget revenues collected by the tax authority because the lower turnover means lower tax liabilities.

The statistically most important variable is the demand of customers, while statistically important are all independent variables because the p-value value is lower than the significance level of $\alpha = 0.05$ at each of them. Therefore, according to the realized and presented data, we accept the Hypothesis raised with a 95% confidence level.

4 The impact of COVID - 19 on the collecting of tax revenues in Kosovo

The impact of coronavirus (COVID-19) is being felt by all businesses around the world. The position of businesses in most sectors has taken a different direction, causing the reduction of transactions and their value with suppliers, customers and, other interest groups. Thus, the total capacity of products and services within the sectors affected by the pandemic situation have been declined, therefore the aggregate supply has suffered the same declining position. On the other hand, consumers may have had the same level of interest in consuming goods and services, but they were not in willing to buy them due to movement restrictions, insecurity, declining revenues as a result of the situation created causing the decrease of purchasing power. The sectors most affected by the pandemic situation were mainly the manufacturing, tourism, and trade sectors. The decrease in the commercial activity of business entities shows immediate implications in terms of

© The Authors, published by EDP Sciences. This is an open access article distributed under the terms of the Creative Commons Attribution License 4.0 (http://creativecommons.org/licenses/by/4.0/).

macroeconomics. Prohibition of economic activity for a certain period, doing economic activity under the influence of reduced demand, reduced revenues as a result of the above factors reflect lower incomes and as a result lower tax liabilities compared to periods of a normal state of business activity.

Thus COVID - 19 in addition to the human impact, had a significant commercial impact that is felt globally as well as an impact on the collection of tax revenues. As in all countries of the world, also the tax authority in Kosovo - Tax Administration of Kosovo was affected by the circumstances and was influenced by the trend of revenue collection as well as increasing the stock of tax liabilities.

The pandemic situation has caused the decrease of business turnover, workers laid off or lost their jobs, the reduction of liquidity which has pushed businesses to finance expenses and liabilities through banking loans. Influenced by the decline in turnover and job losses, tax revenues have marked a negative trend in 2020 compared to previous periods.

Table 6. Tax Revenues 2017 – 2020.

	Tax Revenues	Tax Revenue Planning	Pension Contributions
2017	414,355,294	420,000,000	159,313,875
2018	456,798,495	446,000,000	173,758,536
2019	504,281,866	516,400,000	188,421,119
2020	467,039,833	491,800,000	195,259,485

*Source: Tax Administration of Kosovo.

According to the data above, revenues collected by TAK from 2017 have a positive trend until the period of the pandemic crisis of 2020. The decrease of tax revenues of 7.38% compared to the previous period was influenced by the cessation of business activity (all sectors, excepted grocery stores, pharmacies and bakeries) for 3 months, as well as the decrease in turnover of entities even after the resumption of business activity in pandemic time.

As we can see, revenues from pension contributions which are collected by TAK and managed by KPST (Kosovo Pension Savings); despite the challenging financial situation in which businesses have passed, this account has been increased compared to the previous period which means that the Government decisions to finance the salaries of employees have reflected in the microeconomic terms through maintaining the financial stability of households and the macroeconomic terms by increasing income from pension contributions.

Financing of employees' salaries has also stimulated the business entities to increase transparency, which to benefit from fiscal packages, have presented accurate statements of the number of employees.

The tax account with the highest impact on tax revenues is VAT with an average of 48% of the participation.

© The Authors, published by EDP Sciences. This is an open access article distributed under the terms of the Creative Commons Attribution License 4.0 (http://creativecommons.org/licenses/by/4.0/).

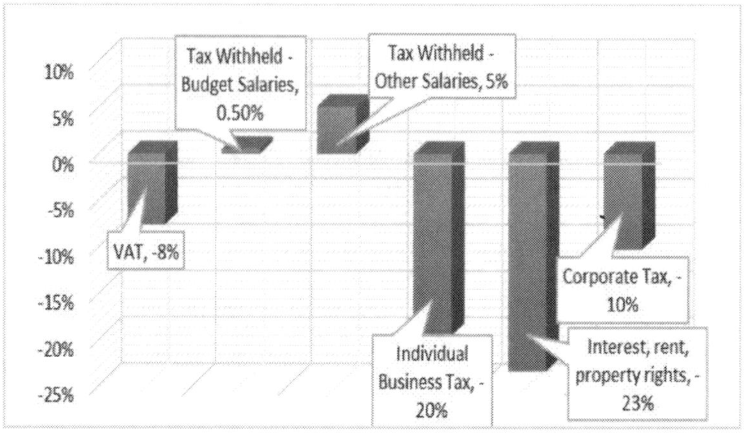

*Source: Tax Administration of Kosovo.

Fig. 1. The trend of tax revenues in 2020 compared to 2019.

Although VAT has the highest impact on tax revenues, during 2020 this account had the smallest decline in revenues to other accounts compared to the previous year, of 8% or expressed in the amount of 18,572,099 million euros, where one of the factors is the decrease of imports (affected by lower demand for goods) compared to 2019 of 6% as trade activity in Kosovo is greatly affected by the import of goods, on the other hand, the level of domestic production has decreased which reflects on the reduction of the supply of goods and this results in a lower level of VAT calculated. Individual business tax and Corporate tax are decreased by 20% and 10% due to reduced business turnover. The positive trend of withheld tax on wages is reflected in fiscal policy assistance in avoiding reducing the number of employees.

Kosovo's economy for 2020 was projected to grow by 4%, but the spread of COVID - 19 in Kosovo also caused a decrease in nominal GDP of 4.02%. The ratio of tax revenues to GDP of the previous year shows an insignificant decrease, in 2020 this share was 9.73% compared to the previous year by 9.77%. The GDP trend in 2020 compared to the previous year was downward by 4.02%, while tax revenues for the same period had a downward trend of 4.39%, as we can see in the chart below.

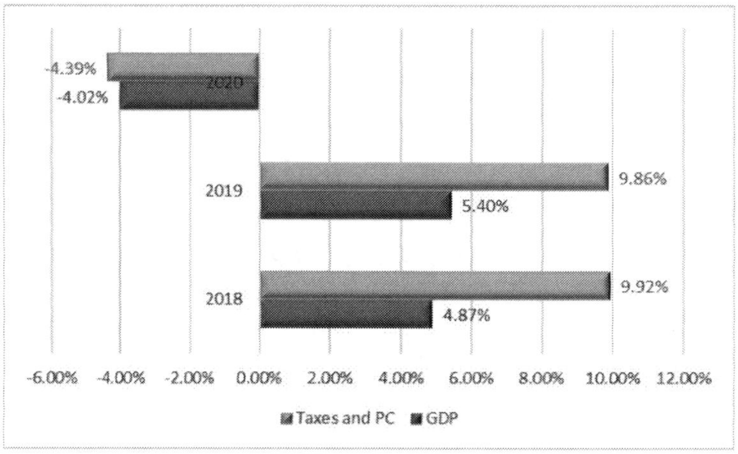

*Souece: Kosovo Agency of Statistics and Tax Administration of Kosovo.

Fig. 2. Tax Revenue Trend plus Pension Contributions and Nominal GDP.

© The Authors, published by EDP Sciences. This is an open access article distributed under the terms of the Creative Commons Attribution License 4.0 (http://creativecommons.org/licenses/by/4.0/).

Closing the 2019 year with a relatively low public debt has provided opportunities for fiscal policy to mitigate the effects of the pandemic, although due to the measures taken, have caused an increase in current expenditures.

5 Conclusions

While Kosovo's economy in 2019 was experiencing a surge of growing, the COVID-19 explosion disrupted it along with many other sectors. The rapid spread of the crisis reduced private consumption, business income, household income, imports, exports, remittances, etc. The impact of economic activity for a limited time (3 months 2020) for all business activities except grocery stores, bakeries and pharmacies, reduced the turnover of businesses and increased difficulties in their solvency capabilities. Thus the Government had an immediate reaction to mitigate the negative financial effect that the pandemic is causing to the vast majority of businesses, especially through the financing of employees' salaries and other financial facilities such as postponement of deadlines for declaration and payment of taxes, postponement of deadlines for following the procedures of blocking or confiscation of assets as a result of non-payment of obligations, remission of administrative penalties created by delays in declaration or payment of basic tax, and others.

Although government spending increased to support funding through fiscal policy, the effects of government measures were on maintaining the financial stability of employees and households as well as guaranteeing the payment of salaries which ensured stable tax withheld revenues and pension contributions, as well as guaranteeing partial financing of rental costs enabled landlords to maintain financial stability and tenants to be freed from liquidity and solvency capabilities. This situation has affected the increase of revenues from tax withheld which includes salaries, rents, and other of 5% compared to the previous period as the accuracy of their declaration was increased in order to benefit from financial assistance. On the other hand, restrictions on the movement and realization of business activities created insecurity of consumers in the consumption of goods and services which from the usual demands they turned into essential consumption requirements.

The econometric model used in this research found that business turnover during the pandemic period was reduced and significantly with a 95% confidence level was influenced by the decrease in consumer demand due to restrictions on activities, incomes decreasing, current consumption uncertainty, and savings trends due to uncertainty about the future. Thus, the impact of global markets all affected by the pandemic situation was also reflected in international import and export transactions on the one hand in the reduction of supply reserves and on the other hand in the lower turnover of businesses.

The restriction of business activity reflected in some aspects in tax revenues, in terms of increasing the debt stock affected by the reduction of liquidity and solvency of businesses, and lower collection of tax revenues compared to previous periods as a result of non-realized regular business activity; in 2020 tax revenues decreased by 7.38% compared to the previous period in which the crisis in Kosovo was not present.

References

1. Eastern Europe Coronavirus Tracker: Preparing for the Worst (2020)
2. The Year of "Tectonic Shifts", **40** (2020)
3. V. Ponkratov, N. Kuznetsov, N. Bashkirova, M. Volkova, M. Alimova, M. Ivleva, L. Vatutina, I. Elyakova, JOP : Technology, Market, and Complexity, **6**, 85 (2020)
4. Kosovo COVID – 19: Emergency Fiscal Measures Package (2020)

© The Authors, published by EDP Sciences. This is an open access article distributed under the terms of the Creative Commons Attribution License 4.0 (http://creativecommons.org/licenses/by/4.0/).

5. International Labour Organization, ILO Monitor 1st Edition, COVID-19 and the world of work: Impact and policy responses, First Edition (2018)
6. COVID – 19 Socio - Economic Impact Assessment (2020)
7. The COVID – 19 Crisis in Serbia, Tackling Coronavirus (COVID – 19) Contributing to a Global Effort (2021)
8. V. Georgieva, O. Svrtinov, T. Georgieva, E. Miteva-Katsarski, B. Koleva, J. of Economits, **5**, 35 (2020)
9. R. Andoh, Int. J. of Multidiscilpinary and Adcanced Technology, **1**, 22 (2020)
10. O. Abodunrin, B. Adesola, Int. J. of Arts, Languages and Business Studies, **4**, 13 (2020)
11. B. McDonald, S. Larson, J. of Public and Nonprofit Affairs, **6**, 377 (2020)
12. C. C. Williams, A. Kayaoglu, The Service Industries J., **40**, 914 (2020)
13. Deloitte Albania, Shock Economic Recovery and Prosperity in Albania, The Effect of COVID – 19 (2020)
14. Business Albania, Impact of the COVID - 19 Pandemic on Enterprises in Albania (2020)
15. Albania Investment Council, Key Findings of the COVID - 19 Business Survey (2020)
16. CIAT/IOTA/OECD, Tax Administration Responses to COVID-19: Measures Taken to Support Taxpayers, OECD (2020)
17. T. Burzanovic, Western Ballkans Regular Economic Report, **17** (2020)
18. Kosovo Chamber of Commerce, COVID - 19 and the Impact on the Economy: Opportunities for Economic Recovery and Transformation (2020)
19. Kosovo Agency of Statistics, Series 3: Economic Statistics - Foreign Trade of Goods (2020)
20. Tax Administration of Kosovo, Annual Report of the Tax Administration of Kosovo (2021)
21. Tax Administration of Kosovo, Annual Report of the Tax Administration of Kosovo (2020)
22. Tax Administration of Kosovo, Annual Report of the Tax Administration of Kosovo (2019)
23. World Bank, www.worldbank.org
24. Tax Administration of Kosovo, www.atk-ks.org
25. Kosovo Agency of Statistics, www.ask.rks-gov.net

© The Authors, published by EDP Sciences. This is an open access article distributed under the terms of the Creative Commons Attribution License 4.0 (http://creativecommons.org/licenses/by/4.0/).

The linkage between innovation and labour market performance: some empirical evidence from the Nordic region

Svetlana A. Balashova*, and *Anastasia A.* Abramova

Peoples Friendship University of Russia (RUDN University), 6 Miklukho-Maklaya St, 117198 Moscow, Russian Federation

Abstract. It is common knowledge that innovations are the drivers of economic growth. However, the direct relationship between innovation and socio-economic performance is questionable. In the article, we examine the linkage between innovation performance, measured by the Summary innovation index, and labour market indicators in Nordic countries, being the leaders of innovation performance and welfare states at the same time. The study shows that employment is growing among people with higher education with the improvement of innovation performance. At the same time, there is a large share of unemployed among low-skilled workers, which also tends to grow. The causality test analysis between Summary Innovation Index and labour statistics by Granger test for panel data does not reveal the causal relationships.

1 Introduction

One of the burning topics of the late 20th century and, in particular, of the entire period of the 21st century is innovation and innovative development. Following the Lisbon Strategy, adopted in 2000, European Union is aimed at achieving global innovation leadership. Each EU country and other European countries have adopted national strategies for innovative development. And although the strategies are not fully implemented and still there are many unsolved issues, the efforts already led to great success, and most European countries are assessed as the top innovatively developed countries in the world. The Nordic countries are highly successful among European countries in terms of innovation performance: Sweden, Denmark and Finland are among the top 10 world leaders. Norway and Iceland are ranked 20 and 21among 130 countries of the world according to the Global Innovation Index 2020. Moreover, the Nordic countries have the lowest wage inequality and, as a group, the lowest poverty rates, as well as the highest job and life satisfaction. All this is also reflected in the level of innovative development and on working life in general [1].

Innovation strategies, predominantly in high-tech sectors, are attractive and successful in industrialized countries mainly [2]. Firms in such countries invest heavily in R&D, contributing to economic growth and international trade, focusing on the modernization of production, methods, and processes in traditional sectors of the economy. One of the main

*Corresponding author Balashova-sa@rudn.ru

© The Authors, published by EDP Sciences. This is an open access article distributed under the terms of the Creative Commons Attribution License 4.0 (http://creativecommons.org/licenses/by/4.0/).

indicators of achieving the goals of innovative development is the share of R&D expenditures in the gross domestic product. By increasing spending on R&D, leadership in the technology race should be ensured, the output of high-tech products should increase, and, as a result, the quality of life should improve. In the long term, R&D spending should support economic growth by increasing productivity and capital. The relationship between innovation capabilities and economic performance are well documented in literature [3].

Innovation development plays a key role in the labour market. On the one hand, new technologies, which are the main features of innovative development, may cause unemployment in the traditional sectors of the economy. On the other hand, these technologies are expected to form high demand in new sectors and even create new sectors [4]. The active introduction of new technologies is causing shifts in employment structure, and wages in some industries can increase significantly [5]. High wages attract highly skilled workers to enterprises. High technology and skilled personnel are expected to increase productivity [6].

The study aims to find empirical evidence that illustrates the relationship between employment growth, the structure of employment and unemployment and the progress in innovative development. We selected the Nordic countries (Denmark, Sweden, Finland, Norway, and Iceland) for the analysis. This region is remarkable in many ways and presents an interesting case study. The countries of this region have close historical, economic and cultural ties. On the other hand, these countries are closely integrated into the world and, above all, the European economy. These countries are united not only by the so-called Scandinavian model of economic and social development, which is based on the idea of the welfare state, but also by the desire of the state and society to develop national innovation systems, create and introduce new technologies, and improve processes in different sectors of the economy. All such features of the Nordic countries cannot but be reflected in the labour market. Therefore, this study analyses the relationship between labour market indicators and innovative performance in the Nordic countries.

2 Methodology and data description

Innovation performance is a complex phenomenon and is traditionally assessed by various indexes. This study uses the Summary Innovation Index (SII) from the EU, which consists of 27 indicators and fully reflects countries' innovative structure [7]. Among these indicators, the most significant ones can be pointed, most responsible for the differences between countries [8]. We consider the period from 2012 to 2019. During this period, the Nordic countries are listed as Innovative Leaders (Sweden (SWE), Denmark (DNK) and Finland (FIN)) and Strong Innovators (Norway (NOR) and Island (ISL)), with an index score higher than or close to the European average (Fig. 1). We can note that Sweden constantly has a high score of SII. Denmark has slightly decreased its performance in the middle of the 2010s but has improved its position in 2019 as well as Finland. Since 2016, Norway demonstrates growth (a leap in 2016 is manly a statistical issue). And only in Iceland there is a decrease in innovative development in recent years.

To characterize the labour market in the Nordic countries, we use such indicators as labour force, total employment (both in the number of employees aged 20–64 and as a percentage of the total population); the number of people employed in the sectors of the economy to determine the most demanded industries; employment and unemployment among workers with different levels of education and skills to assess demand for qualified and not qualified workers. Data are retrieved from European database and OECD database.

© The Authors, published by EDP Sciences. This is an open access article distributed under the terms of the Creative Commons Attribution License 4.0 (http://creativecommons.org/licenses/by/4.0/).

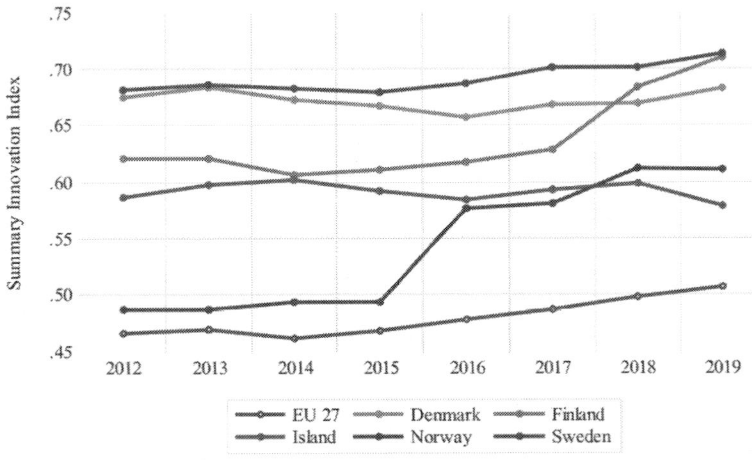

Fig.1. Innovation performance of the Nordic countries

We examine the structure of the employment by the sectors of the economy and assess employment trends. We assess the annual growth rate of employees using the OLS trend regression

$$lnx_t = \alpha + \beta t + u_t \quad (1)$$

Here x_t is the number of employees/labour force in a country at time t ($t = 2011 \div 2019$). α and β are parameters, u_t is an error term.

To assess the direct association between labour market indicators and innovation performance we utilize cross-correlation analysis for panel data between SII and a labour market indicator. Due to short panel (7 years from 2012 to 2019 corresponding to data available for SII and 8 years from 2011 to 2019 for labour indicators), we use only 2 lags for cross-correlation analysis as well as for the Granger causality analysis.

The cross-correlation between the two series x and y are given by

$$Cor_{xy}(l) = \frac{r_{xy}(l)}{\sqrt{r_{xx}(0)}\sqrt{r_{yy}(0)}}$$

$$r_{xy}(l) = \begin{cases} \sum_{t=1}^{T-l} \frac{(x_t - \bar{x})(y_{t+l} - \bar{y})}{T}, & l = 0,1,2 \\ \sum_{t=1}^{T+l} \frac{(x_{t-l} - \bar{x})(y_t - \bar{y})}{T}, & l = 0,-1,-2 \end{cases} \quad (2)$$

This approach can help understand the association between the two variables in dynamics and determine the lag between innovation and labour market indicators.

However, correlation does not necessarily imply causation in any meaningful sense of that word. We use the Granger causality test to address a question of whether series x causes y . y is said to be Granger-caused by x if x helps in the prediction of y , or equivalently if the coefficients on the lagged x 's are statistically significant in the regression (3a)

$$y_t = \alpha_0 + \alpha_1 y_{t-1} + \alpha_2 y_{t-2} + \beta_1 x_{t-1} + \beta_2 x_{t-2} + \varepsilon_t \quad (3a)$$

$$x_t = \alpha_0 + \alpha_1 x_{t-1} + \alpha_2 x_{t-2} + \beta_1 y_{t-1} + \beta_2 y_{t-2} + \varepsilon_t \quad (3b)$$

If the hypothesis $H_0: \beta_1 = \beta_2 = 0$ is not rejected for the second regression (3b), it is said that x is not Granger cause by y. It is worth noting that Granger causality measures

© The Authors, published by EDP Sciences. This is an open access article distributed under the terms of the Creative Commons Attribution License 4.0 (http://creativecommons.org/licenses/by/4.0/).

precedence and information content but does not by itself indicate causality in the more common use of the term.

3 Empirical results and discussion

3.1 Employment trends

Examining the labour market indicators of the five Nordic countries, it can be noted that all countries have a high employment rate which exceeds the average rate for the OECD countries and EU countries: the employment rate of Iceland is 80.3% (it is in first place according to OECD estimates), Sweden - 75.5%, Norway - 74.7%, Denmark - 74.4%, Finland - 72% as of 2019.

The employment policy of the Nordic states is based on the close cooperation of employers' associations and trade unions, the main issue of which was the centralized decision on wages. Trade union participation is about 80% in Iceland and about 70% in Nordic countries. In addition, it is precisely such centralized solutions that have created the image of countries as states with high employment and a small difference in wages. Another interesting government decision was the introduction of a basic income into social security, which stimulated citizens to find employment [9].

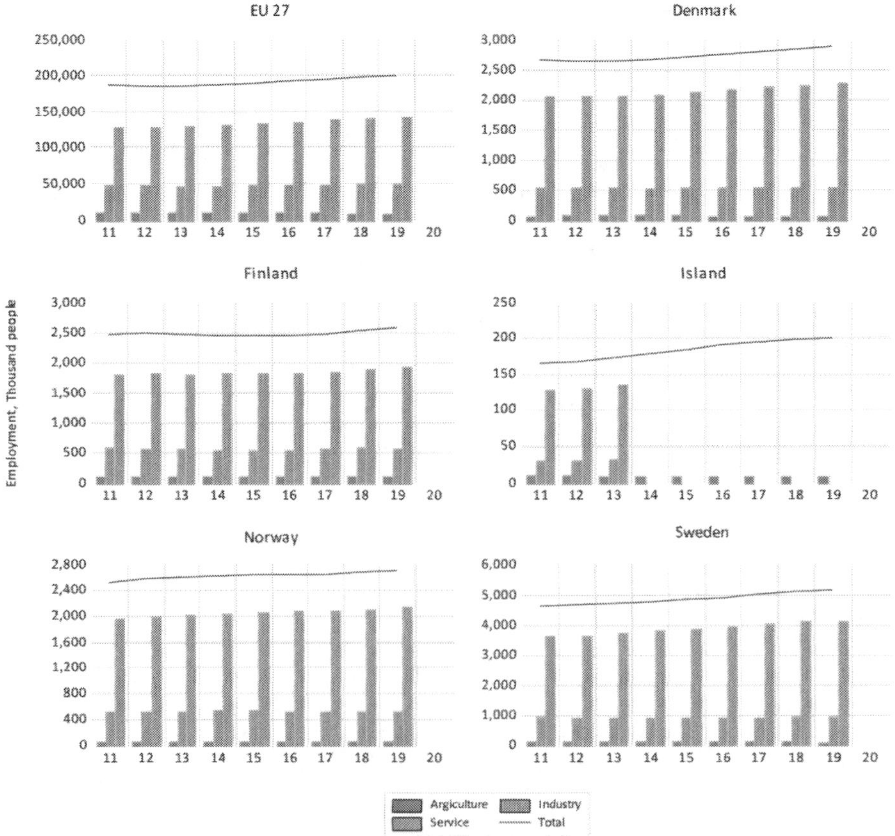

Fig.2. The structure of the employed in the sectors of the economy. Source: Eurostat database

Fig. 2 shows the structure of employed by sectors of the economy for each country and the European Union totally. These statistics demonstrate that countries have relevantly small working force mainly employed in the service sector.

© The Authors, published by EDP Sciences. This is an open access article distributed under the terms of the Creative Commons Attribution License 4.0 (http://creativecommons.org/licenses/by/4.0/).

Simple charts of employment growth suggest the trend rate is very low (see Fig.2). However, Sweden and Denmark overperform other Nordic countries as well as EU in average in terms of the annual growth rate of employees.

Changes in employment breakdown by aggregate sectors are summarized in Table 1. Employment in industry (and in high-tech industry in particular) is not growing in the Nordic countries. Employment in agriculture has a negative trend although the pace is lower than in EU-27. Employment in service has a positive trend, especially in the knowledge-intensive services. Note that labour force was growing in all Nordic countries in the considered period.

In all Nordic countries the growth rate of number of employees in knowledge-intensive service is higher than the growth rate of labour force (see Table 1). We can link it with high pace of innovative development in these countries, which cause the growing demand for high-skilled workers.

Table 1. The annual average percentage change of number of employees and labour force

	Sweden	Denmark	Finland	Norway	Island	EU-27
Total employment	1.4***	1.1***	0.37***	0.7***	2.6***	0.98***
Industry	0.3	0.02	-0.2	0.0	n/d	0.56**
High-tech industry	0.65	-0.3	-1.1	-2.4***	n/d	1.8**
Agriculture	-0.9*	-1.1**	-1.4**	-0.9*	n/d	-2.5***
Service	1.7***	1.5***	0.6***	0.9***	n/d	1.4***
Knowledge-intensive services	2.1***	0.7***	1.0***	1.0***	3.4***	1.4***
Labour force	1.2***	0.7***	0.3***	0.7***	2.0***	

Notes. *** p-value<0.01, p-value<0.05, * p-value<0.10. n/d – unavailable data

Next, we examine the average unemployment rate among different groups of working force categorized by educational attainment. The educational attainment level is coded according to the International Standard Classification of Education (ISCED). In the Fig.3 and after we use the following notations: High skill – levels 5 to 8, Middle skill- levels 3 and 4, Low skill – levels 0 to 2 (see Eurostat database for details). We calculate a simple annual average unemployment rate for each country by calculating the arithmetic mean of the unemployment rate for each of the years from 2012 till 2019. The method is acceptable because the labour force was changing very slowly in the considered period in the countries of interest.

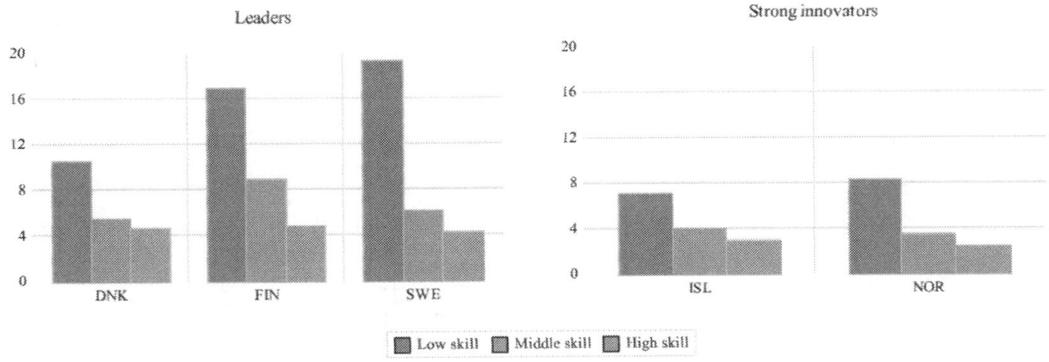

Fig.3. The simple annual average unemployment rate of different working groups

© The Authors, published by EDP Sciences. This is an open access article distributed under the terms of the Creative Commons Attribution License 4.0 (http://creativecommons.org/licenses/by/4.0/).

Fig.3 suggests that for Innovation leaders, the gap in unemployment between the groups of workers with different skill levels are rather large, and the unemployment rate among low-skilled workers is much higher than among the high-skilled. This is where the paradox arises: an increase in the level of innovative development leads to increased unemployment, at least for low-skilled workers in general. Innovative development has a somewhat negative effect on the unemployment rate. Although unemployment itself is decreasing, it has a positive simultaneous association with the index of innovative development. This is understandable, since with active innovations, new jobs are created in different industries, and this takes time. Nevertheless, new jobs cannot replace lost ones [10], therefore, states solve this problem in at least three ways: introduction of basic income [11] or its various analogues [12], a tax on robots, and a reduction in the length of the working week [13].

3.2 Innovations vs. employment for Innovation leaders

Let us look closer for innovation leaders (Sweden, Denmark, and Finland), whose state social policy is aimed at investing in the general welfare, but always taking into account the principles of equal opportunities for the population and the abundance of equal elements of the economic, environmental and social spheres of human life [14]. First, we consider the relationship between employment rate among people with different levels of education (different skills) and the Summary innovation index. The employment rate by the skill levels is calculated by dividing the number of employed people within the age group 20-64 years having attained a specific level of education by the total population of the same age group and with the same educational attainment level.

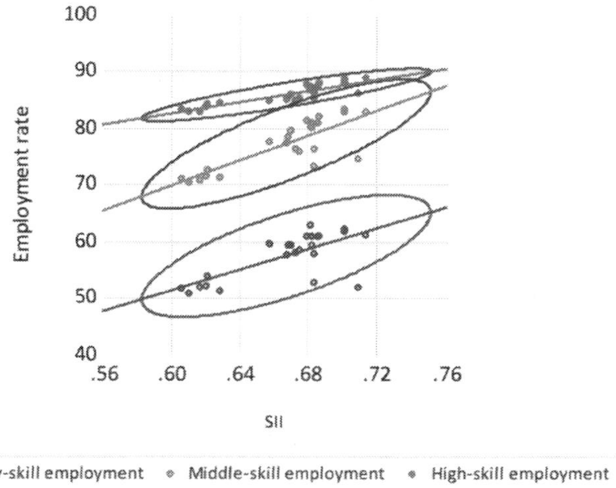

Fig. 4. The linkage between the employment rate among workers with different skill levels and the Summary innovation index for Innovation leaders.

The linkage between the employment rate and the innovation performance is very solid for highly skilled workers (the confidence ellipse is very narrow, see Fig.4.). Since innovative development requires highly qualified specialists and the higher the level of innovative development, the more qualified the employee must be. However, the slope coefficient is lower than for other skilled levels. One of the reasons is that in the Nordic countries, the employment rate of highly educated workers is very high and cannot grow at a high pace. The relationship between innovation performance and employment for middle- and low-skilled workers is not so strong, but the slope coefficient is high. The estimate shows that an increase in SII by 0.01 is associated with a more than 1 pp increase in the employment rate for middle-skill jobs. These findings are in line with the projections of high demand for

© The Authors, published by EDP Sciences. This is an open access article distributed under the terms of the Creative Commons Attribution License 4.0 (http://creativecommons.org/licenses/by/4.0/).

middle-skill jobs in future with increasing technology penetration in all sectors of the economy. On the other hand, it is a result of 'countries place an emphasis on vocational education and training, as well as maintaining a tradition of cooperative social dialogue' [15].

Table 3 shows the results of cross-correlation analysis and causal inference. This table provides the highest correlation coefficient and the corresponding lag (see formula (2)). Interpreting the results, we should consider that SII is based on indicators of previous or even the year before previous.

There is a strong negative association between the share of employment in the agriculture sector of the previous period and SII (-0.78 with l=1). Note that Value added in the agricultural sector as per cent of GDP is about 1.5 pp for the Nordic region.

The correlation between employment in industry and SII is modest but positive, while employment in high- and medium-tech industries is declining with innovation development. The strongest association is observed between employment in knowledge-intensive services and SII. This is a consequence of a large share of service in GDP in these countries and a policy for creating new knowledge-intensive services.

Remarkably, the employment rate is highly correlated with the innovation index for all workers breakdown by skill levels. However, the relationship is even stronger for the high skill workers. And only the high skill workers employment rate revealed to be Granger cause for innovative performance. A high level of employment among skilled workers, the intensity of spending on R&D, support for innovation in society result not only in high values of the innovation development index but also in higher labour and capital productivity than in less innovative European countries [16].

Table 3. Relationship between SII and Labor market indicators

Variable, Y	**Cor (SII_t,Y_{t-l})**	**Granger Causality**
Employment in agriculture, % of employment	-0.78 (l=1)	none
Employment in industry, % of employment	0.55 (l=0)	none
Employment in high- and medium-tech industry, % of employment	-0.58 (l=3)	none
Employment in service, % of employment	0.74 (l=1)	none
Employment in knowledge-intensive service sectors, % of employment	0.77 (l=1)	none
High skill employment rate	0.89 (l=0)	Empl-> SII
Middle skill employment rate	0.81 (l=0)	none
Low skill employment rate	0.71 (l=0)	none

4 Concluding remarks

The article examines the relationship between innovative activity and labour market indicators in the Nordic countries with effective tools and instruments for development.

© The Authors, published by EDP Sciences. This is an open access article distributed under the terms of the Creative Commons Attribution License 4.0 (http://creativecommons.org/licenses/by/4.0/).

Since countries are highly developed, and the service sector prevails, innovative activity has a positive relationship with employment in the service sector and a slightly negative relationship with those employed in the agricultural sector. This is understandable since innovative activity predominates in science-intensive services. The growth rate of employees in knowledge-intensive services is higher than the growth rate of total employment.

From the conducted analysis, we can conclude that innovative development does not directly affect the unemployment rate, at least when the state pays considerable efforts to reduce the negative consequences of high technologies implementation. In the more innovative Sweden, Finland and Denmark, unemployment rates for low-skilled workers are higher than in Norway or Iceland. And significantly higher than among highly and medium-skilled workers. However, the employment rate is positively related to innovation performance for all skill levels.

In order to study the relationship between innovative development, the labour market and state employment policy instruments in more detail, the pool of countries under consideration should be expanded.

References

1. S. Torp, J. Reiersen, Int. J. Environ. Res. Public Health, **17** (2020)
2. D. Ornston, Review of Policy Research, **31**, 5 (2014)
3. S. S. Alam, V. Arumugam, N. G. Mohd Nor, P. Kaliappan, and L. S. Fang, Bus. Manag. Horizons, **1(1),** 59 (2013)
4. V. Matyushok, V. Krasavina, A. Berezin, and J. S. García, Econ. Res. Istraživanja, **27**, 1 (2021)
5. V. Y. Lyashok, T. M. Maleva, M. V. Lopatina, Economic policy, **15(4),** 62 (2020)
6. J. Zysman, *How revolutionary was the revolution? National responses, market transitions and global technology"* (23 Stanford, CA: Stanford Business Books, 2006)
7. European Innovation Scoreboard 2020 (2020)
8. S. A. Balashova, Economics and Mathematical Method, **53(1),** 21 (2017)
9. A. Volkov, World Economy and International Relations, **64(9),** 48 (2020)
10. S. Aronowitz, W. DiFazio, *The Jobless Future: Sci-Tech and the Dogma of Work* (Minneapolis: University of Minnesota Press, 1994)
11. E. Korchak, A. Korchak, The North and the Market: Formation of the Economic Order, **3(6),** 46 (2014)
12. U. Gentilini, M. Grosh, J. Rigolini, R. Yemtsov, World Bank, **337** (2020)
13. A. Hines, Foresight and STI Governance, **13(1),** 19 (2019)
14. P. Frase, Verso Books, **160** (2016)
15. OECD Employment Outlook 2019, OECD (2019)
16. S. Balashova, A. Abramova, Rudn J. Econ., **29(1),** 88 (2021)

© The Authors, published by EDP Sciences. This is an open access article distributed under the terms of the Creative Commons Attribution License 4.0 (http://creativecommons.org/licenses/by/4.0/).

Triad of innovative development: universities-municipalities-business in the development of territories based on cluster policy: case of France

Elena Ponomarenko[1,2,*]

[1]RUDN University, Miklukho-Maklaya Str., 6, 117198 Moscow, Russia
[2]RANEPA, Vernadskogo Pr., 82, 119571 Moscow, Russia

Abstract. Having conducted a comparative analysis of higher education and science reforms in France and in Russia in recent years, we have concluded that with obvious progress in a number of areas of reform, the link between education and science and the implementation of their innovative results in practice is significantly more successful in France. This connection is implemented within the framework of cluster policy, the development of poles of competitiveness and technopolises. They became the basis for promoting a new economic model of development based on the knowledge economy in the early 2000s. A new impetus to the rapid development of clusters in European countries was given by the processes of digitalization, which penetrated into all spheres of activity – research, education, commercialization of ideas, production of new products. The closest connection between science (University and academic science), education (higher educational institutions, their laboratories) and the development of territories (primarily cities) with the help of digital technologies is becoming an everyday reality, the engine of social progress. The growth points that appeared in different regions, expanding within the framework of projects that were selected on a competitive basis and supported by the state, showed the successful result of a targeted policy of supporting innovation, helped to find a way to get the economies of a number of European countries out of the permanent crises of the late 90s, 2008-2009. What is the secret of the success of cluster policy in France? Which role does each of the leading players in the cluster play? Who is the coordinator, connecting link – government institutions? Universities? Research laboratories? Companies? What unites clusters, poles of competitiveness and technology parks? This article is devoted to the answers to these questions.

1 Introduction

As we know from economic theory and confirmed by the practice of recent decades, there are not so many drivers and stimulators of economic growth. This is either a powerful

*Corresponding author: elponamorenko@yandex.ru

© The Authors, published by EDP Sciences. This is an open access article distributed under the terms of the Creative Commons Attribution License 4.0 (http://creativecommons.org/licenses/by/4.0/).

investment in the economy – 30-40% GDP (the example of China, India), growth in exports of products, primarily high – tech (the example of the «Asian tigers» – South Korea, Taiwan, Singapore, Japan), or growth in domestic demand (the example of Russia between 2000-2008 – about 10% per year), with an increase in exports of hydrocarbons.

Export growth and a positive foreign trade balance have always been signs of a «healthy» economy. At the same time, the developed countries of Europe were the first to face the problems of depleted natural resources (their overuse), which are always in demand on world markets, which forced these countries to search for alternative sources of economic development and growth at the very beginning of the years 2000.

The key source was the production and export of high-tech innovative products. This is not just a product (service) of single enterprise, the technological innovation implies the existence of a modern «factories of the future» on the basis of the triple alliance of science and education in the face of universities, government (regions and municipalities), business (from large to small) that implement the process of bringing advanced scientific research to industrial designs, creation of innovative products, competitive not only in domestic market but also in external. Innovations have decisively taken the place of a driver of economic growth in today's globalized economies.

The concept of the «triple helix» was developed in the late 90's by H. Etzkowitz and L. Leydesdorff on the basis of the transition from the dual relationship «state – business» in the industrial economy to the triple – «University-business-state» in the knowledge economy. The new role of universities (in addition to the traditional tasks of teaching and research, it also involves the transfer of technologies and innovations) is also justified in the concept of the «third mission of universities» by Burton R. Clark and in the famous book by J. Wissema «New generation University» [1] [2].

The triple helix, with the support of the state and the local community, gives stability to both national and, most importantly, regional development, and is implemented on the basis of cluster policy. Clusters – are not individual companies or breakthrough research organizations, but a new network phenomenon that has united scientific, educational, and industrial organizations in a single area under the patronage and support of government agencies. Such networks have become able to produce and introduce the latest ideas, patents, and create innovative products that are so necessary to give dynamism to the territorial economy in Europe.

Clusters have united three aspects: research, political and economic. Research – is the connection with universities, as conductors of new knowledge, their research laboratories; political – the attraction of regional (municipal) political elites to the idea of territorial development, economic – finding investors for progressive innovative products.

Digitization processes have given an additional impetus to the rapid development of clusters, poles of competitiveness, and technopoles in European countries. Based on the use of digital technologies, regional growth points are becoming more widespread – smart homes, smart districts, smart cities.

Balanced modern territorial development, including the creation of «smart» cities, is impossible without the participation of educational and scientific institutions located in a particular area. In this regard, it is becoming more relevant to study the new role of universities in the development of territories, their relationship with local government and the business community.

The forms and methods of successful coordination of joint efforts of numerous organizations in clusters, poles of competitiveness in a tough international market competition are the subject of close study of researchers and practitioners of cluster policy, implemented since the 2000s in developed countries: Spain, Germany, Austria, Japan, the United States, and France.

© The Authors, published by EDP Sciences. This is an open access article distributed under the terms of the Creative Commons Attribution License 4.0 (http://creativecommons.org/licenses/by/4.0/).

https://doi.org/10.1051/shsconf/202111401033

Numerous studies have been devoted to the evolution of cluster development. It is based on the ideas of Alfred Marshall about «industrial areas», based on the study of the phenomena of Birmingham and Sheffield in England at the end of the 19th century. Later, there were Italian competition districts, and then – clusters, the theoretical justification for which was given by Michael Porter. Thanks to him, since the late 90s of the twentieth century, clusters have been explained as a geographical concentration of economic activity, production networks generated, on the one hand, by competition that encourages the development of innovations, and on the other hand – by the cooperation of organizations included in the cluster (exchange of information through official and unofficial contacts), which leads to the spread of innovations. Public and private investors play an important role in the implementing of cluster policy, which requires significant investment, according to Porter. Clusters are not necessarily formed with significant state support, but M. Porter justifies state intervention for the development of wide, competitive clusters (Silicon Valley, Boston's Route 128, etc.) [3].

2 Materials and methods

The paper provides qualitative analysis of the development and specifics of cluster policies in France. The methodology is based on the principles of formal logic and systemic analysis. The methods of generalization, analysis and synthesis were also used. The research paper and databases are used as a source of information.

To assess this problem we analyzed implemented policies and other different factors influencing the state of cluster developing.

3 Results and discussion

The success of cluster initiatives has inspired many economic strategies and given impetus to the development of country cluster policies. It is particularly interesting to examine the experience of France's cluster policy, where several hundred cluster initiatives with direct state support have passed a number of stages of development since 2004, have grown into several dozen (about 70 in 2019) poles of competitiveness (PC). Understanding of the importance of updating industrial policy based on investment in innovation, new technologies, by encouraging the transfer of innovation from research laboratories to industrial companies came to France in 1995. However, only since 2004, cluster policy has been developing, which has gone through several stages of evolution.

At the first stage of cluster development (2004-2008), the state was the initiator. The first stage is characterized by strong state support based on competitive selection, serious competition between cluster initiatives (innovative projects in various fields of activity and different geographical regions).

At the first stage, more than 1000 projects participated in the competitive selection process, and 1,5 billion euro of public funds were invested in the selected innovative projects.

Clusters were rapidly gaining strength in the context of Internet penetration in all spheres of life, specializing primarily in everything related to innovation: digitalization, telecommunications, logistics projects, microelectronics, biotechnologies, etc.

Having played a positive role in giving dynamics to territorial development, showing examples of efficiency, clusters gradually concentrated around the poles of competitiveness (PC). The poles of competitiveness are specialized operators, qualified intermediary structures that search for innovations, unification of professional participants in their development and implementation, and investments for the implementation of innovations. Along with the poles of competitiveness, as a form of state support for clusters, there are other mechanisms for the development of territorial cooperation – clusters of enterprises,

© The Authors, published by EDP Sciences. This is an open access article distributed under the terms of the Creative Commons Attribution License 4.0 (http://creativecommons.org/licenses/by/4.0/).

territorial poles of economic cooperation, centers of excellence in rural areas, innovative poles for artisans, mechanisms for partner research, etc.

At the next stage – in 2009-2012 – issues of strategic management of the poles of competitiveness, accelerated implementation of innovative platforms for venture financing, stimulating the growth of small and medium-sized enterprises (SMEs), investment in the ecosystem of innovative development (social, transport, digital, technological, environmental infrastructure) were being resolved. At this second stage, more than 1,5 billion euro were also spent, but already with the involvement of business resources.

Since 2013 to date, poles of competitiveness have been developing from design to implementation factories, strengthening the partnership of participants, creating centers for the transfer of cluster technologies. The movement «for open innovation» is becoming not just an emotional appeal, but an urgent need for mutual disclosure of information in order to achieve effective results. The European strategy «Europe 2020», dedicated to promoting «smart growth» (including bringing research funding to 3% of GDP), claims that success now does not depend on the efforts of a single firm, but is based on the exchange of research results and knowledge among all cluster participants.

At the end of the second stage, it was concluded that the policy of creating poles of competitiveness has reached its goal, 85 technologies with «high potential for industrial development» were identified in 2010 as key points by 2015-2020. These are the aviation industry (Aerospace Valley, Astech, Pegase), railway (iTrans), automobile (iDForCar, Véhicule du Futur, LUTB, Moveo), pharmaceutical industry (Medicen, Lyon Biopôle, Eurobiomed), agriculture, food industry and information-computer technology (Systematic, Images et Réseaux, Minalogic, Cap Digital) . The poles of competitiveness cover both the service sector (finance, logistics) and environmental technologies, as well as the agricultural and agro-industrial sectors.

In 2014, about 8,500 member companies and 1,150 public research and educational institutions joined and collaborated in the 71 poles of competitiveness in France. [4]

Currently, the third stage of cluster development has been completed (2013-2017) and the fourth stage has begun in 2018. Cluster policy instruments and financial flows have changed significantly – innovative products and technologies are initiated and implemented in the current poles of competitiveness with the institutional support of government structures, but often without the financial participation of the state.

Since 2017, 50% of all innovative companies, all companies of the national exchange index CAC 40 and state research structures are included in the poles of competitiveness. 70% of research organizations in the PC have partnerships with small and medium - sized enterprises, 59% - with large enterprises.

In 2019 France ranked first in Europe in terms of the number of new startups, flows of investment in innovative products, and a high share of innovation, surpassing recognized leaders in this area – both Germany and the United Kingdom. At the same time, it is noteworthy that the share of private investment in total investment in innovative products is more than 70%, and the share of public (local) funds is more than 25%, while the share of universities is only about 5%.

By now, France has become one of the leading countries in the field of research and innovation. The key indicators of 2017-2020 are listed below [7]:

1) France is ranked 9th most innovative country in the Innovation Index (Bloomberg 2018), with South Korea leading, followed by Sweden.

The French Alternative Energies and Atomic Energy Commission (CEA) ranks 1st in Europe and 2nd in the world in the 2017 ranking of the most innovative public organizations (Thomson Reuters ranking).

18 French universities are among the 100 most innovative universities in Europe (Thomson Reuters 2018 ranking).

© The Authors, published by EDP Sciences. This is an open access article distributed under the terms of the Creative Commons Attribution License 4.0 (http://creativecommons.org/licenses/by/4.0/).

France ranks 4th in the world for the number of Nobel Prize winners, including the 2018 Prize awarded to French physicist Gerard Moore and 2nd in the world for the number of Fields medals (the highest award in the world in mathematics).

2) In terms of the number of researchers, France ranks third in the European Union, after Germany and Great Britain.

More than 600,000 people are involved in research activities, 58,000 postgraduates (doctoral students); (277,600) are professional researchers, almost a third (27%) of whom are female. Half of the researchers have the right to devote up to 50% of their time to research in business companies.

3) France has the 4th place in the world for patent applications to the European Patent Office. 16,000 patent applications are filed every year: 22% comes from small and medium-sized enterprises, 57% from large companies. More than 2,250 patent applications were filed in 2018 by government research institutions.

4) Institutional research facilities in France.

CNRS (National Center for Scientific Research) is the Europe's top in the ranking of research institutions and ranks #2 in the world (Nature Index 2017). There are 99 research infrastructures in the country.

At the same time, 2,400 companies, supported by government research for 20 years, created more than 38,000 jobs and a capitalization of 12 bln euros.

5) R&D funding is 49.8 bln euro or 2.2% of GDP, which is lower than that of the leading countries.

R&Ds are implemented both in public and private structures. Public research is carried out mainly in universities and national research organizations, in 3200 own research and joint university units. These structures are major participants in international research, have partners in more than 250 locations around the world: CNRS, Institut Pasteur, Cirad, INRA, INRIA, INSERM, CNES, large research universities.

France is an attractive country for private research, offering a certain set of incentive instruments - a tax credit for research, as well as direct assistance in financing private R&D, etc. Allocating almost 0.4% of GDP for public support for private R&D, France ranks 2 -th place among the OECD countries.

Technological research has been developed: development of accelerometers used in airbags (CEA-Leti), inventions of smart cards (R. Moreno), ultra-low consumption electronic components (SOITEC - CEA), development of TGV, Ariane program, many computer data processing languages, used all over the world.

A number of medical innovations are also the result of French research. Over the past 60 years, 127 international innovations have been presented at French university hospital centers

Turning to some intermediate conclusions, we note that the essence of the transition to the poles of competitiveness, in our opinion, is a departure from the state paternalism of each project at all its stages, the rejection of benefits (including tax), traditional for «first wave» clusters and the transition to decentralized initiation of innovative projects, primarily in the most promising areas of activity that are in demand on world markets. Innovative projects of the poles of competitiveness are financed without state financial resources, or with their partial participation.

The differences between the poles of competitiveness and the clusters of the previous stage are:

- firstly, there is an independent project operator (the pole itself becomes a qualified intermediary that unites the efforts of PC participating organizations, facilitating the search for all types of resources), and there are differences in the scale of the PC (on average, one pole of competitiveness unites more than 127 organizations, although most of them are small innovative enterprises, startups);

© The Authors, published by EDP Sciences. This is an open access article distributed under the terms of the Creative Commons Attribution License 4.0 (http://creativecommons.org/licenses/by/4.0/).

- Secondly, there is a rejection of internal competition. Participants develop partnerships and form partner ecosystems. It is more profitable for large business structures to invest small funds in a startup, risking only small investments if the idea fails, or to cooperate with the laboratories of leading universities-participants of the PC, funded by the state in various projects (IDEX, PIA-1, PIA-2), reducing their own financial risks at the stage of basic and applied research;
- Thirdly, the poles of competitiveness operate based on their own or local initiatives, rather than based on priorities determined in advance by the state. The initiative goes «outside-in», and not vice versa, as it was before in case of clusters;
- Fourthly, innovation funding is mainly provided by participants of the pole of competitiveness, as well as attracted extra-budgetary resources. The PC has greater financial autonomy, since a wide variety of sources are involved in the implementation of projects – primarily – venture capital, resources of European investment funds and programs, own funds of the pole participants, as well as state budget sources, especially within the framework of allocated state priorities (aerospace, biomedical and other projects).

Digital technologies give all the processes that take place in the implementation of an innovative project a special cartoon effect, repeatedly speeding up the progress from the idea to its implementation, turn into a powerful factor of production. Based on the use of digital technologies, smart cities are becoming increasingly widespread, as well as many socio-economic projects to localize the production of innovative products and technologies that give impetus to the development of territories where PCs operate.

Next, the most developed stage of cluster development was the promotion of technopolises (technoparks) that were developing in parallel with them. They appeared in the 70s of the twentieth century, marking a new era in the industrial policy of many countries.

Technopolises are divided into two types: multipurpose for the global economy (such as Sophia Antipolis in France, Silicon Valley in the United States) and specialized (biotechnologies, space technologies, aviation, and shipbuilding, cosmetology). Thus, Technopolis Sophia Antipolis is a space of globalization on the Cote d'Azur, which is developing in accordance with global markets. Specialized technopolises are developing along with multipurpose technopolises – their life cycle is much less than that of multipurpose ones. Common technopolises are based on cross-technologies, long cycles, with upward diversification, and they are therefore less sensitive to crises. Technopolises in any specialization are the place of cluster development. [5]

Currently, the International Association of technoparks includes about 400 members representing 70 countries. These are not all existing technology parks. Nowadays, there are more than 700 of them in the world, including 42 % in the United States, 34% in the European Union and 11% in China. The remaining 13 % are accounted in all other countries of the world.[6]

Following the example of California's Silicon Valley, well-known technopolises have appeared in Japan, Malaysia, Singapore, and Sweden. The most famous in France is the science-city – Sophia Antipolis.

4 Conclusions

Here is a generalized experience of Grand Paris and Cote d'Azur universities in the implementation of the results of R&D and the development of the poles of competitiveness (the connection between science, education, territories, megacities)

Grand Paris is an interesting case of a large-scale project in the implementation of which cluster technologies played an important role. The project started in 2007 at the initiative of the then President of France, Nicolas Sarkozy. The idea of the project is to turn Paris and its

© The Authors, published by EDP Sciences. This is an open access article distributed under the terms of the Creative Commons Attribution License 4.0 (http://creativecommons.org/licenses/by/4.0/).

metropolitan area (Ile-de-France) into a modern world-class metropolis of the 21st century, which:

- creates a favorable living environment for its inhabitants, equalizes the level of social and economic development of all included territories, and also stimulates a sustainable growth of the entire agglomeration;

- carefully preserves culture, historical values, actively develops, but at the same time fulfills obligations under the Kyoto Protocol when constructing modern housing and business offices, improves transport accessibility between the center and the periphery, improves the environmental component, preserves the green belt and protects the environment

The most important characteristics of Paris and Ile-de-France are:

- a world-class hyper-centralized agglomeration that has expanded geographically around Paris and has nearly 12 mln inhabitants. Currently it positions itself as one of the leading economic regions in Europe as a whole;
- the capital of France with a limited area of the territory, but with one of the highest urban densities in the world and strong demographic growth;
- high concentration of economic, tourist, cultural and social life in Paris;
- powerful traffic flows from home to work and vice versa, converging in Paris due to the location of the transport network, an urgent need to connect peripheral cities with each other;
- the city ring road that surrounds Paris, making a clear demarcation with the suburbs.

As a result, the overdue such multi-layered and very complex problems had to be solved. The project started in 2007, 10 teams of international level were involved: architects, urban planners, interdisciplinary specialists. In 2010, Parliament adopted Law No. 2010-597 of June 3, 2010 on Grand Paris, which has been successfully implemented to date.

The implementation of such an ambitious large-scale project became possible due to creation and development of competitiveness poles and large clusters as network structures uniting public and private research structures, prestigious universities, promising enterprises and firms in the Ile-de-France agglomeration.

The Grand Paris project is based on the creation of 7 cluster to increase competitiveness, which operate on the basis of CDT - territorial development agreements with local authorities:

1) The Saclay Plateau Reasearch and Innovation: 20 km south-west of Paris, the highest concentration of higher schools in this cluster (HEC, Polytechnique, ENSAE, Centrale, Agro, Telecom Paristech, etc.), prestigious scientific institutions (CEA, CNRS) and since 2014 the several universities have merged to create Paris-Saclay University. In 2027, the RER 18 metro line will be launched for the Saclay cluster.

2) The International Trade and Events Cluster of Roissy – Charles de Gaulle: 30 km north of Paris, the cluseter includes the leading French airport platform Roissy-Charles de Gaulle and the largest European business airport Le Bourget. The following major projects are being implemented in this territory: the international congress center, the European high-speed freight transport project "Carex","Aéropolians", which will ensure the development of an area of 200 hectares, the Europa City retail and leisure project supported by Immochan and the Chinese company Wanda with more than 3 bln euro investment. This cluster will be served by the future Metro line 17 between 2027 and 2030.

3) The Life Sciences Cluster in Villejuif-Evry (Santé Villejuif-Evry): located from Villejuif to Evry, it should promote the emergence of a world-class cluster. It includes 4 engineering schools, 17 CNRS research laboratories and 4 hospitals, the Gustav Russi Institute included and has almost 35,000 students and 8,500 researchers. The Interdisciplinary University's School of Health will strengthen the center in Villejuif. Health R&D activities extend to Evry through Genopole, the first center in Ile-de-France specializing in research in biotherapy and genetics. It brings together more than 70 companies

© The Authors, published by EDP Sciences. This is an open access article distributed under the terms of the Creative Commons Attribution License 4.0 (http://creativecommons.org/licenses/by/4.0/).

from this sector and 21 research laboratories, organized based on the Center of Hospitals of the South of Ile-de-France. The cluster will be served by metro lines 14 and 15 from 2024.

4) Saint-Denis Plaine Commune Creation Cluste: the cluster covers the towns of Saint-Denis, Saint-Ouen, Aubervilliers, Epinay-sur-Seine, La Courneuve, Pierrefitte, Stains, Villetaneuse and L'Ile-Saint-Denis and is located around the Cité du Cinéma (14 film sets and more than 50 television studios), the studios of Plaine Saint Denis, a large university center and large companies. Numerous projects have been launched within the cluster, which will also host the future Olympic village. The cluster is already well served by several RER and metro lines and from 2024 there will be 4 more new GPE lines (14, 15, 16 and 17).

5) La Défense' finance district: Europe's largest business district is the headquarters for many major banks and financial institutions. The business district is undergoing reconstruction and construction of new modern towers. Located close to Paris, it has good transport accessibility, plus there will be a new metro line 15 in 2030.

6) Le Bourget's Airspace cluster: located 13 km north of Paris, it includes towns of Blanc-Mesnil and Bonneuil-en-France. It is Le Bourget Airport that provides the air navigation orientation of this cluster, where the airshow takes place, and a new Airbus Hélicopters facility has been created in the 60,000 m^2 campus in Dugny. Le Bourget will also host some of the Olympic venues. The cluster is served by RER B, and will be reinforced by metro lines 16 and 17 from 2024.

7) The Sustainable City Cluster around Cité Descartes: the cluster, located on the Seine and Marne rivers, starches across the new town of Marne-la-Vallée, from the Cité Descartes to the Val de Europa sector. The Cité Descartes brings together research laboratories and high schools, with about 15,000 students and 3,000 university researchers, and represents 25% of the national research on sustainable urban development. Since the launch of the cluster, the FCBA Institute of Technology (wood sector) and the Efficiency (Urban Energy Efficiency R&Dt Center) have also been established. Disneyland Paris and Pierre & Vacances have joined forces to create the "Nature Town", an ecotourism destination 6 km from Disneyland. Transport - RER A, in Marne la Valais there will be a new line 15 in 2024 and line 16 in 2030.

Technopolis on the Cote d'Azur, created with public and private funds by Pierre Laffitte in 1970-1985, is still one of the most successful example of cluster policy implementation – to date, despite periodic economic crises, its development has been on the rise for more than 35 years, Technopolis «earns» more than the entire Cote d'Azur with all the tourist attractions, hotels, restaurants combined. What is its secret? Why its development is on the rise even in times of traditional economic crises?

In our opinion, the secret of technopark's sustainability is that its founders managed to capture the main promising trend of global development – digital technologies – and attract the world's «trendsetters» in this area to the South of France. The technopark is a home for a number of major European and international organizations operating in the field of digital technologies (for example, the European headquarters of the World Wide Web Consortium (W3C), which develops standards for the Internet, the European Telecommunications Standards Institute (ETSI), The National Institute for Research in Computer Science and Automation INRIA (Institut national de recherche en informatique et en automatique), dozens of divisions of major international companies that work in the most high-tech markets in the world (EURECOM, IBM, etc.).

One of those who introduced the ideas of Technopolis on the Cote d'Azur – an associate of Pierre Laffite – Dominique Fache, who knows Russia well. He worked in our country. As a specialist in cluster policy, Fache offers interesting ideas for creating a technopolis not on the American, but on the French model and in the Russian Crimea. «Crimea can become an innovative center of Russia's development – a technopark of the future, and the crisis can help it in this», says Dominique Fache.

© The Authors, published by EDP Sciences. This is an open access article distributed under the terms of the Creative Commons Attribution License 4.0 (http://creativecommons.org/licenses/by/4.0/).

The successful experience of interaction between educational centers and research laboratories of universities in France with municipalities and business structures in the territories within the framework of the poles of competitiveness demonstrates, in our opinion, a very constructive experience of a non-competitive policy of trilateral cooperation. [8] This policy is based on the logic of partnership, not competition between the structures that make up the cluster.

The French experience testifies to the birth of new forms of industrial policy that successfully implement innovations in a country where the educational and scientific potential is competitive for the world community. It is very important for its implementation in modern Russia.

References

1. H. Etzkowitz, Tech. An. & Str. Man., **20**, 653 (2008)
2. J. G. Wissema, Delft: Ed. El. Pub, 252 (2009)
3. M. E. Porter, Bost.: Harv. Bus. Sch. Pr., 485 (1998)
4. L. Renou, Thés.: Sci. pol. (2016)
5. D. Bretones, Vie & Sciences de l'Entreprise, **185/186**, 189 (2010)
6. V. Toporkov, Gazeta.ru (2014)
7. Vers une loi de programmation pluriannuelle de la Recherche, https://cache.media.enseignementsup-recherche.gouv.fr
8. E. Ponomarenko, A. Oganesyan, V. Teslenko, Intern J of Econ Pol in Emerg Econom **2(4),** 391 (2019)

© The Authors, published by EDP Sciences. This is an open access article distributed under the terms of the Creative Commons Attribution License 4.0 (http://creativecommons.org/licenses/by/4.0/).

AUTHOR INDEX